The production of new potato varieties: technological advances

The production of new potato varieties: technological advances

Edited by

G. J. JELLIS
Principal Scientific Officer
Plant Breeding Institute, Cambridge

D. E. RICHARDSON
National Institute of Agricultural Botany

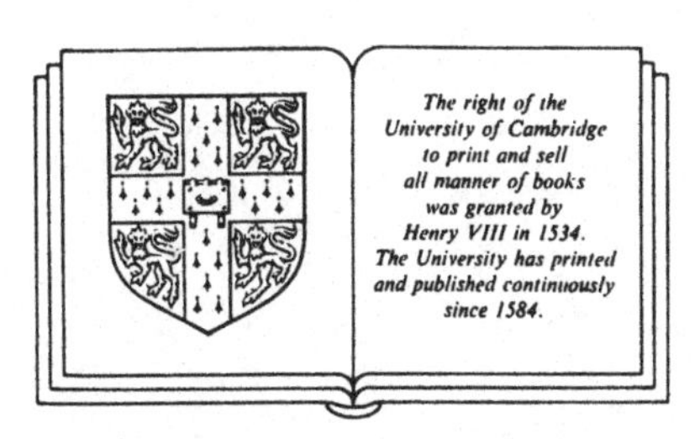

CAMBRIDGE UNIVERSITY PRESS

Cambridge

New York   New Rochelle

Melbourne   Sydney

CAMBRIDGE UNIVERSITY PRESS
Cambridge, New York, Melbourne, Madrid, Cape Town, Singapore, São Paulo

Cambridge University Press
The Edinburgh Building, Cambridge CB2 8RU, UK

Published in the United States of America by Cambridge University Press, New York

www.cambridge.org
Information on this title: www.cambridge.org/9780521324588

First published 1987
This digitally printed version 2008

A catalogue record for this publication is available from the British Library

ISBN 978-0-521-32458-8 hardback
ISBN 978-0-521-06378-4 paperback

CONTENTS

CONTRIBUTORS

N.P. BATTY
John Innes Institute, Norwich NR4 7UH, UK.

W. BATZ
Bundessortenamt, Hannover, D-3163 Sehnde 8, Federal Republic of Germany.

A.G.B. BEEKMAN
Foundation for Agricultural Plant Breeding (SPV), P.O. Box 117, 6700 AC Wageningen, The Netherlands.

A. BLAU
Max-Planck-Institut für Züchtungsforschung, 5000 Köln 30, Federal Republic of Germany.

J. BORYS
The Research Centre for Cultivar Testing, 63-022 Slupia Wielka, Poland.

R.E. BOULTON
Plant Breeding Institute, Cambridge, CB2 2LQ, UK.

J. BROWN
Scottish Crop Research Institute, Pentlandfield, Midlothian, EH25 9RF, UK.

B. COLIN
Laboratoire d'Amélioration des Plantes, ADAR, 91405 Orsay, France.

L. CONCILIO
Conzorzio Provinciale per la Valorizzasione delle Produzioni Agricole, Mario Neri, Imola, Italy.

D.L. CORSINI
University of Idaho, Research and Extension Center, Aberdeen, ID 83210, USA.

S.B. CURRELL
Plant Breeding Institute, Cambridge, CB2 2LQ, UK.

M.F.B. DALE
Scottish Crop Research Institute, Pentlandfield, Midlothian, EH25 9RF, UK.

Y. DATTEE
Laboratoire d'Amélioration des Plantes, 91405 Orsay, France.

S.C. DEBNATH
Biologische Bundesanstalt für Land und Forstwirtschaft, Institut für Resistenzgenetik D-8059 Grünbach, Federal Republic of Germany.

T.J. DIXON
National Institute of Agricultural Botany, Cambridge, CB3 OLE, UK.

B. DOROZHKIN
Siberian Research Institute of Agriculture, Omsk, USSR.

J.M. DUNNETT
Caithness Potato Breeders Ltd, Canisbay (by Wick), Caithness, UK.

J.M. DUNWELL
John Innes Institute, Norwich NR4 7UH, UK.

P. EEKES
Max-Planck-Institut für Züchtungsforschung, 5000 Köln 30, Federal Republic of Germany.

R.N. ESTRADA
Instituto Colombiano Agropecuario (ICA), Apartado Aéreo 151123 ElDorado, Bogotá, Colombia.

S.J. FLACK
National Institute of Agricultural Botany, Cambridge, CB3 OLE, UK.

R.B. FLAVELL
PLant Breeding Institute, Cambridge, CB2 2LQ, UK.

N.E. FOLDØ
Danish Potato Breeding Foundation, 7184 Vandel, Denmark.

B. FOROUGHI-WEHR
Biologische Bundesanstalt für Land und Forstwirtschaft, Institut für Resistenzgenetik D-8059 Grünback, Federal Republic of Germany.

L. FRUSCIANTE
Cattedra di Genetica Agraria, Università di Napoli, Portici, Italy.

P.T. GANS
National Institute of Agricultural Botany, Cambridge, CB3 OLE, UK.

D.R. GLENDINNING
Scottish Crop Research Institute, Pentlandfield, Midlothian, EH25 9RF, UK.

G.M. GURR
National Institute of Agricultural Botany, Cambridge, CB3 OLE, UK.

C.P. HAMPSON
Potato Marketing Board, Broadfield House, 4 Between Towns Road, Cowley, Oxford, OX4 3NA, UK.

S.A. HERMUNSTAD
University of Wisconsin, Madison, WI 53706, USA.

J.G.TH. HERMSEN
Agricultural University, Wageningen, The Netherlands.

M.K. IMAM
Faculty of Agriculture, Assiut University, Assiut, Egypt.

M.T. JACKSON
University of Birmingham, B15 2TT, UK.

J. JAKUBIEC
University of Agriculture, 02-766, Warsaw, Poland.

G.J. JELLIS
Plant Breeding Institute, Cambridge, CB2 2LQ, UK.

M.G.K. JONES
Rothamsted Experimental Station, Harpenden, Herts, AL5 2JG, UK.

E. JONGEDIJK
Agricultural University, 6700 AY Wageningen, The Netherlands.

S.P. KERR
National Institute of Agricultural Botany, Cambridge, CB3 OLE, UK.

C.N.D. LACEY
Plant Breeding Institute, Cambridge, CB2 2LQ, UK.

F. LAMMIN
Laboratoire d'Amelioration des Plantes A.D.A.R., 91405 Orsay, France.

A. LEONE
Centro Studio Miglioramento Genetico Ortaggi-C.N.R., Portici, Italy.

D. LEVY
The Volcani Center, ARO, Bet Dagan 50 250, Israel.

J. LOGEMANN
Max-Planck-Institut für Züchtungsforschung, 5000 Köln 30, Federal Republic of Germany.

J.P. VAN LOON
Hettemer Zonen B.V., Randweg 25, Emmeloord, The Netherlands.

K.M. LOUWES
Foundation for Agricultural Plant Breeding, SVP, Wageningen, The Netherlands.

G.R. MACKAY
Scottish Crop Research Institute, Pentlandfield, Midlothian, EH25 9RF, UK.

M.W. MARTIN
Agricultural Research Service, US Department of Agriculture, Irrigated Agriculture Research & Extension Center, Prosser, WA 99350, USA.

L. MARTINETTI
Institute of Agronomy, University of Milan, Italy.

M.F. MASSON
Germicopa Research & Creation, 29119 Chateauneuf du Faou, France.

H.A. MENDOZA
The International Potato Center, P.O. Box 5969, Lima, Peru.

M. MUNZERT
Bayerische Landesanstalt für Bodenkultur und Pflanzenbau, Freising, Federal Republic of Germany.

A.E.F. NEELE
Foundation for Agricultural Plant Breeding, SVP, Wageningen, The Netherlands.

G. OOMS
Rothamsted Experimental Station, Harpenden, Herts, AL5 2JQ, UK.

J.J. PAVEK
University of Idaho, Research and Extension Center, Aberdeen, ID 83210, USA.

A. PAWLAK
Potato Breeding Station Zamarte, 89-655 Ogorzeliny, Poland.

S.J. PELOQUIN
University of Wisconsin, Madison, WI 53706, USA.

P. PERENNEC
INRA - Station d'Amélioration de la Pomme de terre et des Plantes à Bulbes, BP 5 - 29207 Landerneau, France.

R.L. PLAISTED
Cornell University, Ithaca, New York, NY 14853, USA.

D.E. RICHARDSON
National Institute of Agricultural Botany, Cambridge, CB3 OLE, UK.

S. ROSAHL
Max-Planck-Institut für Züchtungsforschung, 5000 Köln 30, Federal Republic of Germany.

R. SANCHEZ-SERRANO
Max-Planck-Institut für Züchtungsforschung, 5000 Köln 30, Federal Republic of Germany.

J. SCHELL
Max-Planck-Institut für Züchtungsforschung, 5000 Köln 30, Federal Republic of Germany.

R. SCHMIDT
Max-Planck-Institut für Züchtungsforschung, 5000 Köln 30, Federal Republic of Germany.

M. SCHOLTZ
Institut für Kartoffelzüchtung, 2591 Bohlendorf, German Democratic Republic.

R. SCHUCHMANN
Max-Planck-Institut für Züchtungsforschung, 5000 Köln 30, Federal Republic of Germany.

A.M. SQUIRE
Plant Breeding Institute, Cambridge, CB2 2LQ, UK.

N.C. STARLING
PLant Breeding Institute, Cambridge, CB2 2LQ, UK.

R.M. STOREY
Potato Marketing Board, Broadfield House, 4 Between Towns Road, Cowley, Oxford, OX4 3NA, UK.

K.M. SWIEZYNSKI
Institute for Potato Research, Młochow, 05-832, Rozalin, Poland.

M. TALBOT
Agriculture & Food Research Council Unit of Statistics, University of Edinburgh, EH9 3JZ, UK.

A.J. THOMSON
Plant Breeding Institute, Cambridge, CB2 2LQ, UK.

I. WASILEWICZ
Institute for Potato Research, 05-832 Rozalin, Młochow, Poland.

G. WENZEL
Biologische Bundesanstalt für Land un Forstwirtshaft
Institut für Resistengenetik D-8059 Grünbach, Federal Republic of Germany.

J. WHITE
National Institute of Agricultural Botany, Cambridge, CB3 OLE, UK.

L. WILLMITZER
Max-Planck-Institut für Züchtungsforschung, 5000 Köln 30, Federal Republic of Germany.

P. WOOSTER
National Institute of Agricultural Botany, Cambridge, CB3 OLE, UK.

K. VAN DER WOUDE
Government Institute for Research on Varieties of Cultivated Plants, (RIVRO), PO Box 32, 6700 AA Wageningen, The Netherlands.

XU PEI WEN
Institute of Vegetable Research, Shandong Academy of Agricultural Sciences, Jinan, China.

E. ZIMNOCH-GUZOWSKA
Institute for Potato Research, 05-832 Rozalin, Młochow, Poland.

PREFACE

It has become a tradition that the section Potatoes of the European Association for Research on Plant Breeding, EUCARPIA, and the section Breeding and Varietal Assessment of the European Association for Potato Research, EAPR, hold their section meetings simultaneously, thereby benefiting mutually from the experience of experts from both Associations.

The present proceedings 'The Production of New Potato Varieties Technological Advances', constitute however the very first publication *in extenso* of papers presented at joint meetings of the sections.

The book contains, among other items, papers presented at the meeting held at Cambridge, England, between December 16th - 20th, 1985, under the main theme 'The development and Identification of Superior Potato Genotypes - Limitations and Prospects for the Future'. It is a coherent publication, offering non-participants as well as participants in the section meetings a presentation of such important aspects as the current strategies employed in the breeding of new potato varieties, the achievements to date, and future prospects for varietal improvements.

The European Association for Research on Plant Breeding and the European Association for Potato Research welcome this initiative on the part of the sections, and hope that readers of the book will give it the appreciation it deserves.

J. Bijanowski
President of the European
Association for Research
on Plant Breeding, EUCARPIA

N.E. Foldø
President of the European
Association for Potato
Research, EAPR

EDITORS' NOTE AND ACKNOWLEDGEMENTS

As indicated in the Preface, this book is based on the EAPR/EUCARPIA Breeding & Variety Assessment Meeting in 1985. A restriction on the size of the book has meant that the most space has been allocated to papers from invited speakers at the Meeting. This does not mean that the shorter contributions are of lesser importance.

In the arrangement of papers we have attempted to follow a logical sequence, but there is often an overlap of interest between the different sections of the book. Cross referencing partly helps to overcome this problem. Four papers are included which are additional to those presented at the Meeting. These aim to cover some additional aspects and to give a general background to the past, present and future state of breeding and variety assessment in potatoes (Foldø, Jackson, Jellis & Richardson, Thomson).

We thank all contributors for their excellent cooperation in compiling this book, particularly those who have been willing to communicate in a language not their own.

We would also like to thank our colleagues at the Plant Breeding Institute and the National Institute of Agricultural Botany for help in organizing the Meeting on which this book is based. In particular we are greatly indebted to both Mrs Susan Jellis for preparing the Index and to Mrs Sheila Tassell and Miss Ann McDonnell for help with the secretarial work.

G.J. Jellis
Plant Breeding Institute,
Cambridge.

D.E. Richardson,
National Institute of
Agricultural Botany,
Cambridge.

INTRODUCTION

J.G.Th. Hermsen, Chairman of EUCARPIA: Section "Potatoes"
K.M. Swiezynski, Chairman of EAPR: Section "Breeding and Variety Assessment"

The potato is one of the world's most important food crops, being surpassed in total production only by wheat, corn and rice. Therefore, advances in potato breeding may greatly contribute to the world's food supply.

Potato breeders are expected to produce improved cultivars that give high yields of high quality tubers. Furthermore, resistance is required to diseases and pests during growth and storage, to stress conditions and to mechanical damage. Finally, specific properties of the tubers are required for various processing industries.

A rich source of genetic variation is available in existing cultivars, in primitive forms and in wild relatives of potato. The main problem faced by potato breeders is how to exploit most efficiently this large genetic variation.

In the last 75 years there has been a considerable expansion of world potato breeding and associated research. Before World War I breeders used to grow no more than several thousand seedlings, and a few years of basic selection were sufficient to put a new cultivar on the market. Research associated with potato breeding was very limited. Nowadays breeders in many countries grow several hundreds of thousands of seedlings. The breeding cycle is usually 10-12 years and breeding often involves sophisticated selection procedures. Governments in many countries have organized an elaborate varietal assessment system to make sure that only cultivars of satisfactory quality are being released. In addition, many sophisticated research centres make available to breeders new genetic variation, new breeding methods and improved selection techniques.

With such advances one might expect considerable progress in breeding and a quick replacement of old cultivars by better ones. Although significant results have been achieved in potato breeding, e.g. improvement of tuber quality and disease and pest resistance, very old cultivars are still predominant in some countries with advanced agriculture. Striking examples are Russet Burbank introduced in 1876 in the USA and Bintje introduced in 1910 in the Netherlands; King Edward introduced in 1902 in Great Britain has only recently declined in popularity. The question may be raised as to why so much recent effort has yielded relatively limited results. The conference at Cambridge, on which this volume is based, aimed at elucidating the answer to this question.

In invited papers, breeding strategy and varietal assessment in various countries was presented, and several specialists summarized their experience with various new breeding procedures which they have introduced or developed. These papers were supplemented by numerous others related to all aspects of potato breeding. The opinions of the conference participants on various matters were sought in a general discussion.

The following considerations may be helpful in evaluating the present situation.

1. Potato breeding is at a transition stage. The expansion of activities during the past few decades has not produced satisfactory results, probably because insufficient attention has been paid to some basic problems in potato breeding. The potato is a highly heterozygous autotetraploid crop plant. High heterozygosity is needed for high vigour and tuber yield, but leads to segregation for too many important characters in breeding progenies (Simmonds 1969), thus decreasing the probability of detecting superior genotypes. In addition, many characters of the potato are highly sensitive to environmental influence, especially in early generations. At this stage, satisfactory selection techniques are still lacking (Brown, this volume).

In order to increase the frequency of desired genotypes in breeding progenies, superior parents with high breeding values are needed. However, the production of such parents is a long term and tedious job. Breeding at the diploid level simplifies the genetic segregation patterns and may thus render breeding more efficient if it can be associated with effective techniques for intact transfer of selected diploid genotypes to tetraploid progeny. Such techniques exist but more research is needed on their manipulation in a breeding programme. The concept of introducing homozygous or multiplex loci for certain simply inherited traits, and thereby reducing the number of characters to be selected for in the early generations, deserves careful consideration. An increase in research into these problems is needed. International cooperation should be promoted.

2. The selection in segregating breeding progenies and the selection of advanced clones in varietal assessment should be regarded as one process

with one objective - to eliminate defective genotypes and to introduce those that are superior to those now in production. Consistent selection criteria should be utilized throughout the whole selection process and the limited possibilities for successful selection at each stage should be duly considered. It is necessary to bear in mind that until now the final evaluation of a potato cultivar may only be obtained after testing it in practice.
3. New developments in tissue culture techniques and in genetic engineering of the potato are presented in some interesting and critical papers. Breeders must be prepared to try out any new development which is likely to make their work more efficient. However, new procedures often are not very useful to breeders because they do not help to solve the most important difficulties which are inherent in potato breeding.

REFERENCE

Simmonds, N.W. (1969) Prospects of potato improvement. Ann. Rep. Scott. Pl. Breed. Stat., 1968-1969, pp.18-37.

THE DEVELOPMENT OF POTATO VARIETIES IN EUROPE

G.J. Jellis and D.E. Richardson

HISTORY

The potato was introduced into Europe from South America sometime between 1565 and 1573. It was first grown in Spain but by the early part of the seventeenth century it was found in the botanical gardens of many European states. This was largely due to the efforts of the botanist Charles d'Ecluse, or Clusius, who received two tubers and a fruit from the Prefect of Mons, Philippe de Sivry in 1588, multiplied the tubers and distributed them to a number of friends. The first botanical description of the potato was published by the Swiss botanist, Caspar Bauhin in 1596. Bauhin also gave the potato its Latin binomial, *Solanum tuberosum*, although he later added *esculentum*. A second description was published in England, where the potato was probably introduced between 1588 and 1593, by Gerard in his 1597 "Herball".

The original introductions were almost certainly from the Andean regions of Peru or Columbia and were of the subspecies *andigena* (hereafter called Andigena potatoes), with a short-day photoperiodic response. Under European conditions they would have produced many stolons but poor yields of late-developing tubers. During the following two centuries, seedlings from these original introductions were selected for yield and earliness, to give rise to clones well adapted to longer summer days. This selection may have been made in hybrids and selfs of only two original introductions (Salaman 1926). By the middle of the eighteenth century the potato had become a universally cultivated field crop in Europe. A major problem was "Curl", a degenerating condition which meant that varieties had a limited life. To overcome this, seedlings were raised from self-set potato fruit and so there were large numbers of varieties. Towards the end of the century a number of distinct varieties came into use, some of which were specified for human consumption and others for animals.

Two events of major importance in the history of potato breeding occurred in the middle of the nineteenth century. The late blight (Phyto-phthora infestans) epidemics of 1845-46 demonstrated the lack of blight resistance in contemporary varieties. This was not surprising as previous selection had been done in the absence of blight. New varieties were developed with some degree of resistance. These included Paterson's Victoria, Champion and Magnum Bonum, all of which feature in the ancestry of many modern varieties (Figure 1).

Figure 1. Three UK potato varieties grown extensively in the nineteenth century; Champion (top left, introduced 1876), Magnum Bonum (top right, introduced 1876) and Lumper (bottom, introduced prior to 1810 and widely grown in Ireland until 1846, when its failure due to late blight caused the famine).

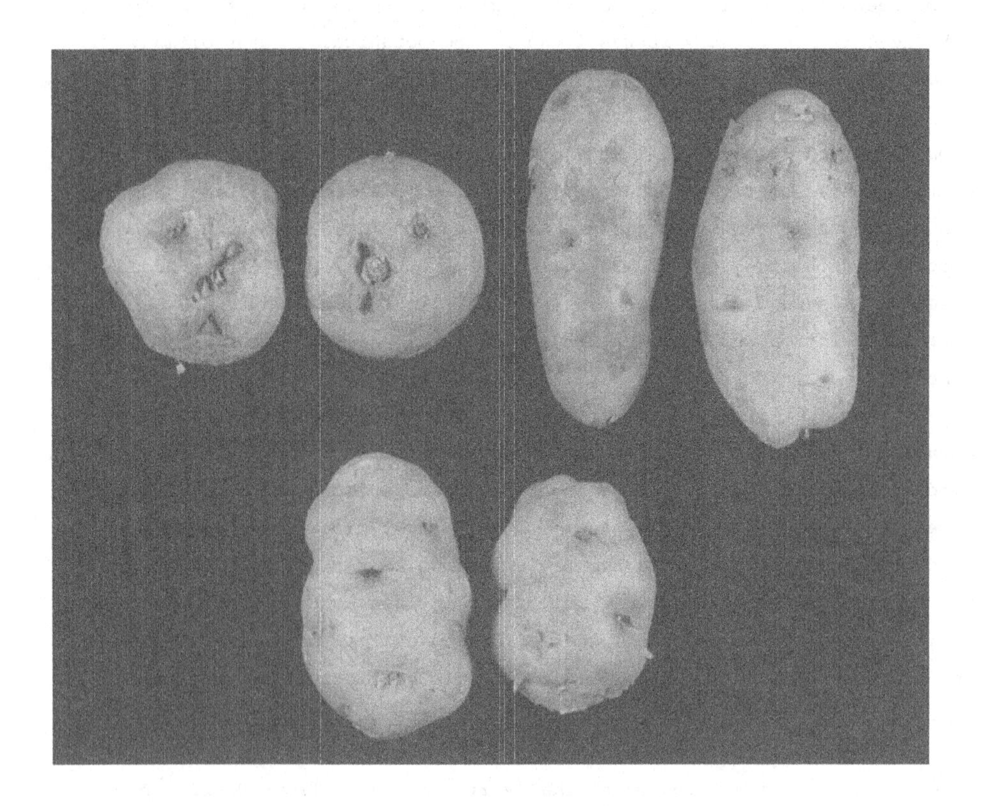

The second event was the introduction of Rough Purple Chili by the Rev. Chauncey Goodrich of New York in 1851. He obtained it as one of a small collection of tubers via the US Consul in Panama. Goodrich assumed that it was a Chilean variety and since it was well adapted to the long days of New York, this may have been the case (Hawkes 1967). From Rough Purple Chili Goodrich raised a number of selfed seedlings including Garnet Chili, which was widely grown in the USA for many years. Among the varieties derived from Goodrich's stocks were Beauty of Hebron, Russet Burbank and Early Rose. The last named variety, in particular, has been very important in European breeding programmes.

Another source of breeding material in the last century was an Andigena clone known as Daber which was introduced into Germany in about 1830 and gave rise to many new varieties, including President.

The next major stimulus to potato breeding was the spread of wart disease (*Synchytrium endobioticum*) in the early part of this century. The recognition of field immunity in a few varieties, including Snowflake, led to the breeding of a whole range of wart-immune varieties. Good blight resistance was still being sought, and we can imagine Salaman's delight when he found field immunity, first in *S. edinense* and later in *S. demissum* (Salaman 1985). By 1914 he had produced a series of families segregating for resistance, and a programme which was to occupy the minds of many potato breeders, particularly in the UK, Germany and the USSR, for the next half a century had been initiated. The first signs that this major gene resistance was not durable came in 1932 but it was another 30 years before breeding for this type of resistance was abandoned. Nowadays, breeding programmes world-wide are aimed at introducing race non-specific resistance into new varieties. This is not proving to be an easy task; the resistance is controlled by a number of genes and is often associated with undesirable traits such as late maturity.

Degeneration of potato stocks, or "Curl", has already been mentioned. During the early part of this century it was discovered that this phenomenon was caused by a number of viruses. Sources of resistance were identified and breeding programmes initiated. Virus resistance is now a major objective in many breeding programmes, using many different *Solanum* species as sources of resistance.

Another objective common to many present-day breeding programmes is resistance to potato cyst nematodes (*Globodera* spp.). Breeding work was made possible by Ellenby (1952) who found resistance to *G. rostochiensis* in five accessions of Andigena and also in *S. vernei*. Populations of cyst nematodes which multiplied on clones incorporating this resistance were found in many European countries. This led to the recognition of a range of pathotypes and eventually to two separate species being described, *G. rostochiensis* and *G. pallida*, each with pathotypes. Resistance to pathotypes of *G. pallida* has been found in accessions of Andigena and *S. vernei* and also in other *Solanum* species. An international scheme for identifying and classifying potato cyst nematodes was proposed by Kort *et al.* (1977) and this is presently being modified.

Potatoes have many uses and programmes have been designed specifically to breed for the requirements of the processing and starch industries. Some European countries have a large export market for seed potatoes and have breeding programmes geared to producing varieties suitable for foreign markets.

As will quickly be appreciated in the following chapters, potato breeding technology is rapidly developing, providing new opportunities for crop improvement. Early generation screening techniques are leading to more rapid identification of the superior genotypes. Breeding at the diploid level and exploiting unreduced gametes are providing ways of introducing new sources of variation and transferring them efficiently (Hermundstad & Peloquin, this volume). Just around the corner is the promise of directed variation by genetic manipulation and protoplast fusion. Already we are reaping some of the benefits of molecular biology in the form of efficient virus and viroid detection techniques (Boulton *et al.*, this volume; Flavell, this volume). Tissue culture has resulted in the rapid propagation of breeding material, and the production of variation through regenerating calluses has led to the exploitation of somaclonal variation (Jones, this volume; Thomson, this volume).

Finally, there is the whole new concept of breeding true potato seed (TPS). The best way of producing such seed is still being investigated (Jackson, this volume) but already TPS is having a major impact in some developing countries.

The "humble potato" has come a long way in the past 400 years, but its future looks just as exciting as its past.

For more detailed and fascinating accounts of the origins of the potato in Europe, the reader is referred to Salaman's book "The History and Social Influence of the Potato", recently reprinted (Salaman 1985), and Hawkes' 1966 Royal Horticultural Society Masters Memorial Lecture "The History of the Potato" (Hawkes 1967).

PLANT BREEDERS' RIGHTS

In Britain during the early part of this century, there was a multiplicity of "new" varieties - bogus selections of existing varieties masquerading under new names. Stocks with the new names usually fetched a higher price than those with the original name. This led the National Institute of Agricultural Botany to set up its Potato Synonym Committee in 1919, when over 200 synonyms of one variety, Up to Date, were found. This Committee continued in operation until the introduction of the distinctness regulations for the Plant Breeders' Rights (PBR) in the 1960s.

The concept of PBR is an extension of the ownership of an invention protected by patent legislation. The aim is to provide a fair return on the long term investment required to breed and assess a new variety; this normally takes about 12 years for potatoes. Protection on all propagating material is given for a minimum period of 15 years from the date of registration; this has been extended for potatoes to 30 years in some countries. Royalties are collected on the basis of the area of potatoes grown for seed certification.

The owner of a variety may transfer the proprietary rights on his variety to individuals or companies for the purpose of growing and selling seed material. He has the legal right to exclude others from reproducing, selling, exporting or importing seed of this variety. There are, however, no restrictions on anyone growing a variety exclusively for their own use, or for selling it for consumption, either fresh or processed.

Plant variety production is internationally controlled by the Union pour la Protection des Obtentions Vegetales (UPOV), which was established in 1961. Membership extends to most European countries, Australia, New Zealand, Israel, Japan, South Africa and the United States. Its aim is to promote progress by stimulating plant breeding and facilitating the interchange of varieties between countries. The benefit of UPOV membership is the availability of reciprocal rights, which avoids duplication of effort by both breeders and government authorities. UPOV has produced test guidelines including 107 taxonomic characters which may be used to establish distinctness in potato varieties (Anon. 1974).

Until now PBR has provided protection only for the variety, which is the end product of plant breeding. In future, with the introduction of genetic engineering there is justification for some form of patent protection of certain technological or microbiological processes used to produce a new variety. The legal implications of this concept have been discussed by Lange (1985).

The general consensus of opinion is that PBR have been advantageous in encouraging investment in private breeding and in increasing the numbers of improved varieties. It has been suggested, however, that PBR are responsible for the recent interest of large multinational chemical and oil companies in taking over seed and breeding establishments, and that such companies may restrict access to varieties with improved disease and pest resistance, but this remains unproven.

It has also been suggested that PBR may cause a reduction of government support for state breeding, which may reduce cooperation and exchange of information between the state and private breeders. In most European countries this is not a problem since the state breeders tend to concentrate on doing basic research and providing the private breeders with basic breeding material.

COOPERATION

There is a good spirit of international cooperation in both potato breeding and variety assessment. This is fostered by such organisations as the European Association for Potato Research (EAPR), European Association for Research in Plant Breeding (EUCARPIA), Potato Association of America (PAA), International Potato Center (CIP) and the International Board for Plant Genetic Resources (IBPGR).

In order to help with the collection, conservation, documentation, evaluation, exchange and use of germplasm, guidelines on descriptors have been produced (IBPGR 1977). These include over 100 "passport", taxonomic, agronomic, disease and quality characteristics. More recently a shorter minimum list of descriptors has been produced by the Commission of the European Communities Agricultural Research Programme on Plant Productivity in collaboration with IBPGR (CEC 1985).

MERIT TESTING

Within the European Economic Community (EEC) a new variety can only be marketed if it is included on the Common Catalogue. This is a list of all varieties included on the National List of all member states. A derogation order can be obtained by any member state to prevent a variety being grown in their country if it presents a particular disease or pest hazard.

In order to comply with the EEC Directive for the inclusion of a variety on a National List it must be distinct, uniform and stable (DUS) as for PBR, and have value for cultivation and use (VCU). The testing systems in several countries are described in this volume.

The need for VCU testing to be a statutory requirement has sometimes been challenged, but few agree that the success or failure of a variety should be dictated by market forces alone. Some form of independent merit testing is generally considered necessary for advisory purposes and for protection against exaggerated or misleading claims. However, potato experts agree that it is difficult to predict from trials the future of some potato varieties and that commercial experience may also be required.

Fifty years ago the Scottish potato breeder of the famous "Arran" varieties, Donald MacKelvie (1937), wrote "No one can foretell what will be the verdict of the practical grower on a new variety of potato, or its future in the market, even if it is successful in trials". To a lesser extent this is still true today, and few varieties get beyond the "promising" stage.

REFERENCES

Anon. (1984). Test Guidelines for Potato (TG/23/2). Geneva: International Union for the Protection of New Varieties of Plants.

Ellenby, C. (1952). Resistance to the potato root eelworm, Heterodera rostochiensis Wollenweber. Nature, 170, 1016.

Hawkes, J.G. (1967). The history of the potato. Masters Memorial Lecture, 1966. J.R. Hort. Soc., 92, 207-24 and 249-302.

Huaman, Z., Williams, J.T., Salhuana, W. and Vincent, L. (Eds.) (1977). Descriptors for the Cultivated Potato. Rome: International Board for Plant Genetic Resources.

Kort, J., Ross, H., Rumpenhorst, H.J. and Stone, A.R. (1977). An international scheme for identifying and classifying pathotypes of potato cyst-nematodes Globodera rostochiensis and G. pallida. Nematologica, 23, 333-9.

Lange, P. (1985). The nature of Plant Breeders' Rights (plant variety protection law) and their demarcation from patentable inventions. Pl. Var. Prot., 44, 17-27.

MacKay, G.R., Hijink, M.J. & Mix, G. (Eds.) (1985). Minimum List of Characters of European Communities. Rome: International Board for Plant Genetic Resources.

Salaman, R.N. (1926). Potato Varieties. Cambridge: Cambridge University Press.

Salaman, R.N. (1985). The History and Social Influence of the Potato. Revised Impression ed. by J.G. Hawkes. Cambridge: Cambridge University Press.

GENETIC RESOURCES

GENETIC RESOURCES: THEIR PRESERVATION AND UTILIZATION

N.E. Foldø

Centres of diversity ("gene-centres") are areas where wild crop ancestors developed and adapted themselves throughout the ages to what were to become their natural habitats. In these areas, in order to survive as a crop species, coevolution between parasites and host plants and between the environment and plant species in general, took place and cultivation of the better adapted native cultivars commenced (Hawkes 1971).

The rich genetic heritage of the gene pool in these areas is of immense importance to plant scientists and plant breeders who are concerned with the incorporation of resistance to pests and diseases, the improvement of quality characters, and a wider environmental adaptability of cultivated varieties.

Plant extinction is a serious threat in many areas of diversity, due among other things to the introduction of new varieties possessing a narrow genetic base (Hawkes 1979), thus lacking the adaptability of the original flora. Fortunately this "genetic erosion" has attracted considerable interest, particularly throughout the 1960s and 1970s, not only from plant geneticists and taxonomists, but also from governments and official bodies. Increasing concern has resulted in greater efforts to preserve this "treasure of nature".

Inseparably linked to the preservation of this genetic diversity should be its utilization for the continued improvement of cultivatable varieties: without preservation its utilization is impossible, and without utilization preservation becomes meaningless.

GENETIC VARIABILITY

The genus *Solanum*, to which the cultivated potato belongs, contain more than 2000 species (Hawkes 1978), of which close to 180 are tuberiferous, and new species are still being found and described (Hawkes & Hjerting 1983, 1985; Ochoa 1983, a,b).

The total gene pool available for research and breeding purposes consists of, in order of increasing "improvement":

- wild species, including interspecific hybrids
- primitive, edible cultivars and their hybrids; often tolerant, non-degenerating, naturally adapted cultivars
- landraces, the varieties of prescientific agriculture
- materials used in ongoing breeding programmes; often interspecific and intervarietal hybrids
- advanced varieties of present or past cultivation.

Generally speaking, the following are of particular value for breeding purposes: the wild species and primitive cultivars on account of their resistance characteristics, landraces for their quality characteristics, and hybrids of existing breeding programmes and advanced varieties for their agronomic characteristics. Use should be made, however, of as broad a genetic base as practicable, depending on the "crossability" of the particular species and hybrids, the inheritance of the particular property to be incorporated, and the number of backcross generations necessary for upgrading the "species characteristics" to varieties of cultivatable value in order to broaden the genetically based adaptability of modern varieties.

Genetic variability in primitive cultivars and wild species

The centres of diversity of the potato are found in North, Central and South America, namely in Argentina, Bolivia, Brazil, Chile, Colombia, Costa Rica, Ecuador, Guatemala, Mexico, Panama, Paraguay, Peru, SW United States, Uruguay and Venezuela (Dambroth & Schittenhelm 1984; Hawkes 1978).

The centres of diversity are generally grouped into two overlapping geographical areas: the Andean region consisting of the South American gene pool, where such series as ACAULIA, CONICIBACCATA, COMMERSONIANA, MEGISTACROLOBA occur, and the Central American region consisting of the Mexican gene pool, where BULBOCASTANA, DEMISSA, LONGIPEDICELLATA, PINNATISECTA and POLYADENIA are found (Hawkes 1958b, 1978; the taxonomy and nomenclature used is in accordance with Hawkes 1963, 1978). The only series with centres of origin outside these areas are ETUBEROSA, JUGLANDIFOLIA (non-tuberiferous) and PIURANA (Hawkes 1958b). The primitive cultivars *S. ajanhuiri*, *S. goniocalyx*, *S. phureja*, *S. stenotomum* (diploids) *S. chaucha*, *S. juzepczukii* (triploids) *S. tuberosum* ssp. *tuberosum*, *S. tuberosum* ssp. *andigena* (tetraploids) and *S. curtilobum* (pentaploid) are all found in the

Andean region (Hawkes 1978).

The climatic and ecological characteristics of the natural habitats of primitive cultivars and wild species are of greater importance as background information to their utilization than is knowledge of their geographical distribution. As a result of the overall environmental selection pressures, biotypes have evolved where the indigenous plant species have adapted themselves through coevolution to the pests, diseases and environmental stresses present within a particular zone. Thus, as was realized very early in the history of potato breeding (Hawkes 1958a), late blight (*Phytophthora infestans*) resistant species may be found within the Mexican gene pool, where the late blight fungus has later been found to reproduce sexually. Frost resistance may be found in species capable of growing at altitudes above *c.* 3500 metres e.g. *S. acaule*, *S. ajanhuiri*, *S. boliviense*, *S. curtilobum*, *S. etuberosum* (nontuberiferous) *S. juzepczukii*, *S. pumilum*, *S. raphanifolium* and *S. sanctae-rosea* (Ochoa 1955; Ross & Rowe 1969). In dry and warm climate areas drought and heat tolerant (*S. chacoense*, *S. commersonii*, *S. gourlayi*, *S. ochoae*, *S. spegazzinii*, *S. tarijense*) as well as scurf and nematode resistant species may be found (Hondelmann & Soest 1980; Ross 1960).

Although resistance to pests and fungal, bacterial and viral potato diseases, as well as agronomic (environmental) and quality characters, may be found in diverse cultivars and species, it is necessary to refer to the evaluation data of specific accessions of a collection of germplasm to judge the potential value and possible use of a particular accession.

Late blight resistance has been derived from Mexican sources in particular. As pointed out by Soest, Schöber and Tazelaar (Soest *et al.* 1984), however, Andean sources might also be of benefit. Race specific and race nonspecific resistance may be found, however, in germplasm of both sources. Race specific resistance is particularly found within the species *S. cardiophyllum*, *S. demissum*, *S. edinense*, *S. stoloniferum* and *S. verrucosum* (Frandsen 1958; Lehmann 1937). A large number of investigations have been made on resistance to late blight. In most breeding programmes emphasis is now on race nonspecific resistance (also called field, horizontal or general resistance although these terms are not strictly synonymous), which may be found in species like *S. berthaultii*, *S. chacoense*, *S. circaeifolium*, *S. demissum*, *S. microdontum*, *S. pinnatisectum*, *S. stoloniferum* and *S. vernei* (Soest 1983a; Soest *et al.* 1984). Resistance to early blight, caused by *Alternaria solani*, may be found within the species *S. bulbocastanum*, *S.*

chacoense and *S. tarijense* (Frandsen 1958).

Resistance to potato wart (*Synchytrium endobioticum*) is, among others, found in the species *S. acaule*, *S. berthaultii*, *S. boliviense*, *S. gourlayi*, *S. spegazzinii*, *S. sucrense* and *S. vernei* (Soest 1983a). Present day potato breeding generally incorporates wart resistance from existing varieties, and despite the existence of different races (Hampson & Proudfoot 1974; Winkelmann 1953), attention is only given to the most common race, 1 (Dahlem D_1), in most breeding programmes.

Resistance to other fungal diseases such as *Rhizoctonia* canker, *Fusarium* wilt and *Fusarium* dry rot, *Verticillium* wilt, gangrene (*Phoma* rot), *Cercospora* leaf blotch, have traditionally been scarcely considered in breeding programmes. Reliable sources of resistance to *Rhizoctonia* have not been reported, although some resistance may exist in the species *S. demissum* and *S. suaveolens* (non-tuberiferous)(Focke 1955; Frandsen 1958). However, differences in disease expression of cultivated varieties have been reported, which has encouraged some screening for varietal differences in susceptibility (Frank *et al.* 1976; Richter & Scheider 1954). The same may be said about the other plurivorous pathogens causing wilts and rots (*Fusarium* spp., *Verticillium* spp. and *Phoma* spp.), where minor or major differences in varietal resistances have been discovered and utilized in resistance screening, but generally without systematic breeding approaches in order to obtain the desired character (Ayers 1952; Boyd 1952; Busch 1966; Jellis 1975; Jellis & Starling 1983; Langerfeld 1979; McLean 1952; Wellving 1976). However, resistance to *Fusarium* dry rot occurs in *S. sparsipilum*, *S. tarijense* and *S. tuberosum* ssp. *andigena* (Soest 1983a), to *Fusarium* wilt in *S. acaule*, *S. kurtzianum* and *S. spegazzinii*, and to *Verticillium* wilt in *S. chacoense* and *S. kurtzianum* (Webb & Buck 1955).

In the case of bacterial diseases such as blackleg (*Erwinia carotovora*) and common scab (*Streptomyces* spp.) resistance is generally inherited from cultivated varieties, though scab resistance is reported to exist in *S. chacoense*, *S. commersonii*, *S. jamesii* and *S. tuberosum* ssp. *andigena* (Reddick 1939) and resistance to blackleg in *S. acaule*, *S. chacoense*, *S. microdontum*, *S. vernei* and others as well as *S. tuberosum* ssp. *andigena* (Soest 1983a). Ring rot (*Corynebacterium sepedonicum*) and bacterial wilt (brown rot) are the subject of attempted control by means of legislative measures and agronomic practices, but resistance is known to occur within the species *S. chacoense*, *S. demissum*, *S. jamesii* for ring rot (Frandsen 1958), and *S. chacoense*, *S. microdontum*, *S. phureja* and *S.*

sparsipilum for bacterial wilt (Anon. 1985; French & de Lindo 1982; Thurston & Lozano 1968).

Resistance to the virus diseases of the potato has been reviewed extensively by Baerecke (1958), Bagnall (1972) and Ross (1958b). Resistance to leaf roll (potato leafroll virus) is evident in the species S. acaule (Brown et al. 1984), S. demissum, S. stoloniferum, S. tuberosum ssp. andigena (Stelzner 1950), and in the nontuberiferous species S. brevidens (Jones 1979) and S. etuberosum (Rizvi 1983). Resistance to the potato viruses X, Y and M (PVX, PVY and PVM) occurs within several species: extreme resistance to PVX is found in S. acaule, S. berthaultii, S. brevicaule, S. tarijense and S. tuberosum ssp. andigena (Schultz & Raleigh 1933; Soest 1983a; Soest & Hondelmann 1983), hypersensitivity (field resistance) to PVY occurs in S. demissum, S. phureja and S. stoloniferum and extreme resistance to PVY is found in S. acaule, S. chacoense, S. gourlayi and S. stoloniferum (Ross 1958b; Soest 1983a). Resistance to PVM is found in S. gourlayi (Dziewonska & Ostrowska 1978), and S. guerreroense, S. hjertingii and S. multidissectum are reported to be resistant to potato spindle tuber viroid (Bagnall 1972).

In most potato breeding programmes in Europe, breeding for resistance to the cyst nematodes Globodera rostochiensis and Globodera pallida is probably given priority in the context of resistance breeding. Since resistance to these widespread (Evans & Stone 1977) pests was found in S. tuberosum ssp. andigena in 1948 and later by Ellenby (1948, 1952, 1954) in the almost historical CPC-clones 1673, 1685, 1692 and 1695, efforts to incorporate resistance into new varieties have been in progress. When Quevedo and coworkers (Quevedo et al. 1956) in 1956 demonstrated the presence of pathotypes in Peru and Dunnett (1957) the year after reported the presence of a population of what is now known as pathotypes Pa3 (P_5A) of Globodera pallida (Canto-Saenz & de Scurrah 1977; Kort et al. 1977), and Stone in 1972 (Stone 1972) described the second species Globodera pallida, the need for breeding for resistance to specific pathotypes became obvious. The search for resistance in collections is very intensive, and resistance has been found in several species including S. berthaultii, S. gourlayi, S. kurtzianum, S. leptophyes, S. multidissectum, S. oplocense, S. spegazzinii, S. sucrense and S. tuberosum ssp. andigena (Soest 1983a; Soest & Hondelmann 1983). Differences in screening and testing procedures, together with differences in the nematode populations used and problems with correct pathotype designation have, however, caused discrepancies among the results obtained.

Resistance to root-knot nematodes (*Meloidogyne* spp.) has been known to occur within several species such as *S. bulbocastanum*, *S. cardiophyllum*, *S. chacoense*, *S. hjertingii*, *S. kurtzianum*, *S. microdontum* and *S. tuberosum* ssp. *andigena* (Hoyman 1974; Jatala & Mendoza 1978).

Breeding for resistance to the Colorado beetle (*Leptinotarsa decemlineata*) is generally not given high priority but several species are resistant, including *S. chacoense*, *S. commersonii*, *S. jamesii*, *S. pinnatisectum* and *S. polyadenium* (Torka 1958), of which *S. chacoense* is the most important one (Ross 1960).

Resistance to aphids, which indirectly control the aphid-transmitted virus diseases, based on the presence of glandular hairs, has been reported among others for *S. berthaultii* (Gibson 1979; Gibson & Turner 1977), and for *S. neocardenasii* (Hawkes & Hjerting 1983), and resistance to potato tuber moth (*Phthorimaea operculella*) is found in *S. stenotomum* and *S. tuberosum* ssp. *andigena* (Raman & Palacios 1983).

The genotypic influence on agronomic (environmental) and quality characters of the potatoes has been reviewed by Holden (1981), Howard (1974), Keller & Baumgartner (1982) and Rudorf & Baerecke (1958). Particularly worthy of mention in relation to practical use may be the reports of frost tolerance in *S. acaule* (Dearborn 1969), of high protein content in *S. phureja* (Desborough & Wieser 1972), of the storability of *S. phureja* hybrids at low temperatures without enhanced accumulation of reducing sugars (Lauer & Shaw 1970), and of high starch content in *S. vernei* derivatives (Hermsen 1977).

Genetic variability in cultivated varieties

Since the initial introduction of the potato to Europe and North America as cultivars of *S. tuberosum* ssp. *tuberosum* from Chile, and *S. tuberosum* ssp. *andigena* from Peru/Bolivia (Salaman 1926), genetic variation has been present within the cultivated varieties, and several selected varieties - the majority of which must probably be regarded as landraces - appeared in general cultivation (McIntosh 1927). Extensive reviews of these older varieties, many of which may be considered "base varieties" for subsequent and even present-day potato breeding, have been prepared by Ross (1958a) and Salaman (1926). Among the more important of these varieties are Rough Purple Chili, giving rise to Garnet Chili, from which many well-known varieties originated (Earlaine, Epicure, Frühe Rosen (syn. Early Rose and

Tidlig Rosen (Denmark)), Oberarnbacher Frühe, Russet Burbank), Ackersegen, Gineke, Sebago (race nonspecific resistance to late blight); Jubel, Hindenburg, Saphir, Snowdrop (syn. Witchhill; wart resistant); Jubel, Menominee and Ontario (common scab resistant), Aquila, Bona, Flava and Shamrock (leafroll resistant); Bountiful, Saco, Saphir, Tawa, Villareola (PVX resistant); Bison, Fanal, Frühbote, Maritta, Sieglinde (PVY resistant) and others (Baerecke 1958; Frandsen 1958; Ross 1958a,b, 1979; Salaman 1926).

Breeding for specific resistance characters is often initiated by sudden problems such as outbreaks of diseases and pests or the discovery of specific resistance. Thus attempts to incorporate resistance to late blight from S. demissum were commenced after the disastrous blight epidemics of the 1840s (Hawkes 1958a), and the recognition and spread of wart disease around 1910 initiated breeding for resistance to this disease (Salaman 1926). Demonstration of resistance to cyst nematodes in the 1950s gave rise to efforts in breeding for resistance against these pests (Ellenby 1952, 1954; Dunnett 1957), and more recently, serious gangrene problems in Europe in the 1970s initiated screening of varieties for resistance to this complex disease (Jellis 1975; Wellving 1976) at many breeding stations. In spite of the extremely diverse genetic variability available, the total gene pool has contributed only slightly to the widening of the genetic basis of cultivated varieties, probably due to the need for many generations of backcrosses to obtain modern varieties fulfilling the demands of modern agriculture, industry and consumers (see Glendinning, this volume).

The 1985 issue of the 'Index of European Potato Varieties' (Stegemann & Schnick 1985) lists 627 cultivars. Though as many as 391 (62%) have primitive cultivars or species other than S. tuberosum ssp. tuberosum in their pedigrees, only 10 other primitive cultivars or species are listed viz: S. acaule (30 varieties), S. demissum (336), S. microdontum (1), S. phureja (29), S. microdontum ssp. gigantophyllum (syn. S. simplicifolium) (1), S. sparsipilum (1), S. spegazzinii (7), S. stoloniferum (32), S. tuberosum ssp. andigena (300), S. vernei (36), and Chiloense (29), a Chilean form of S. tuberosum ssp. tuberosum. Of the varieties listed, 431 (69%) are reported to possess resistance to potato wart, 22 varieties of which are resistant to more races than race 1 (Dahlem D_1), 165 (26%) of the varieties are resistant to cyst nematodes (25 with resistance to more pathotypes than pathotype Ro1, including 5 varieties reported to have resistance to Globodera pallida), and 280 (45%) varieties with resistance to late blight, of which 165 varieties are reported to possess race nonspecific resistance. It is reasonable to believe, however, that a much wider range of genetic resources are in current use in various breeding programmes.

GENE PRESERVATION

Access to genetic resources is a prerequisite for their utilization in the production of new varieties. Not only preservation of endangered material is desirable however, but also the evaluation and exploitation of its potential. It is generally impossible within the scope of a practical breeding institute to keep more than a small working collection of material, thus it must rely on other sources when the need arises (Hawkes 1971).

This has been realized for a long time, and the worldwide network of gene banks has been extended continuously so that it now comprises about 50 institutions holding different crop genetic resources (Anon. 1983; Damania & Williams 1980), which cooperate with international organizations such as the Food and Agriculture Organization of the United Nations (FAO), the International Board for Plant Genetic Resources (IBPGR), founded in 1974 by the Consultative Group on International Agricultural Research (CGIAR) to coordinate existing national, regional and crop specific international conservation efforts, the European Cooperative Programme for Conservation and Exchange of Crop Genetic Resources (ECP/GR) founded in 1980, and the consultative committees of EUCARPIA (Anon. 1981, 1984; Hawkes 1985; Kåhre 1985.

The major institutions holding genetic resources of the potato are (Damania & Williams 1980):

- Estación Experimental Regional Agropecuaria (INTA), Balcarce, Argentina
- EMBRAPA-UEPAE de Brasília, Brazil
- Institut für Kartoffelforschung, Gross-Lüsewitz, D.R. Germany
- Dutch-German Potato Collection, Braunschweig, F.R. Germany
- International Potato Center (CIP), Lima, Peru (WPC -World Potato Collection; IBPGR-base collection)
- Commonwealth Potato Collection (CPC), Pentlandfield, Scotland
- Inter-Regional Potato Introduction Station (IR-1), Sturgeon-Bay, Wisconsin, USA
- N.I. Vavilov Institute of Plant Industry, Leningrad, USSR.

The overall duties of gene banks for each particular crop comprise:

- exploration and collection
- identification and documentation
- preservation and recovery

- evaluation
- distribution.

Methods of storage

Once collected, identified and documented, the material needs to be stored in such a way that it can be made available to breeders and other users on request.

The material may be maintained sexually as true seed and pollen or asexually as clones through tissue culture or through traditonal propagation as tubers. Preservation of germplasm through sexual propagation as true seed is less laborious and costly than the preservation of vegetatively propagated material. In addition it is easier to maintain the material free of pathogens this way, as only a few virus diseases are known to be seed-transmitted, and to distribute it with little trouble from quarantine restrictions. Furthermore true seeds and pollen with a low moisture content can be kept at low temperatures for many years (Towill 1983, 1984). Preservation as true seeds and pollen is only practicable, however, when it is the total genetic diversity which needs to be maintained.

Since the potato is highly heterozygous it is advisable to maintain the exact genotype of the primitive cultivars, some of the highly selected clones, and breeding selections and varieties. For this purpose preservation as tissue culture by *in vitro* methods, with or without pathogen elimination and with careful evaluation regarding possible alterations in karyotype, has attracted some attention as an alternative to traditional propagation as tubers. Long term storage, for as long as 2-3 years without transplanting, and distribution of vegetatively propagated material may be achieved through shoot-tip (meristem) cultures, cultures of nodal cuttings or *in vitro* tubers (micro-tubers), and storage over a prolonged period may become possible by cryo-preservation (Estrada *et al.* 1983; Grout & Henshaw 1978; Pett & Thieme 1981; Roca *et al.* 1979).

UTILIZATION OF THE GENE POOL

When one discusses the utilization of the gene pool, it is just as important to be aware of possible undesirable characters as it is to know of desirable characters in a specific source. These include the late blight susceptibility of accessions of *S. acaule*, *S. chacoense*, *S. phureja* and *S. vernei*, the early blight susceptibility of *S. demissum*, the potato wart susceptibility of *S. demissum*, *S. stoloniferum* and *S. verrucosum*

(Frandsen 1958; Soest et al. 1984) and the high glycoalkaloid content in S. chacoense (Zitnak & Johnston 1970).

Description and evaluation

It is necessary to provide sufficient description and screening or evaluation data on desirable as well as undesirable characters for the materials contained in gene banks in order to attract interest from potential users (Soest 1983a). Thus descriptor lists, containing not only index and characterization data but detailed and broad evaluation data, with information on the methods used to obtain these, is necessary to stimulate and facilitate the assimilation and communication of plant genetic resources information, and to turn "gene bank collections" into "working collections". For the practical breeder, expression of disease resistance characters on, for instance, the 1-9 scale is often desirable in addition to a code for a particular form of resistance or susceptibility, such as hypersensitivity (H), immunity (I), resistance (R), susceptibility (S) and tolerance (T), and to information on characters such as resistance to specific races, pathotypes or virulence groups of potato wart and cyst nematodes. Detailed information and retrieval systems, to a great extent based on common descriptors recommended by IBPGR, have been prepared for most gene bank collections (Huaman et al. 1977; Seidewitz 1976; Soest & Seidewitz 1981).

A natural progression of the work of most gene banks is further evaluation, or pre-utilization ("pre-breeding") of the material collected and preserved into a semi-finished state, which will require a greater or lesser number of crosses and/or comprehensive selection for a marketable variety to emerge. This undoubtedly leads to a closer association between gene banks and gene bank users and the efficient utilization of genetic resources materials. As stated at the first 'Planning Conference on Utilization of Genetic Resources' in 1974 (Anon. 1974) "breeders do not so much lack genetic resources as the ability to manage efficiently the existing resources". As mentioned by Ochoa and Schmiediche (1983), breeding efforts could concentrate on the species most closely related to the cultivated potato, to facilitate crossing and minimize the number of backcrosses necessary.

Utilization of genetic resources may be speeded-up by the building up of "resistance columns" (breeding populations with single or combined resistances; Anon. 1974; Ochoa & Schmiediche 1983), by the formation of new gene pools (Glendinning 1975; Simmonds 1969), by population breeding (Anon.

1980; Mendoza & Rowe 1977), and by the use of cell and tissue culture techniques to increase the efficiency of specific steps in the breeding efforts (Hermsen 1983). Particularly noteworthy are: 1) the possibilities for heterosis in *Solanum tuberosum* ssp. *tuberosum* x *S. tuberosum* ssp. *andigena* hybrids (Cubillos & Plaisted 1976; Mendoza & Haynes 1974; Tarn & Tai 1977) and *S. demissum* derivatives (Hermsen 1977), 2) the overcoming of crossing barriers by means of bridging species (Hermsen & Ramanna 1973; Soest 1983b), the hybridization of *S. etuberosum* (nontuberiferous) and *S. pinnatisectum* for leafroll resistance (Hermsen & Taylor 1979), 3) attempts to establish protoclonal variation of known varieties such as Bintje, King Edward, Maris Bard, Russet Burbank and others (Jellis *et al.* 1984; Secor & Shephard 1981; Thomas *et al.* 1982), 4) the variety Yukon Gold, originating from diploid (*S. phureja*) x dihaploid (Katahdin) breeding (Johnston & Rowberry 1981) and 5) the frost tolerant, tetraploid hybrid Acaphu, a cross between tetraploid *S. acaule* and diploid *S. phureja* using unreduced gametes of the latter (Estrada 1984) and the endosperm balance number (EBN) hypothesis (Johnston *et al.* 1980).

International transfer of genetic resources

The mode of storage of potato material in gene banks and other collections is related to the subsequent transfer of the material. A prerequisite for unimpeded international use of genetic resources is that the material can be freely exchanged between countries without constituting a risk for the dissemination of pests and diseases. The concept of pests and diseases of the potato in relation to the international transfer of genetic resources has been reviewed by French, Jatala and Turkensteen (French *et al.* 1977) and Shepard (1977).

Absolutely pathogen-free material is not needed for the preservation *per se* of the material in collections, as a certain degree of recovery and indexing of possible infections normally must be carried out at intervals and prior to the shipping of any material. For restricted use of such material for breeding (pollination) purposes, attention need only be paid to the very few seed-transmitted diseases such as potato spindle tuber viroid (Hunter *et al.* 1969), Andean potato latent virus (Jones & Fribourg 1977) and potato virus T (PVT; Jones 1982; Salazar & Harrison 1978).

Because of the risks of disseminating harmful pests and diseases, international transfer of plant material is very often restricted by import regulations which prevent the introduction of new germplasm, particularly of vegetatively propagated crops. For the European Common Market (EEC) countries, import of potatoes is regulated by the EEC Council Directive 77/93/EEC of 1977 on protective measures against the introduction into the Member States of organisms harmful to plants and plant products. Derogations from this directive are permitted in respect of potato breeding material according to conditions detailed in the Commission Decision 80/862/EEC of 1980, which authorizes Member States to import quantities of true seed, tissue culture, rooted plants, unrooted cuttings or tubers having regard to the available quarantine testing facilities, as the material introduced must be subjected to official post-entry quarantine testing. Tests executed on each unit of the material include those for Andean potato latent virus, Andean potato mottle virus, potato black ringspot virus, potato spindle tuber viroid, potato virus T, non-European strains of PVA, PVM, PVS, PVX, PVY, leafroll, and *Corynebacterium sepedonicum*. In addition, if symptoms of any disease are observed during visual examination, the causal organism(s) are isolated and identified.

Listed in order of increasing risk of infection by pathogens, potato germplasm is usually available as true seeds, *in vitro* plantlets (nodal cuttings or shoot-tip cultures), rooted cuttings, *in vitro* tubers ("micro-tubers") or tubers. An increasingly popular method of transferring potato germplasm is in the form of *in vitro* plantlets and in connection with their establishment and preparation for shipment, modern and accurate techniques for disease indexing and elimination are available, including the nucleic acid spot hybridization technique (NASH) for the detection of potato spindle tuber viroid, ELISA technique and latex agglutination test (LAT) for the detection of viruses and the fluorescent antibody techniques (IFAS and FAS) for the detection of bacteria and other pathogens (Diener *et al*. 1983; Slack & French 1983). These methods, when combined with specified certification, should facilitate freer access of material to nationally approved plant breeding and research institutes.

ACKNOWLEDGEMENTS

The author is indebted to J.P. Hjerting, Copenhagen Botanic Garden, Denmark, and M. Umaerus, Swedish University of Agricultural Sciences, Uppsala, Sweden for critical comments on the manuscript, and to J. Hockenhull, The Royal Veterinary and Agricultural University, Copenhagen, Denmark for linguistic correction.

REFERENCES

Anon. (1974). International Potato Center. Report of the Planning Conference on Utilization of the Genetic Resources of the Potato. CIP, Lima, Peru.

Anon. (1980). Report of the Planning Conference. Utilization of the Genetic Resources of the Potato. III. International Potato Center (CIP), Lima, Peru.

Anon. (1981). Consultative Group on International Agricultural Research. Crop Genetic Resources. International Board for Plant Genetic Resources. Rome, Italy.

Anon. (1983). European Cooperative Programme on Conservation and Exchange of Crop Genetic Resources. Directory of European Institutions holding Crop Genetic Resources Collections. FAO, Rome.

Anon. (1984). The ECP/GR. An Introduction to the European Cooperative Programme for the Conservation and Exchange of Crop Genetic Resources. UNDP-IBPGR, Rome, Italy.

Anon. (1985). International Potato Center. Annual Report CIP 1984. Lima, Peru.

Ayers, G.W. (1952). Studies on Verticillium wilt. Am. Potato J., 29, 201-5.

Baerecke, M.-L. (1958). 2. Blattrollresistenzzüchtung. In Kartoffel. Handbuch der Pflanzenzüchtung. 2. Aufl., III. Band, ed. W. Rudorf et al., pp. 97-106.

Bagnall, R.H. (1972). Resistance to potato viruses M, S, X and the spindle tuber virus in tuber-bearing Solanum species. Am. Potato J., 49, 342-8.

Boyd, A.E.W. (1952). Dry rot disease of the potato. VI. Varietal differences in tuber susceptibility obtained by injection and riddle-abrasion methods. Ann. Appl. Biol., 39, 339-50.

Brown, C.R., Salazar, L., Ochoa, C., Chavez, R., Schilde-Rentschler, L. & Lizarraga, C. (1984). Ploidy manipulation of a new source of resistance to PLRV from Solanum acaule. EAPR Abstracts of Conference Papers. 9th Trienn. Conf. of the European Association for Potato Research. Interlaken, pp. 288-9.

Busch, L.V. (1966). Susceptibility of potato varieties to Ontario isolates of Verticillium albo-atrum. Am. Potato J., 43, 439-42.

Canto-Saenz, M. & de Scurrah, M. (1977). Races of the potato cyst nematode in the Andean region and a new system of classification. Nematologica, 23, 340-9.

Cubillos, A.G. & Plaisted, R.L. (1976). Heterosis for yield in hybrids between S. tuberosum ssp. tuberosum and S. tuberosum ssp. andigena. Am. Potato J., 53, 143-50.

Damania, A.B. & Williams, J.T. (1980). Directory of Germplasm Collections. II. Root Crops. International Board for Plant Genetic Resources, IBPGR Secretariat, Rome.

Dambroth, M. & Schittenhelm, S. (1984). Nutzbarmachung von Wildarten und Primitivformen der Kartoffel für die praktische Pflanzenzüchtung. Der Kartoffelbau, 4, 145-9.

Dearborn, C.H. (1969). Alaska Frostless, an inherently frost resistant potato variety. Am. Potato J., 46, 1-4.

Desborough, S. & Wieser, C.J. (1972). Protein comparisons in selected Phureja-Haploid Tuberosum families. Am. Potato J., 49, 227-33.

Diener, T.O., Salazar, L.F., Owens, R.A. & Cross, D.E. (1983). New potato spindle tuber test: implications for the future. In: Research For the Potato in the Year 2000, ed. W.J. Hooker, pp. 71-6. Lima: International Potato Center

Dunnett, J.M. (1957). Variation in pathogenicity of the potato root eelworm (Heterodera rostochiensis Woll.,) and its significance in potato breeding. Euphytica, 6, 77-89.

Dziewonska, M.A. & Ostrowska, K. (1978). Resistance to potato virus M in certain wild potato species. Potato Res., 21, 129-31.

Ellenby, C. (1948). Resistance to the potato-root eelworm. Nature, 162, 704.

Ellenby, C. (1952). Resistance to the potato eelworm, Heterodera rostochiensis Wollenweber. Nature, 170, 1016.

Ellenby, C. (1954). Tuber forming species and varieties of the genus Solanum tested for resistance to the potato-root eelworm. Euphytica, 3, 195-202.

Estrada, R., Schilde-Rentschler, L. & Espinoza, N. (1983). In vitro storage of potato germplasm. In: Research for the Potato in the Year 2000, ed. W.J. Hooker, pp. 80-1. Lima: International Potato Center.

Estrada, R.N. (1984). Acaphu: a tetraploid, fertile breeding line, selected from an S. acaule x S. phureja cross. Am. Potato J., 61, 1-7.

Evans, K. & Stone, A.R. (1977). A review of the distribution and biology of the potato cyst-nematodes Globodera rostochiensis and G. pallida. Pest Articles and News Summaries, 23, 178-89.

Focke, R. (1955). Rhizoctonia-Resistenzprüfung an Sämlinger einiger Wild- und Kulturkartoffeln. Der Züchter, 25, 138-40.

Frandsen, N.O. (1958). IV. Grundlagen und Methoden der Züchtung. A. Resistenzeigenschaften und ihre Vererbung. 1. Resistenzzüchtung gegen pilzliche und bakterielle Krankheiten der Kartoffel. In Kartoffel. Handbuch der Pflanzenzüchtung. 2. Aufl., III. Band, ed. W. Rudorf et al., pp. 71-97.

Frank, J.A., Leach, S.S. & Webb, R.E. (1976). Evaluation of potato clone reaction to Rhizoctonia solani. Plant Dis. Reptr., 60, 910-2.

French, E.R. et al. (1977). Potato (Solanum spp.): Fungi, bacteria, and nematodes. In Plant Health and Quarantine in International Transfer of Genetic Resources, ed. W.B. Hewitt & L. Chiarappa, pp. 225-31. Cleveland, Ohio: CRC Press.

French, E.R. & de Lindo, L. (1982). Resistance to Pseudomonas solanacearum in potato; specificity and temperature sensitivity. Phytopathology, 72, 1408-12.

Gibson, R.W. & Turner, R.H. (1977). Insect trapping hairs on potato plants. Pest Articles and News Summaries, 23, 22-7.

Gibson, R.W. (1979). The geographical distribution, inheritance and pest-resisting properties of sticky-tipped foliar hairs on potato species. Potato Res., 22, 223-36.

Glendinning, D.R. (1975). Neo-Tuberosum: new potato breeding material. 1. The origin, composition, and development of the Tuberosum and Neo-Tuberosum gene pools. Potato Res., 18, 256-61.

Grout, B.W. & Henshaw, G.G. (1978). Freeze preservation of potato shoot-tip cultures. Annals of Botany, 42, 1227-9.

Hampson, M.C. & Proudfoot, K.G. (1974). Potato wart disease, its introduction to North America, distribution and control problems in Newfoundland. FAO Plant Prot. Bull., 22, 53-64.

Hawkes, J.G. (1958a). Significance of wild species and primitive forms for potato breeding. Euphytica, 7, 257-70.

Hawkes, J.G. (1958b). I. Taxonomy, cytology and crossability. In Kartoffel. Handbuch der Pflanzenzüchtung. 2. Aufl., III. Band, ed. W. Rudorf et al., pp.1-43.

Hawkes, J.G. (1963). A revision of the tuber-bearing Solanums. Records Scottish Plant Breeding Station, pp. 76-181.

Hawkes, J.G. (1971). Conservation of plant genetic resources. Outlook on Agriculture, 6, 248-53.

Hawkes, J.G. (1978). Biosystematics of the potato. In The Potato Crop: The Scientific Basis of Improvement, ed. P.M. Harris, pp. 15-69. London: Chapman and Hall.
Hawkes, J.G. (1979). Genetic poverty of the potato in Europe. Proc. Conf. Broadening Genetic Base of Crops, Wageningen, 1978, pp. 19-27. Wageningen:Pudoc.
Hawkes, J.G. (1985). IBPGR 10th Anniversary, Leuven, Belgium. Eucarpia's collaboration with IBPGR on European genetic resources. Eucarpia Bull., 14, 54-6.
Hawkes, J.G. & Hjerting, J.P. (1983). New tuber-bearing Solanum taxa from Bolivia and northern Argentina. Botanical Journal of the Linnean Society, 86, 405-17.
Hawkes, J.G. & Hjerting, J.P. (1985). Two new wild potato species from Bolivia. Botanical Journal of the Linnean Society, 90, 105-12.
Hermsen, J.G.Th. (1977). Incorporation of new germ plasm: Wild species. In Planning Conference on the Utilization of Genetic Resources of the Potato. II. pp. 91-108. The International Potato Center (CIP), Lima, Peru.
Hermsen, J.G.Th. (1983). New approaches to breeding for the potato for the year 2000. In: Research for the Potato in the Year 2000, ed. W.J. Hooker, pp. 29-32. Lima: International Potato Center.
Hermsen, J.G.Th. & Ramanna, M.S. (1973). Double-bridge hybrids of Solanum bulbocastanum and cultivars of Solanum tuberosum. Euphytica, 22, 457-66.
Hermsen, J.G.Th. & Taylor, L.M. (1979). Successful hybridization of non-tuberous S. etuberosum Lind. and tuber-bearing S. pinnatisectum Dun. Euphytica, 28, 1-7.
Holden, J.H.W. (1981). The Contribution of breeding to the improvement of potato quality. Survey Papers, pp. 37-53. 8th Triennial Conference of the European Association for Potato Research. Munchen, Bundesrepublik Deutschland.
Hondelmann, W. & van Soest, L.J.M. (1980). Expedition in Bolivien. Der Kartoffelbau, 31, 230-2.
Howard, H.W. (1974). Factors influencing the quality of ware potatoes. I. The genotype. Potato Res., 17, 490-511.
Hoyman, W.G. (1974). Reaction of Solanum tuberosum and Solanum species to Meloidogyne hapla. Am. Potato J., 51, 281-6.
Huaman, Z., Williams, J.T., Salhuana, W. & Vincent, L. (1977). Descriptors for the Cultivated Potato, and for the maintenance and distribution of germplasm collections. Rome: International Board for Plant Genetic Resources.
Hunter, D.E., Darling, H.M. & Beale, W.L. (1969). Seed transmission of potato spindle tuber virus. Am. Potato J., 46, 247-50.
Jatala, P. & Mendoza, H. (1978). A review of the research program on the development of resistance to the root-knot nematodes Meloidogyne spp. In Developments in the Control of Nematode Pests of Potato. Report of the 2nd nematode planning conference, pp. 58-65. Lima: International Potato Center.
Jellis, G.J. (1975). The susceptibility of potato tuber tissues to infection by Phoma exigua var. foveata. Potato Res., 18, 116-9.
Jellis, G.J., Gunn, R.E. & Boulton, R.E. (1984). Variation in disease resistance among potato somaclones. EAPR Abstracts of Conference Papers. 9th Triennial Conference of the European Association for Potato Research. Interlaken, pp. 380-1.
Jellis, G.J. & Starling, N.C. (1983). Resistance to powdery dry rot (Fusarium sulphureum) in potato tubers. Potato Res., 26, 295-301.

Johnston, S.A., den Nijs, T.P.M., Peloquin, S.J. & Hanneman Jr., R.E. (1980). The significance of genic balance to endosperm development in interspecific crosses. Theor. Appl. Genet., 57, 5-9.
Johnston, G.R. & Rowberry, R.G. (1981). Yukon Gold: a new yellow-fleshed, medium-early, high quality table and french-fry cultivar. Am. Potato J., 58, 241-4.
Jones, R.A.C. (1979). Resistance to potato leafroll virus in Solanum brevidens. Potato Res., 22, 149-52.
Jones, R.A.C. (1982). Tests for transmission of four potato viruses through potato true seed. Ann. Appl. Biol., 100, 315-20.
Jones, R.A.C. & Fribourg, C.E. (1977). Beetle, contact and potato true seed transmission of Andean potato latent virus. Ann. Appl. Biol., 86, 123-8.
Keller, E.R. & Baumgartner, M. (1982). Beeinflussung von Qualitätseigenschaften durch Genotyp und Unwelt. Der Kartoffelbau, 33, 12-5.
Kort, J., Ross, H., Rumpenhorst, H.J. & Stone, A.R. (1977). An international scheme for identifying and classifying of pathogens of potato cyst-nematodes Globodera rostochiensis and G. pallida. Nematologica, 23, 333-9.
Kåhre, L. (1985). IBPGR - genebank collaborator with special emphasis on developing countries. Sveriges Utsädesförenings Tidsskrift, 95, 25-30.
Langerfeld, E. (1979). Prüfung des Resistenzverhaltens von Kartoffelsorten gegenüber Fusarium coeruleum (Lib.) Sacc. Potato Res., 22, 107-22.
Lauer, F. & Shaw, R. (1970). A possible genetic source for chipping potatoes from 40°F. Am. Potato J., 47, 275-8.
Lehmann, H. (1937). Das heutige Ausgangsmaterial für die Züchtung Phytophthora-wiederstandsfähiger Kartoffeln. Der Züchter, 9, 29-35.
McIntosh, T.P. (1927). The Potato: its History, Varieties, Culture and Diseases. Edinburgh: Oliver and Boyd.
McLean, J.G. (1952). Results of testing lines and varieties of potatoes for field resistance to Verticillium wilt. Phytopathology, 42, 26.
Mendoza, H.A. & Haynes, F.L. (1974). Genetic basis of heterosis for yield in the autotetraploid potato. Theor. Appl. Gen., 45, 21-5.
Mendoza, H. & Rowe, P.R. (1977). Strategy for potato population breeding for adaptation to the lowland tropics. Am. Potato J., 54, 488.
Ochoa, C. (1955). Species of Solanum (Tuberarium) of South America. Present taxonomic status and species used in plant breeding with special reference to Peru. Phytopathology, 45, 247-50.
Ochoa, C. (1983a). Solanum nemorosum, a new hexaploid tuber-bearing species in Peru. Am. Potato J., 60, 389-92.
Ochoa, C. (1983b). Solanum bombycinum, a new tuber-bearing tetraploid species from Bolivia. Am. Potato J., 60, 849-52.
Ochoa, C. & Schmiediche, P. (1983). Systemic exploitation and utilization of wild potato germ plasm. In: Research for the Potato in the Year 2000, ed. W.J. Hooker, pp. 142-4. Lima: International Potato Center.
Pett, B. & Thieme, R. (1981). Untersuchungen zur Depothaltung eines Kartoffelsortimentes in vitro. Potato Res., 24, 105-10.
Quevedo, D.A. et al. (1956). Estudios de resistencia a la 'Anguilula dorada' de la papa. Inf. Estac. Agric. La Molina, 347, 10-5.
Raman, K.V. & Palacios, M. (1983). Approaches to integrated control of Phthorimaea operculella. In: Research for the Potato in the Year 2000, ed. W.J. Hooker, pp. 155-6. Lima: International Potato Center.

Reddick, D. (1939). Scab immunity. Am. Potato J., 16, 71-6.
Richter, H. & Schneider, R. (1954). Untersuchungen zur Rhizoctonia-Anfälligkeit der Kartoffelsorten II. Der Züchter, 24, 264-71.
Rizvi, S.A.H. (1983). Extreme resistance to potato leafroll virus (PLRV) in seedlings of Solanum etuberosum x S. pinnatisectum (EP) with 4x chromosomes. In: Research for the Potato in the Year 2000, ed. W.J. Hooker, p. 162. Lima: International Potato Center.
Rocha, W.M., Bryan, J.E. & Roca, M.R. (1979). Tissue culture for the international transfer of potato genetic resources. Am. Potato J., 56, 1-10.
Ross, H. (1958a). II. Ausgangsmaterial für die Züchtung. In Kartoffel. Handbuch der Pflanzenzüchtung. 2. Aufl., III. Band, ed. W. Rudorf et al. pp. 43-59.
Ross, H. (1958b). 3. Resistenzzüchtung gegen die Mosaik- und andere Viren der Kartoffel. In Kartoffel. Handbuch der Pflanzenzüchtung. 2. Aufl., III. Band, ed. W. Rudorf et al., pp. 106-25.
Ross, H. (1960). Über die Zugehörigkeit der Knollentragenden Solanum-Arten zu den pflanzengeographischen Formationen Südamerikas und damit verbundene Resistenzfragen. Z. Pflanzenzüchtung, 43, 217-40.
Ross, H. (1979). Wild species and primitive cultivars as ancestors of potato varieties. Proc. Conf. Broadening Genetic Base of Crops, Wageningen, 1978, pp. 237-45. Wageningen:Pudoc.
Ross, R.W. & Rowe, P.R. (1969). Utilizing the frost resistance of diploid Solanum species. Am. Potato J., 46, 5-13.
Rudorf, W. & Baerecke, M.-L. (1958). C. Variabilität der Wertmerkmale und ihre züchterische Nutzung. In Kartoffel. Handbuch der Pflanzenzüchtung. 2. Aufl., III. Band, ed. W. Rudorf et al., pp. 138-56.
Salaman, R.N. (1926). Potato Varieties. Cambridge: Cambridge University Press.
Salazar, L.F. & Harrison, B.D. (1978). Host range, purification and properties of potato virus T. Ann. Appl. Biol., 89, 223-35.
Schultz, E.S. & Raleigh, W.P. (1933). Resistance of potato to latent mosaic. Phytopathology, 23, 32.
Secor, G.A. & Shepard, J.F. (1981). Variability of protoplast-derived potato clones. Crop Science, 21, 102-5.
Seidewitz, L. (1976). Thesaurus for the international standardisation of gene bank documentation. Forschungsanstalt für Landwirtschaft, Braunschweig-Völkenrode, F.R. Germany.
Shepard, J.F. (1977). Potato (Solanum tuberosum L.): Virus and mycoplasma-like diseases. In Plant Health and Quarantine in International Transfer of Genetic Resources, ed. Hewitt, W.B. & L. Chiarappa, pp. 233-40. Cleveland, Ohio: CRC Press.
Simmonds, N.W. (1969). Prospects for potato improvement. Rep. Scott. Pl. Breed. Stn. 1968, pp. 18-38.
Slack, S.A. & French, E.R. (1983). New disease elimination techniques in seed production programs. In: Research for the Potato in the Year 2000, ed. W.J. Hooker, pp. 25-8. Lima: International Potato Center.
Soest, van L.J.M. (1983a). Evaluation and distribution of important properties in the German-Netherlands potato collection. Potato Res., 26, 109-21.
Soest, van L.J.M. (1983b). Interspecific hybridization with the allotetraploid tuber-bearing Solanum species S. hjertingii and S. fendleri. Genetika, 15, 257-68.

Soest, van L.J.M., Schöber, B. & Tazelaar, M.F. (1984). Resistance to Phytophthora infestans in tuber-bearing species of Solanum and its geographical distribution. Potato Res., 27, 393-411.
Soest, van L.J.M. & Hondelmann, W. (1983). Taxonomische und Resistenz-Untersuchungen an Kartoffel-Wildarten und -Primitivformen der deutsch-niederländischen Sammelreise in Bolivien 1980. Landbauforschung Völkenrode, 33, 11-23.
Soest, van L.J.M. & Seidewitz. L. (1981). Evaluation data on tuber-bearing Solanum species. Inst. für Pflanzenbau und Pflanzensüchtung der FAL-Stichting voor Plantenveredeling (SVP).
Stegemann, H. & Schnick, D. (1985). Index 1985 of European Potato Varieties. National Registers, Characteristics, Genetic Data. Biologische Bundesanstalt für Land- und Forstwirtschaft. Institut für Biochemie, Braunschweig.
Stelzner, G. (1950). Virusresistenz der Wildkartoffeln. Z. Pflanzenzüchtung, 29, 135-58.
Stone, A.R. (1972). Heterodera pallida n.sp. (Nematoda:Heteroderidae), a second species of potato cyst-nematode. Nematologica, 18, 591-606.
Tarn, T.R. & Tai, G.C.C. (1977). Heterosis and variation of yield components in F_1 hybrids between group tuberosum and group andigena potatoes. Crop Sci., 17, 517-21.
Thomas, E., Bright, S.W.S., Franklin, J., Lancaster, V.A. & Miflin, B.J. (1982). Variation among protoplast-derived potato plants (Solanum tuberosum cv. "Maris Bard"). Theor. Appl. Genet., 62, 65-8.
Thurston, H.D. & Lozano, J.C. (1968). Resistance to bacterial wilt of potatoes in Colombian clones of Solanum phureja. Am. Potato J., 45, 51-5.
Torka, M. (1958). 4. Resistenzzüchtung gegen den Kartoffelkäfer. In Kartoffel. Handbuch der Pflanzenzüchtung. 2. Aufl., III. Band, ed. W. Rudorf et al., pp. 125-8.
Towill, L.E. (1983). Longevity of true seed from tuber-bearing and closely related non-tuber-bearing Solanum species. Am. Potato J., 60, 75-83.
Towill, L.E. (1984). Seed set with potato pollen stored at low temperatures. Am. Potato J., 61, 569-75.
Webb, R.E. & Buck, R.W. (1955). Reaction of some Solanum species to virus Y and Verticillium albo-atrum. Nat. Pot. Breed. Progr. 25. Ann. Rep. 1954, pp. 22-4.
Wellving, Å. (1976). Studies on the resistance of potato to storage rots caused by Phoma exigua var. foveata and Fusarium spp. The Swedish Seed Association, Svalöv, Sweden.
Winkelmann, A. (1953). Weitere Fundstellen von Biotypen des Kartoffelkrebserregers in Westdeutschland. Nachrichtenbl. Deutsch. Pflanzenschutzd. (Braunschw.), 5, 173-5.
Zitnak, A. & Johnston, G.R. (1970). Glycoalkaloid content of B5141-6 Potatoes. Am. Potato J., 47, 256-60.

GENE POOL OF MODERN POTATO VARIETIES

D.R. Glendinning

The gene pool contained in varieties grown early this century was small. It was derived from a few 16th to 18th century introductions, their genetic contributions eroded by selection and disease epidemics, plus a very few 19th century introductions (Glendinning 1983).

The pedigrees of 21 varieties issued from the Scottish Plant Breeding Station between 1934 and 1974 (Anon. 1977) have been studied. *Solanum rybinii*, *S. tuberosum* ssp. *andigena*, *S. demissum* and *S. salamanii* (a *demissum-andigena* hybrid) were used in the programme and, while the pedigrees of old varieties or imported breeding lines used in it were often incomplete, some involved the 19th century introductions Daber or Rough Purple Chili or 20th century introductions *S. rybinii*, *S. demissum* or Villaroela. One old variety, Majestic, might have another introduction in its parentage.

The genotypic constitutions of the 21 varieties were estimated. Some examples are shown in Table 1. The combined contributions of 20th century introductions never exceed 24%, and those of 19th + 20th century introductions never exceed 44%; if the suggestion regarding the parentage of Majestic is discounted the maximum is 31%. Most genes in all the varieties

Table 1. Genotypic constitutions (%) of some varieties.

	P. Crown	P. Dell	P. Falcon	P. Javelin	P. Squire	Croft
Old intros*	85	61	56	73	73	61
Daber	t	-	-	t	t	t
R.P. Chili	2	9	9	6	5	8
?(M)**	-	8	18	2	4	7
Villaroela	12	-	-	12	6	3
S. andigena	-	-	-	6	-	3
S. rybinii	-	9	12	-	5	8
S. demissum	-	12	5	-	6	9

* 16th-18th century introductions.
** Putative introduction in the parentage of Majestic
t = trace P = Pentland

trace to 16th-18th century introductions.

Coancestry analyses (Mendoza & Haynes 1974) were used to quantify relationships between, and inbreeding levels of, the varieties (Table 2). In the first analysis the old varieties used in our programme were assumed to be unrelated to each other and noninbred. This showed that little inbreeding had been involved and, while most of the issued varieties were related, few of the relationships were as close as for full sibs. In the second analysis such as is known of the pedigrees of the old varieties was included; this made little difference.

In the third analysis it was assumed that ancestors of unknown breeding were related to each other, being derived from a gene pool to which four 16th-18th century introductions had contributed unequally. This gave very different results. Almost half of the 210 pair-comparisons between the 21 varieties indicated relationships closer than for full sibs, many much closer, and the varieties appeared much more inbred. While these assumptions are arbitrary they probably resemble the truth as regards the population which survived the 19th century blight epidemics. As that population is involved in all breeding programmes, all modern varieties will be inter-related and somewhat inbred. The modern gene pool is still very limited.

The development of germplasm from primitive cultivars to near-commercial standards (Glendinning 1979) before crossing with the currently-utilized gene pool offers prospects of varieties with 50% new genes, and of major reductions in inter-relationships and inbreeding levels. Issue of many such varieties, each with a different parent, could broaden the gene pool considerably.

Table 2. Coancestry analyses

a) No. of coancestries (of 210) falling in the ranges:

	0	<.03	<.06	<.09	<.12	<.15	>.15
Analysis 1	24	58	64	34	13	14	3
Analysis 2	3	59	70	38	21	13	6
Analysis 3	-	-	6	43	61	49	51

(Standards Full sibs or parent-offspring .125; half sibs or grandparent-grandchild .063, first cousins .031)

b) No. of varieties (of 21) with inbreeding coefficients:

F =	0	<.04	<.08	<.12	<.16	<.20
Analysis 1	7	6	8	-	-	-
Analysis 2	-	12	8	1	-	-
Analysis 3	-	-	5	7	7	2

REFERENCES

Anon. (1977). (Pedigrees of SPBS varieties). Ann. Rep. Scott. Pl. Breed. Stat., 1976-77, pp. 88-97.

Glendinning, D.R. (1979). Enriching the potato gene-pool using primitive cultivars. In Proc. Conf. Broadening the Genet. Base Crops, Wageningen 1978, pp. 39-45. Wageningen:Pudoc.

Glendinning, D.R. (1983). Potato introductions and breeding up to the early 20th century. New Phytol., 94, 479-505.

Mendoza, H.A. & Haynes, F.L. (1974). Genetic relationship among potato cultivars grown in the United States. HortScience, 9, 328-30.

BREEDING STRATEGIES

POTATO BREEDING STRATEGY IN THE GERMAN DEMOCRATIC REPUBLIC

M. Scholtz

In the German Democratic Republic (GDR) the potato is used as a high quality food for human consumption, as raw material in industry, and as fodder. Table 1 gives a survey of the development of the potato crop area and *per capita* consumption from 1970 to 1984. Potatoes are produced on specialized farms with an average potato crop area of 400 ha each. They are grown predominantly on sandy to loamy soil sometimes containing a large number of stones in the topsoil. Potatoes for specific purposes are grown in different areas of the GDR. Breeding of new varieties and clonal selection, as well as multiplication of high quality potatoes (Grades S, SE and E), are concentrated in the north where there is a low degree of virus infection. Elite and certified potatoes are multiplied in the central and southern areas of the country. Ware potatoes are produced in all regions. The consumer prefers a ware potato which becomes mealy to firm-fleshed on boiling and which is suitable for processing; special varieties for processing are not required. In view of this production system and the consumer requirements, potato breeding has the following main aims (Kleinhempel *et al*. 1983), some of which are dealt with in more detail in this paper:

1) High yields in all maturity groups
2) High proportion of marketable tubers
3) Suitability for ware, starch or processing

Table 1. Potato crop area and *per capita* consumption during the period 1970-1984

Area (1000 ha)	1970	1980	1984
Ware	470	339	318
Seed	167	151	143
Starch	30	30	28
Total	667	519	489
% total arable area	14.4	10.8	10.3
Per capita consumption (Kg)	145	143	143

4) High relative resistance to the economically most important pathogens:
   a) Viruses: leafroll virus (PLRV), PVX, PVY, PVM, PVA
   b) Potato wart disease (*Synchytrium endobioticum*)
   c) Late blight on tuber and leaves (*Phytophthora infestans*)
   d) Fusarium dry rot (*Fusarium* spp.)
   e) Erwinia soft rot (*Erwinia carotovora*)
5) Resistance to *Globodera rostochiensis* (RO1) and *Globodera pallida* (Pa3)
6) Low susceptibility to external damage and black spot.

Potato breeding in the GDR is carried out by two national institutes which have several breeding stations. Every year about 1 000 000 seedlings are raised in greenhouses from 4*x* x 4*x* crosses (see Table 2). In addition to their own breeding material, the breeders obtain from the Division for Basic Material of the Institute of Potato Research Gross-Lüsewitz, tetraploid parents with specific characteristics which originate from a dihaploid programme. These genotypes have been pretested so that information is available on many characters and thus crosses for specific combinations of characters are possible.

Success in breeding can only be achieved if there is an efficient testing system. Therefore, the best clones from the D-clone stage onwards are subject to pretests and main tests for yield performance and other characters in different places over 3 years.

Testing for the various types of resistance starts with the harvest of the A-clones, with special attention being attached to virus resistance. The clones are tested over several years under field conditions at various sites subject to heavy virus infections. Table 3 shows results

Table 2. Breeding and testing scheme for potato varieties

Year	Generation	Multiplication	Test
1.	seedling-pot	1 plant)	maturity, type
2.	seedling-tuber/field	1 plant)	tuber shape, quality,
3.	A-clones	10-14 plants)	yield, resistance
4.	B-clones	80-120 plants)	etc. by breeder
5.	C-clones	600-800 plants)	
6.	D-clones		pretest year 1
7.	E-clones		pretest year 2
8.	F-clones		pretest year 3
9.	multiplication		main test year 1
10.	multiplication		main test year 2
11.	multiplication		main test year 3
12.	admission to national trials	500-1000 ha	

from an international collection of varieties in the years 1978/1979. It is obvious that in such a location (Bernburg) the majority of the varieties are 100% infected by severe viruses (PLRV, PVY, PVA, PVM) after 3 years.

Table 4 gives an outline of the progress obtained in breeding for relative virus resistance. Environmental conditions during production and the relatively high labour requirement for selection work make it necessary to continuously improve the relative resistance of the varieties to viruses but not to neglect the other breeding aims. Breeding work in includes not only relative virus resistance but also extreme virus resistance and hypersensitivity.

Much of the potato production in the GDR is now fully mechanized, including storage, handling and marketing. Thus, the potato tuber is exposed to a high risk of damage during the entire production and transport process, so that only genotypes with low susceptibility to damage and mature tubers have good keeping quality. For this reason, the percentage of varieties in

Table 3. Results of degeneration trials at Bernburg 1978/1979 (International collection)

Variety	Maturity group	Severe viruses % infection	
		1978 (after 2 years)	1979 (after 3 years)
Astilla	Early/very early	5	60
Marion	Early/very early	10	100
Linzer Fruhe	Early/very early	40	100
Isabell	Early/very early	28	100
Adretta	Mid-early	5	50
Amsel	Mid-early	7	75
Galina	Mid-early	0	16
Omega	Mid-early	2	60
Pola	Mid-early	5	100
Werta	Mid-early	7	100
Nicola	Mid-early	20	100
Roxane	Mid-early	52	100
Pioner	Mid-early	25	100
Norchip	Mid-early	30	100
Bake King	Mid-early	70	100
Culpa	Mid-early	2	100
Schwalbe	Mid-late/late	5	25
Libelle	Mid-late/late	2	30
Moni	Mid-late/late	5	35
Cores	Mid-late/late	17	100
Belladonna	Mid-late/late	23	100
Paterson's Victoria	Mid-late/late	27	100

maturity group (MG) 3 increased considerably and that of varieties in MG 4 declined over the last few years; late-maturing genotypes have been of no interest since 1977 (Table 5 and Altenburg & Schuman 1983). At present the number of varieties in MG 4 is slightly increasing, particularly among starch varieties.

The breeding clones undergoing pretest 1 also reflect this change (Table 6). Shortening of the growth period needs, of course, to be considered in relation to yield. The ensuing technological benefits will be accepted by the growers only if the yielding capacity of the plants does not decline, or declines only to a small extent during a shorter growth period. Table 7 shows the results of an evaluation of standard varieties in the main test for a period of 30 years. i.e. the tuber yield in relation to the growth

Table 4. Virus resistance of potato varieties (1970-1984)

Virus resistance	Number of varieties and % total					
	1970	%	1980	%	1984	%
High	13	39	12	60	17	74
Medium	15	46	6	30	6	26
Low	5	15	2	10	-	-
Total	33		30		23	

Table 5. Changes in the composition of maturity groups in the period 1970-1984

MG Maturity	Growing season (days)	Share of potato crop area 1970 %	1980 %	1984 %
1 Very early	115	18.0	14.8	8.9
2. Early	122	3.2	1.7	9.6
3. Mid-early	135	41.8	74.9	68.8
4. Mid-late	145*	31.0	8.4	12.7
5. Late	146	3.0	0.0	0.0

* Starch varieties up to 150 days

Table 6. Breeding clones in the first year of pretest (1970-1984)

Year	Maturity groups (% total)			
	1 and 2 (Early)	3	4	5 (Late)
1970	17.5	28.7	43.8	10.0
1980	30.7	48.3	21.0	0.0
1984	19.1	49.4	31.5	0.0

period. It is obvious that as a result of selection (1961-1970) tuber yield does not differ significantly in MG 3, 4 and 5, although the growth period of MG4 and 5 is 12 to 24 days longer than that of MG 3. It would appear that an important genetic improvement has been achieved here.

For a long time, selection of varieties with resistance to Globodera rostochiensis (pathotype RO_1) has been one of the aims of the breeding programme. Due to crop rotation it has not been as important as in some other countries. In recent years it has been possible, after backcrossing several times, to produce varieties that equal the nonresistant varieties in many characters.

Table 8 shows the number of nematode(N)-resistant varieties and their share of the potato crop area as well as the percentage of the N-resistant clones in pretest 1. Since 1980 the proportion of N-resistant varieties and clones has increased significantly. Our aim is to cultivate N-resistant varieties on contaminated areas every second cropping cycle. We hope the multiplication of G. pallida (Pa_3) is not enhanced by this approach. Stelter (1983) demonstrated that the cultivation of RO_1-resistant varieties promoted the mutiplication of Pa_3 more than the cultivation of varieties that were not resistant to RO_1, and observed a variety-specific behaviour.

Table 7. Tuber yield of MG 1,2,4 & 5 as a percentage of that of MG 3 (1951-1980) (Mean of standard varieties in main test)

Year	Maturity Group							
	3		1 and 2		4		5	
	Growth period (days)	%	Growth period (days)	%	Growth period (days)	%	Growth period (days)	%
1951-1960	128	100	111	88	146	111	157	116
1961-1970	133	100	115	96	145	103	159	104
1971-1980	135	100	118	88	150	103	159	103

Table 8. Resistance of potato varieties and clones to nematodes (RO_1) in pretest 1

Year	Number of varieties	Share of potato crop area	% nematode-resistant clones
1970	2	8	21.5
1980	5	10	42.0
1984	8	25	47.8

This and other results indicate that, as we see it, a selection for only nonresistant genotypes is not advisable under our conditions. Selection for genotypes resistant to Pa_3 has been included in the breeding programme for some years. As the resistance and phenotypic ratios are not as well defined for G. pallida as for G. rostochiensis, it is not to be expected that this aim will be realised within a short period of time, unless, by the application of unconventional breeding methods, the breeders obtain suitable material after only a few backcrosses. This and other problems require close cooperation between breeding researchers and breeders to achieve further improvements.

Finally, the breeding scheme (Table 2) shows clearly that seed potatoes should be available for an area of about 1200 ha when a new variety is approved. This is necessary if the new variety is to cover 3-5% of the total potato crop area in the future. With respect to varieties in MG 1 and 2 and starch varieties having a share of less than 3% in the total potato crop area, it is sufficient to provide seed potatoes for about 500 ha. As we assume that any new variety represents a genetic improvement, we are endeavouring to pass on the benefit of this improvement to practical potato growing as quickly as possible.

REFERENCES

Anon. (1971, 1981). Statistisches Jahrbuch der DDR Staatsverlag der DDR, Berlin.

Altenburg, A. & Schumann, G. (1983). Nutzung des Kartoffelsortimentes der DDR nach Gebrauchswert und Reifezeit. Feldwirtschaft, 24, 291-4.

Kleinhempel, D., Oertel, H. & Scholtz, M. (1983). Beitrag der Züchtung für die Intensivierung der Kartoffelproduktion. Feldwirtschaft, 24, 267-90.

Scholtz, M. (1984). Ergebnisse und Probleme bei der Züchtung von Sorten mit extremer Resistenz gegenuber dem Kartoffel-X-und-Y-Virus. Abstracts, 9th Trienn. Conf. EAPR, Interlaken, pp. 247-8.

Stelter, H. (1983). Zur Konkurrenz von Globodera rostochiensis, Pathotyp 1 mit G. pallida, Pathotyp 77 an Kartoffeln unterschiedlicher Resistenzeigen-schafton. Arch. Phytopathol. u. Pflanzenschutz, Berlin, 19, 381-9.

POTATO BREEDING STRATEGY IN THE FEDERAL REPUBLIC OF GERMANY

M. Munzert

INTRODUCTION

In West Germany potato breeding is carried out by private breeders. Nearly 20 breeding stations and breeders' co-operatives are concerned with the breeding of new varieties and clonal selection. In addition to this, four public institutes are working on breeding research to supply the private breeders with basic material. Separate research projects are also handled by university institutes.

The work of the potato breeders is documented in the national list of potato varieties published annually. In the 1985 edition 127 varieties are included. Five of these are for export use only and are not given detailed character descriptions. One hundred and fourteen varieties originate from German breeders; the remainder mainly come from Holland.

The multitude of breeders and varieties makes it impossible to define a uniform potato breeding strategy for West Germany or to give a complete survey of all the differences between the selection systems. In this paper only the main objectives and techniques are presented.

BREEDING OBJECTIVES

In the last 5 years, 42% of the total potato crop (7 million tonnes) has been used as ware potatoes, 29% in the processing industry and 18% for starch production, distillation and fodder (Graf & Menz 1984). Therefore all the breeding stations are producing new ware and starch varieties with intensive selection for processing quality.

Most of the ware varieties belong to the mid-early maturity group. On the other hand most starch variety breeding is within the mid-late group. Varieties suitable for processing are mostly early or mid-early. These differences in maturity for different breeding objectives correspond with the requirements of farmers for damage resistant ware varieties which mature during the first half of September and for highly efficient starch varieties which may be later maturing.

RESISTANCE TO NEMATODES

Resistance to Globodera rostochiensis, pathotype Rol, is an absolute requirement in all breeding programmes. As a rule, lack of resistance to Rol results in immediate rejection of a clone, irrespective of the other characters. As shown in Table 1, there are only two out of 52 varieties released since 1980 which are susceptible; one of these originated abroad. The percentage of resistant varieties continues to increase; at present 76 varieties or 62% are Rol resistant.

Although resistance to all pathotypes of G. rostochiensis, Rol-Ro5, is not a general requirement, most breeders have limited screening programmes. At present 10 varieties with this type of resistance are available; only one has been bred abroad.

Efforts to breed for resistance to G. pallida have, until recently, been unsuccessful because of insufficient genetic sources of resistance. The only variety partially resistant to pathotype Pa2 originates from Holland.

VIRUS RESISTANCE

A new approach to breeding for resistance to virus diseases was necessary when, in the mid 1950s, a calamity caused by potato virus Y (PVY) occurred. Since that time resistance to potato leafroll virus (PLRV), PVY and potato virus A (PVA) has had a high priority in the selection process. A variety with a susceptibility rating 7 (high) to 9 (very high) has very little chance of success. Advances during the last 10 years with regard to resistance to PLRV, PVA and PVY are shown in Table 2. Still more important is the fact that the number of varieties with combined high virus resistance is also high. At present there are 44 varieties which have a score of 1-3 (very high resistance) for PLRV, PVY and PVA.

Table 1. Grouping the national list of potato varieties according to resistance to Globodera rostochiensis

	No. of varieties			
	Susceptible	Rol	Rol-5	Total
Before 1970	23	1	0	24
1970-1974	13	8	0	21
1975-1979	8	14	3	25
1980-1985	2	43	7	52
Total	46	66	10	122

For some years attention has been given to potato virus M (PVM). This was necessary because of the introduction of gene Ry from *S. stoloniferum*, which gave resistance to PVY but also gave increased susceptibility to PVM (Munzert & Scheidt 1984). The losses caused by PVM are similar to those caused by PLRV and PVY (Hunnius 1976).

LATE BLIGHT RESISTANCE

Farmers require varieties which generally need no more than three to four spray operations with fungicides to control late blight (*Phytophthora infestans*). Table 3 shows that the majority of varieties are classified as moderately resistant (4-6) and only four varieties are classified as more susceptible than this. Likewise, varieties with high resistance (1-3) are becoming less frequent. Thus it is obvious that a strategy of "field resistance" is being followed; with varieties needing only a few sprays and allowing a degree of infestation which, hopefully, will not favour the development of new races. In the variety list 82 out of 112 varieties do not possess R-genes (Bundessortenamt 1984).

Table 2. Grouping the national list of potato varieties according to virus resistance

		No. of varieties						
Period	Score	PLRV	PVY	PVA	PLRV +PVY*	PLRV +PVA*	PVY +PVA*	PLRV+ PVY+PVA*
Before 1970	1-3	11	14	21	8	10	12	7
	4-6	7	7	2	8	8	9	9
	7-9	6	3	1	8	6	3	8
1970-1974	1-3	7	9	19	2	7	9	2
	4-6	9	8	2	11	9	8	11
	7-9	5	4	0	8	5	4	8
1975-1979	1-3	10	18	22	8	8	16	7
	4-6	12	7	3	14	14	9	15
	7-9	3	0	0	3	3	0	3
1980-1985	1-3	34	41	51	28	34	41	28
	4-6	14	11	1	20	14	11	20
	7-9	4	0	0	4	4	0	4
Total	1-3	62	82	113	46	59	78	44
	4-6	42	33	8	53	45	37	55
	7-9	18	7	1	23	18	7	23

*In the ranges 4-6 and 7-9, the number of varieties indicated are those with the relevant susceptibility score for at least one virus in the combination

SCAB RESISTANCE

Although common scab (Streptomyces spp.) is an unacceptable defect on ware and processing potatoes, moderate resistance is also desirable for starch varieties. Only 11 out of 122 varieties are in the score range 7-9 (susceptible). Since a high proportion of the potato acreage is situated on sandy soil, where scab incidence tends to be high, this is a very good achievement.

RESISTANCE TO DAMAGE

In the early 1970s damage resistance became an important objective, especially for ware and processing varieties (Fuchs 1971). Among 99 varieties tested for damage resistance 31 have a score between 1-3 (high resistance) and only eight are very susceptible (Table 3). The positive correlation between late maturity, starch content, tuber size and damage susceptibility suggests an impediment for improvement of damage susceptibility; but this difficulty is not insurmountable. Only seven out of 55 varieties released since 1975 are highly susceptible. In contrast to this 17 varieties have been scored as highly resistant (score 1-3). In future damage resistance combined with rot resistance will be even more important.

Table 3. Grouping the national list of potato varieties according to different characters

Period	Score	No. of varieties Late blight	Common scab	Damage	Yield	Flavour score	no
Before 1970	1-3	8	10	6	1	1-2	2
	4-6	14	12	16	20	3-4	13
	7-9	2	2	1	3	5	4
1970-1974	1-3	2	11	8	0	1-2	1
	4-6	18	9	13	18	3-4	15
	7-9	1	1	0	3	5	1
1975-1979	1-3	6	10	7	0	1-2	3
	4-6	19	13	17	23	3-4	13
	7-9	0	2	1	2	5	2
1980-1985	1-3	16	14	10	0	1-2	0
	4-6	35	32	14	30	3-4	31
	7-9	1	6	6	22	5	7
Total	1-3	32	45	31	1	1-2	6
	4-6	86	66	60	91	3-4	72
	7-9	4	11	8	30	5	14

QUALITY

Good cooking quality is regarded as an important varietal attribute. A variety for human consumption is required to have at least a flavour score of 5 (medium). In main trials good genotypes have frequently been rejected because of this limitation. Most varieties on the national list have scores of 3 or 4, and therefore improvements are still possible and desirable. The introduction of new sources of disease resistance mentioned above has probably slowed down advances in this sphere, but Scheidt & Hunnius (1978) demonstrated that with systematic selection it was possible to combine good quality with high yield and disease resistance.

The relatively recent objective, breeding for processing quality, was at first orientated towards finding genotypes with crisp quality. Since then chips (French fries) and dehydrated products have become more important, and these aspects have been given more consideration. There are now 21 varieties suitable for crisps, 13 for chips and 15 for dehydration but the number of varieties actually being used is substantially smaller.

YIELD

The difficult problem of combining resistance to leaf and tuber diseases and good quality has not allowed tuber and starch yield to be given the highest priority during selection. As Table 3 shows, most varieties released before 1979 have average yielding ability, but since the beginning of the 1980s a turning point can be observed: more and more varieties released are showing yielding ability far above average (Table 3). This is because a number of *Solanum* species have been introduced into the breeding programmes, including *S. demissum*, *S. tuberosum* ssp. *andigena*, *S. stoloniferum*, *S. verrucosum*, *S. spegazzinii* and *S. vernei*, for disease resistance. Both Howard (1970) and Scheidt & Munzert (1986) have demonstrated that the introduction of new species can be beneficial to yield.

CHOICE OF PARENTS

Potato breeding for the ware market is done exclusively at the tetraploid level. Only state institutes are interested in using haploids for scientific studies. A few breeders have a small haploid programme but it will still be some time before doubled haploids are produced. An important question in breeding stragegy was how to handle the *H1* gene introduced for Rol resistance. Rothacker (1958) advised the breeding of clones multiplex for *H1* in order to save time doing resistance testing. The West German potato breeding programmes succeeded without using this strategy - probably because of the strict test criteria of the Bundessortenamt. As Pfeffer & Steinbach (1982) and Swiezynski (1982) confirmed, multiplex resistance produced by selfing or by crossing seedlings which were resistant but which had many poor characters from S. tuberosum ssp. andigena, resulted in poor progeny and this could not be easily rectified by back crossing. A better method is first to produce simplex resistant genotypes with a high level of valuable characters, and subsequently to intercross these. For example, at our institute the first Andigena x Andigena crosses were made in 1965 when these Andigena parents had the same good characters as susceptible Tuberosum parents. At that time the offspring of all Andigena crosses resulted in more than 80% Ro1 resistant seedlings. In the near future it will be possible to save time spent on the rootball test at the beginning of the selection process.

Scheidt & Hunnius (1978), in their work on consumer quality, answered the question of which breeding strategy to choose for polygenically inherited quality characters. They showed that offspring only contain improved genotypes if both of the parents have good quality. Crosses with wild species - in order to introduce new resistance - have to some extent been a setback in breeding for quality.

It can be concluded that the careful choice of parents is one of the most important components of any breeding strategy. A comprehensive knowledge of the characters of the parents and careful planning of the crossing programmes are important considerations.

REFERENCES

Bundessortenamt (1984). Beschreibende Sortenliste 1985, Kartoffeln. dtv-Buchverlage Frankfurt a.M.

Fuchs, G. (1971). Untersuchungen zur Vollernteverträglichkeit der Kartoffel. Diss. Techn. Univ. München-Weihenstephan.

Graf, G. & Menz, M. (1984). ZMP-Bilanz Kartoffeln 83/84. ZMP GmbH Bonn-Bad Godesberg.

Howard, H.W. (1970). Genetics of the Potato Solanum tuberosum. London: Logos Press.

Hunnius, W. (1976). Zum Problem der Interferenzen zwischen verschiedenen Virusarten der Kartoffel. Bayer. Landw. Jb., 53, 525-64.

Munzert, M. & Scheidt, M. (1984). Sortenresistenz - Teil des Integrierten Kartoffelbaus. Der Kartoffelbau, 36, 38-40.

Pfeffer, C. & Steinbach, P. (1982). Die Züchtung von Kartoffelgenotypen mit multiplex Resistenz gegen Globodera rostochiensis Pathotyp 1. Arch. Züchtungsforsch., Berlin, 12, 287-95.

Rothacker, D. (1958). Beiträge zur Resistenzzüchtung gegen den Kartoffelnematoden (Heterodera rostochiensis Wollenweber). III Untersuchungen Uber den Einflu unterschiedlicher Kreuzungspartner auf die Ausbildung verschiedener Knolleneigenschaften bei Kartoffelkreuzungen, zugleich ein Beitrag zur Züchtungsmethodik. Der

Züchter, 28, 133-43.

Scheidt, M. & Hunnius, W. (1978). Ein Beitrag zur züchterischen Verbesserung der Qualität von Speisekartoffeln. Bayer, Landw. Jb., 55, 793-816.

Scheidt, M. & Munzert, M. (1984). Resistenzzüchtung gegen den Kartoffelnematoden (*Globodera rostochiensis*) - Entwicklung und Stand der Arbeiten an der Landesanstalt. Bayer. Landw. Jahrb., 63, (in press).

Swiezynski, K.M. (1982). Parental line breeding in potatoes. Acta biologica Jugoslavica, 15, 243-56.

POTATO BREEDING STRATEGY IN THE NETHERLANDS

J.P. van Loon

INTRODUCTION

From the title, the reader of this chapter may expect to get the formula for the Dutch breeders' success. However, there never has been a specific breeding strategy formulated by the fundamental breeders although, in the past, possibilities for future work have been published (Dorst 1963; Anon. 1980). The viewpoint put forward in this paper is that of a practical breeder, and so runs the risk of being one-sided.

The paper has been divided into two sections:

a. the past, because this has led to current achievements,

b. the future, what is desirable and what new opportunities will there be?

THE PAST

Potato breeding in the Netherlands has now been practised for about a century. In that period breeding has developed into a fully-fledged profession. Nevertheless, the basic principles of breeding are still the same - creating variation and selection.

The beginning in the Netherlands

There was no obvious date when breeding commenced. In several countries, during the whole of the last century, naturally produced true seed was grown to prevent degeneration. This slowly evolved into purposeful crossing (Sneep 1968). In the Netherlands the work of G. Veenhuizen in 1888 is considered to be the start of potato breeding. His great influence is described by De Haan (1958).

Breeding incentives

Diseases have, from the beginning, provided a strong incentive. The late blight (*Phytophthora infestans*) epidemic in Ireland in 1845 probably gave the first impetus to breeding; this indirectly affected work in the

Netherlands (Zingstra 1983). In 1915 wart disease (Synchytrium endobioticum) was successfully controlled by a combination of resistance and cultivation measures. Oortwijn Botjes (1924) was the first to emphasize that progress can only be made if good test methods are available. For a long time, remarkably enough, no consideration was given to virus diseases although it was appreciated that healthy seed was important (Dorst 1963).

The potato cyst nematode (Globodera spp.) has provided a strong incentive. Twelve years after discovering the source of resistance, Solanum tuberosum ssp. andigena, CPC 1673, the first resistant varieties appeared on the Dutch list of varieties (1963).

Mutation breeding, either by using bud mutants (Dorst 1924) or irradiation (Dorst 1957), has been considered. Nevertheless it has hardly found acceptance in spite of comprehensive research by Van Harten (1978). The introduction of desired characteristics from wild and primitive species is not new. Veenhuizen (1924) reported introductions as early as 1896. It was not until after the Second World War that resistance was specifically looked for in wild species (Dorst 1963), initially to late blight. Breeding for quality has only received special attention since 1976 (Van Loon 1985); this resulted in cooperation between breeders and the processing industry.

The enthusiasm of some potato breeders has always been a tremendous incentive, with Geert Veenhuizen as an obvious example (De Haan 1958). Thus it is evident that there has been no question of a strategy; we may speak rather of a vague objective to which adjustments had to be made in response to changes in cultivation, processing and use.

Broadening the objectives

At first, breeding activities were only aimed at the Dutch market. It was Prof. Dr. Ir. J.C. Dorst who, working as a practical breeder in the 1930s, took the initiative to concentrate on the export trade. Since then there has been a tripartite objective: potatoes for home consumption, starch potatoes and seed potatoes for export. The breeding stations have begun to specialize in one of these markets, while also paying attention to the other sectors. In the last few years there has been an increasing tendency to combine some objectives, like ware potatoes for different countries with potatoes for the french fried industry, or starch potatoes with potatoes for the crisp industry.

Incentives from outside bodies

Both Dorst (1957) and Hogen Esch (1957) give a good overview of the advisory services, material support, financial support and other measures which promote breeding. It all started when, in 1912, the Institute for Plant Breeding (IVP) was established. Today this Institute, as a department of the Agricultural University, enjoys world-wide fame. Since 1934, NAK (General Netherlands Inspection Service) has encouraged the breeding work with premiums and a system for recompensing the breeder. The breeding work was controlled by The Plant Breeders Decree of 1941 and the Seeds and Planting Materials Act of 1967. These measures led to a large increase in the number of private potato breeders and also led to the establishment of the Commission for the Furtherance of Breeding and Investigation into New Potato Varieties (COA) in 1938 to provide starting material and advice. Since that time potato breeding has expanded enormously (Table 1), partly because of large scale selection in early generations.

The Foundation of Plant Breeding (SVP), which evolved out of IVP and was established in 1948, further developed the breeding research and became a bridge between science and practice. The right to monopolize varieties, implicit in the law of 1967, offered the possibility of total private ownership. It made the work economically more attractive for the breeder, his agent/representative and the grower. The growing of potatoes and seed potatoes intensified (Table 2). In the region where starch potatoes are grown, a foundation (TBM) was set up in 1975 to provide measures to protect potato cultivation. This foundation gives a premium to breeders who have promising new varieties in trial that possess characteristics such as resistance to potato cyst nematode, so important to that region.

Table 1. Number of breeders and total number of seedlings produced in selected years between 1934-1982 (Zingstra 1983)

Year	Number of breeders	Number of seedlings
1934	17	10 000
1940	75	50 000
1948	193	110 000
1956	243	578 000
1964	201	682 000
1972	183	792 000
1980	180	1 063 000
1982	187	1 195 000

THE PRESENT

Present organization

The organization of breeding is part of the strategy and is based on a tripartite system:

a. Good basic scientific research, mainly at the Department of Plant Breeding of the Agricultural University and the Foundation of Plant Breeding.

b. Breeding stations with adequate scientific knowledge.

c. Qualified selectors.

Governmental institutions are not allowed to breed varieties. This takes place at private breeding stations, partly organized into co-operatives. In Table 3 the breeding stations are classified according to seedling production.

The small and medium-sized stations are mainly associated with marketing companies who test and commercialize the varieties. These contacts are continually strengthened, so that the small breeder also receives breeding material and is advised and guided by the large breeder and/or the marketing company.

Objectives

As a whole, breeders devote about 30% of their resources to breeding starch potatoes and about 70% to breeding varieties of ware potatoes for export. To give more precise objectives is impossible, because during the long period of development of a variety the requirements are continually

Table 2. Total area of potatoes and area of seed potatoes in selected years between 1966 - 1984

Year	Total area of potatoes (ha)	Area of seed potatoes (ha)
1966	130 400	21 473
1972	148 500	19 559
1978	161 700	30 810
1984	160 600	31 451

Table 3. Classification of the breeding stations in 1982 according to seedling production (Zingstra 1983)

Number of breeding stations	Number of seedlings per year
137	< 2 000
38	2 000 - 15 000
9	15 000 - 70 000
3	> 70 000

being adjusted. The main objective is thus fixed afterwards and is still best described in the words of Dorst (1957): "Would I as a farmer like to grow that variety on a large scale?" Apart from this, breeders take the opportunity to select for particular characters when it is possible or necessary.

Achievement of objectives

Till now genetic variation has been obtained almost exclusively by crossing and the normal selection methods are well known, although they differ from one breeder to another. Sneep (1970) summarized the five steps in breeding as follows:

- To look for or to create genetic variation.
- To combine useful genotypes within the obtained variation.
- To select within this variation.
- To "build" the selected material into a variety.
- To maintain the variety.

A good education as a plant breeder is a prerequisite for success.

In addition to the above five steps, the newly obtained variety should be properly introduced onto the market and commercialized. To date, most breeders have virtually exclusively restricted themselves to selection within the variation obtained from crossing at random or by experience, because they lack information about the inheritance of the characters of both parents. The choice of parents, however, must be done very carefully and be based on their characters. The parental lines of the Foundation for Plant Breeding are used in special programmes; for obtaining nematode resistance, for example.

Very important factors in the success of breeding in the Netherlands are the very flexible and short lines of communication between breeders and researchers, between breeders and assessing authorities, between breeders and inspection services and between breeders and the trade. There are hardly any obstacles to consultation and exchange of information. When there are common problems, a workshop can be organized, or even a foundation can be established, within a very short time, where all parties concerned are represented and the problems can be tackled quickly and adequately. On this basis NAK and the Netherlands Potato Consultative Institute (NIVAA) have functioned for many years, and also more recently the foundation for rapid *in vitro* multiplication of potatoes (SBSA).

In field trials, the emphasis is on large-scale early generation tests, if possible with regard to specific characters, such as resistance to cyst nematode and starch content. Furthermore, the tests are done at locations where the future variety may be grown. In the past 15 years such development has taken place through commercial enterprises, particularly for export varieties. The breeding material is, on average, tested abroad from the fifth year after crossing. The introduction of promising new varieties is closely connected with this. It is done in an atmosphere of very flexible teamwork between breeders, researchers and the trade. We may safely say: "The world is our trialfield". Of vital importance in this respect is the system of varietal assessment and the introduction of breeding material onto the market, as described by V.d. Woude (1985). An indispensable part of introducing a new variety is the advice given to growers on its cultivation. The information is given to the growers associations and, where possible and necessary, to the customers via the commercial enterprises.

Problems

The biggest problem in a breeding strategy is the complex behaviour of the autotetraploid potato; with tetrasomic inheritance and multiple allelism as basic determinants of performance. The following remarks have been made over the years. Veenhuizen (1924) said about the inheritance: "As for the answer, we are left completely in the dark". Dorst (1941) stated: "Next to nothing is known about the genetic structure of our varieties", and Hermsen (1979): "Intercrossing highly heterozygous parents, implies random combinations of gametes from two highly heterogeneous gametal populations and thus is the main reason for slow progress". Still, in spite of that, progress has been made.

FUTURE PROSPECTS

In formulating a strategy, the future also has to be considered. The strategy is strongly determined by the place one occupies in breeding. In this respect I distinguish three levels: classical breeding procedures at the tetraploid level, breeding at the diploid level and genetic engineering techniques.

Classical procedures

Trip (1979) expects a successful continuation of the existing methods for the next few decades. In the near future the number of characters considered will be increased (Anon. 1980). Progress will be strongly dependent on adequate test methods; more efficient large-scale tests are needed. The necessary genetic variation will continue to be sought for in wild and primitive Solanum species. For example, breeding for resistance to blackleg (Erwinia carotovora) is possible for the practical breeder because it appears that greater variation will become available, but the feasibility of such a project depends on a test method that a breeder is able to apply. Breeding for resistance to frost is still not practicable; the best sources occur among wild Solanum species and testing is as yet rather difficult. The selection methods used so far have resulted in progress for simply inherited characters, but hardly so for polygenically inherited characters. Moreover, the effectiveness of the selection - particularly in earlier generations - is rather low because of the low heritability of many characters. A study of how to increase the efficiency of selection has been instigated at the SVP.

Breeding at the diploid level

This was started in 1961 by IVP/SVP (Van Suchtelen 1966). In the past 25 years, interest has varied, but today it shows a sharp increase and the larger practical breeders are considering expanding small-scale programmes or initiating new ones. There is great potential but understandable reservation, because of the investment required and because of the fact that in the end one has to return to the tetraploid level. The possibilities for using diploids have increased because of genotypes with the ability to produce unreduced gametes (Neele & Louwes 1985). Also the diploid is very useful in the study of the breeding value of a parental line. At the same time one is able to gain more knowledge of polygenic inheritance, heterozygosity, additive effects and multiple allelism, in short, of the genetics of the potato. Of course the diploid potato is very useful for crossing with many wild diploid species and as it produces unreduced gametes, it opens the way to producing potatoes from true seed (TPS). And finally, practical breeders are already considering somatic hybridization, and so we arrive at the third level.

Genetic engineering

The first two levels, continuation and extension of the conventional methods and the use of diploids can be used by the practical breeder. The third level, genetic engineering, is still at the stage where techniques and fundamental processes are being studied. The breeder is interested in the practical application of these techniques and the value they may have in improving the efficiency of current breeding methods.

It is hard to predict the future contribution of genetic engineering *in vitro* to potato breeding. The main contribution may be expected to be the creation of new genetic variation. It should be emphasized, however, that the laborious testing of newly created material along classical lines will remain indispensable. Genetic engineering aims at obtaining genetic variation asexually. The method nearest to conventional breeding is the creation of somaclonal variation which is being done in the Netherlands at ITAL. Sceptics think that this is a modern variation of mutation breeding. One point is certain, to test somaclones is more laborious than to create them. Closely related to this is selection *in vitro*. If usable methods of selection for interesting characters can be found, this way of working can be quickly put into practice. Genetic transformation is not yet ready for practical application. At several Dutch universities and institutes, research on methodology is in progress. Somatic hybridization creates more possibilities. However fusion with the aim of developing varieties requires a large programme at the diploid level, experience with large-scale fusion and good selection procedures for testing the agricultural value.

The experienced breeder has great reservations concerning these developments as well as about the investments required. He also knows, as no one else, that disturbance, by whatever method, of a well-balanced genetic whole will result in little gain.

Will genetic engineering enable us to selectively remove the deficiencies inherent in all varieties? Perhaps, but most likely not before the year 2000. Related to genetic engineering is the technique of micropropagation. In practice, this rapid *in vitro* multiplication method is being used on a large scale for the production of high-grade basic seed. Another application in the Netherlands is storing germplasm *in vitro* centrally through RIVRO.

CONCLUSION

Describing a general breeding strategy in the Netherlands is impossible. The breeding of varieties is entirely in the hands of a great number of private breeders who each have their own strategy. When asked for information, they describe their strategy in vague terms. However, without exception, they are keen to meet the ever changing demands of the users of Dutch varieties throughout the world. During the work they continuously adjust the breeding objective. The same thing can be said of the breeding methods. The application of large-scale early-generation tests for nearly all characters is very important. Every breeder examines new techniques critically, for economic reasons, but when an improvement can be demonstrated, he does not hesitate to adjust the working method quickly and efficiently, either individually or cooperatively. A further improvement of conventional methods can be expected in the near future.

This is what we mean by our general breeding strategy: to be highly alert to all new possibilities for improving our complex trade and to act accordingly.

ACKNOWLEDGEMENTS

The author wishes to thank Prof. J.G.Th. Hermsen and several colleagues for critical comments on the manuscript.

REFERENCES

Anon. (1963-1985). Beschrijvende Rassenlijst voor Landbouwgewassen. Jaarlijks uitgegeven bij Leiter-Nijpels B.V., Maastricht.

Anon. (1980). Ontwikkelingsvisie Aardappelen 1982-1986. DLO, Wageningen: Nederlandse Aardappelassociatie.

Dorst, J.C. (1924). Knopmutaties bij de aardappelplant. In: Voordrachten op den tweede aardappeldag, Wageningen, 2 en 3 juli 1924, pp. 3-14.

Dorst, J.C. (1941). De Plantenveredeling op brederen grondslag. Wageningen: H. Veenman.

Dorst, J.C. (1957). Een kwarteeuw plantenveredeling. In Tussen ras en gewas, pp. 121-32. Nederlandse Algemene Keuringsdienst voor Landbouwzaaizaden en Aardappelpootgoed (NAK), Wageningen.

Dorst, J.C. (1963). Een blik terug, een blik vooruit. Landbouwk. Tijdschrift, 75, 344-51.

Haan, H. de (1958). Geert Veenhuizen (1857-1930), The pioneer of potato-breeding in the Netherlands. Euphytica, 7, 31-7.

Harten, A.M. van (1978). Mutation breeding techniques and behaviour of irradiated shoot apices of potato. Thesis, Landbouwhogeschool Wageningen. Wageningen: Centre for Agricultural Publishing and Documentation.

Hermsen, J.G.Th. (1979). New approaches in potato breeding. In Plant Breeding Perspectives. ed. J. Sneep et al., pp. 153-9. Wageningen: Centre for Agricultural Publishing and Documentation.

Hogen Esch, J.A. (1957). De bevordering van het kweken en het onderzoek van nieuwe aardappelrassen. In Tussen Ras en Gewas, pp. 153-60. Wageningen: Nederlandse Algemene Keuringsdienst voor Landbouwzaaizaden en Aardappelpootgoed (NAK).

Loon, J.P. van (1985). Vele aardappelrassen, nog steeds een oude methode. Zaadbelangen, 39, 86-9.

Neele, A.E.F. & K.M. Louwes (1985). Kweken met diploide aardappels: toekomst voor de aardappelveredeling. De Pootaardappelwereld, 38, 10-4.

Oortwijn Botjes, J. (1924). Aardappelwratziekte. In: Voordrachten op den Tweede Aardappeldag, Wageningen, 2 en 3 juli 1924, pp. 67-83.

Sneep, J. (1968). Caput: Geschiedenis en Wetten. Landbouwhogeschool Wageningen.
Sneep, J. (1970). Kandidaatscolleges Plantenveredeling. Landbouwhogeschool Wageningen.
Suchtelen, N.J. van (1966). Haploide aardappelen. In: Ontwikkelingen in de Plantenveredeling, ed. W. Lange et al., pp. 127-30. Wageningen: Stichting voor Plantenveredeling.
Trip, J. (1979). Potatoes, conventional methods. In: Plant Breeding Perspectives. Ed. J. Sneep et al., pp. 143-53. Wageningen: Centre for Agricultural Publishing and Documentation.
Veenhuizen, G. (1924). Het kweeken van nieuwe aardappelvariëteiten. In: Voordrachten op den Tweede Aardappeldag, Wageningen, 2 en 3 juli 1924, pp. 53-66.
Woude, K. v.d. (1985). Het kweken en het onderzoek van nieuwe aardappelrassen. Mededeling RIVRO no. 83.
Zingstra, H. (1983). Vijftig jaar bevordering van het aardappelkweken en het onderzoek van aardappelrassen. Wageningen: COA.

POTATO BREEDING STRATEGY IN POLAND

K.M. Swiezynski

INTRODUCTION

"Breeding strategy" is defined as the determination of breeding objectives and the means necessary to reach these objectives. In Poland the potato is a field crop of major economic importance (Swiezynski 1982). To improve it an efficient breeding strategy is required. After a short description of present-day potato breeding in Poland, some possible ways of increasing its efficiency will be considered.

THE OBJECTIVES AND STRUCTURE OF POTATO BREEDING IN POLAND

Potatoes are grown in Poland under various conditions and the crop is utilized in various ways. It follows that there must be a number of breeding objectives. Recently Polish breeders agreed that it is desirable to improve the following characters: table quality, dry matter content, resistance to viruses, resistance to blight (*Phytophthora infestans*), resistance to cyst nematode (Globodera spp.), resistance to storage diseases and adaptation to light, water-deficient soils. The breeding work and associated activities may be grouped as follows:

Potato collection Cultivars and breeding lines as well as wild and primitive cultivated species are being maintained and evaluated. It is important to make available to breeders new cultivars developed in countries with extensive potato breeding programmes.

Parental line breeding Parental lines with multiple resistance to pathogens and outstanding in some agronomic characters are being supplied to breeders (Swiezynski 1983, 1984a). Every year each breeder receives a list of potential lines and every 5 years the results of parental-line breeding are summarized (Roguski 1971, Kapsa 1977, Swiezynski 1984a). The first parental lines were supplied to breeders in 1967. At present *ca.* 30% of breeding material in Poland originates from these parental lines and the first cultivars derived from them have been developed.

Breeding new cultivars Each year over 700 000 first year seedlings are being being grown in nine state breeding stations. The best of them are being evaluated in a cycle of 12 years (Swiezynski 1984b). In the last 15 years (1970-1984), in total 52 new potato cultivars have been introduced. In 1984 these occupied 83% of the certified seed area.

Evaluation of breeding clones and cultivars All advanced breeding clones are evaluated at the Research Centre for Testing Varieties in Slupia Wielka (COBORU) using a statutory trialling system. The results are integrated and evaluated so that the Ministry of Agriculture can decide which cultivars should be added to the approved list. (Borys, this volume).

Associated research Most research is being done at the Institute for Potato Research. Some work is done also at the Agricultural Universities, in COBORU and in the breeding stations.

FACTORS INFLUENCING THE EFFICIENCY OF BREEDING WORK

The efficiency of the breeding work increases when breeding objectives are chosen wisely and are achieved more quickly or with less effort. It has generally been noticed that established potato cultivars are only slowly being replaced by new ones (Howard 1978). As there is much breeding effort, we must conclude that in general its efficiency is low. The reasons are fairly well understood (Swiezynski 1983). Three selected problems which seem to be of general interest will be discussed here.

Difficulties with the identification of superior potato genotypes

Let us consider the problem in general terms. The relative value of clones may be measured using three criteria:

1. yield
2. quality for specific purposes
3. economy of production or management of the crop.

To decide that a clone is better than existing cultivars, it is necessary:

1. to define conditions for which it may be suitable
2. to choose cultivars with which it should be compared

3. to determine properly all important characters of the clone and of the cultivars
4. to ascribe a proper weighting to all differences between the clone and the cultivars.

These requirements are difficult to meet. It is also impossible to avoid subjective judgement of the importance of differences in individual characters. The difficulties are not limited to the potato (Simmonds 1981). Attempts have been made to obtain synthetic results, ascribing fixed values to individual deviations (Perennec 1968, Domanski 1974), but this procedure does not eliminate the basic difficulties.

In some countries there is a tendency to overestimate the capabilities of varietal assessment. It may be expected that the recognition of its limitations and utilization of clearly defined and, as far as possible, uniform criteria for selection throughout the whole breeding and assessment work would result in an increase in breeding efficiency in countries in which the introduction of new cultivars is being strictly regulated.

The problem of low frequency of desirable recombinants in breeding progenies

Simmonds (1969) demonstrated that due to probability laws it is impossible to select efficiently for the many important characters for which breeding progenies segregate. The obvious conclusion is that we should attempt to produce parents, in the progeny of which desirable recombinants are more frequent.

Three procedures seem to be of special interest. Hougas & Peloquin (1958) noticed that diploid potatoes might be utilized to produce homozygotes. Mok & Peloquin (1975) demonstrated that diploids producing unreduced gametes of a certain type may transfer their favourable gene combinations to the progeny in increased frequency. Toxopeus (1953) postulated the development of tetraploid parents multiplex for desired, dominant genes.

The utilization of these suggestions in the production of parental lines is likely to result in:

a) reduced numbers of characters considered in selection, due to homozygosity, and

b) increased frequency of desired recombinants in the progeny due to homozygosity, increased frequency of desired genes and utilization of a suitable type of unreduced gametes. Let us call parents of this type second generation parental lines (SGPL).

The strategy for the future

Let us call breeding work in which parents are selected on their phenotype or on a general evaluation of their progeny, traditional breeding (TB). All, or nearly all, potato cultivars are being produced in this way. TB includes both the production of new cultivars and the production of parental lines carrying some favourable genes or gene combinations. There is a chance to increase the efficiency of TB by better selection of breeding objectives, improvement of breeding techniques, etc. However, these activities will not reduce the main difficulty associated with the rare appearance of desired recombinants in the progeny. This difficulty may be reduced by the development of SGPL.

There are apparently no major difficulties in obtaining SGPL, but the job is laborious and time consuming. According to published information their production has been started in Great Britain (Mackay 1984), in the German Democratic Republic (Pfeffer & Steinbach 1982) and in Poland (Swiezynski 1984c). It would be advantageous to compare the results obtained in individual countries. International cooperation in this area should be promoted because a big and long-term effort is needed to obtain SGPL and the results are expected to be beneficial to all potato breeders who utilize these lines.

We may also attempt to evaluate in the above terms the various semi-conventional and unconventional breeding methods which are now being developed extensively. When they reach the stage where practical application may be considered, it would seem desirable to relate them to TB or to the development of SGPL. In this way it may be easier to define how they may be applied to increase the efficiency of potato breeding.

REFERENCES

Domanski, L. (1979). An application of qualimetry to the assessment of the breeding lines of table potatoes. Ziemniak, 39-67.

Hougas, R.W. & Peloquin, S.J. (1958). The potential of potato haploids in breeding and genetic research. Am. Potato J., 35, 701-7.

Howard, H.W. (1978). The production of new varieties. In The Potato Crop, ed. P.M. Harris, pp. 607-44. London: Chapman & Hall.

Kapsa, E. (1977). Genetics and breeding of potatoes. Zesz. Probl. Post. Nauk Roln. 191, PWN Warszawa.

Mackay, G.R. (1984). Potato breeding. Ann. Rep. Scott. Crop. Res. Inst., 1983, pp. 62-79.

Mok, D.W.S. & Peloquin, S.J. (1975). Breeding value of 2n pollen diplandroids in tetraploid x diploid crosses in potato. Theor. Appl. Genet., 45, 21-5.

Perennec, P. (1968). Les modalites de l'inscription au catalogue francais des varieties de pomme de terre. Pomme de terre Franc., 327, 8-15.

Pfeffer, C. & Steinbach, P. (1982). Die Züchtung von Kartoffelngenotypen mit multiplex Resistenz gegen Globodera rostochiensis. Arch. Züchtungsforsch., 12, 287-95.

Roguski, K. (1971). Problems of potato breeding. Zesz. Probl. Post. Nauk. Roln. 118, PWN, Warszawa.

Simmonds, N.W. (1969). Prospects of potato improvement. Ann. Rep. Scott. Plant Breed. Sta., 1968-1969, pp. 18-37.

Simmonds, N.W. (1981). Principles of Crop Improvement. London: Longman.

Swiezynski, K.M. (1982). The production of potatoes and some other crops in Polish agriculture. Biul. Inst. Ziemn. 27, 5-17.

Swiezynski, K.M. (1983). Parental line breeding in potatoes. Genetica (Yug.), 15, 243-56.

Swiezynski, K.M. (1984a). Problems of potato breeding. Zesz. Probl. Post. Nauk. Roln, 273, PWN Warszawa.

Swiezynski, K.M. (1984b). Early generation selection methods used in Polish potato breeding. Am. Potato J. 61, 385-94.

Swiezynski, K.M. (1984c). An attempt to evaluate the importance of breeding potatoes at the 24-chromosome level. Abstr. Conf. Papers 9th Trienn. Conf. EAPR. Interlaken, pp. 47-8.

Toxopeus, H.J. (1953). On the significance of multiplex parental material in breeding for resistance to some diseases in the potato. Euphytica, 2, 139-46.

POTATO BREEDING STRATEGY IN THE UNITED KINGDOM

G.R. Mackay

INTRODUCTION: HISTORICAL BACKGROUND

At the turn of this century, potato cultivars were produced by interested farmers and enthusiastic amateurs. If a common breeding strategy could be perceived, it was to replace existing cultivars, whose stocks had "degenerated" during the normal course of vegetative reproduction over a number of years in agriculture (Robb 1921, 1948). As with other crop species, potato breeding gained impetus with the increase in understanding of the science of heredity and the rediscovery of Mendel's laws of inheritance. With potatoes, in particular, the practical implications of the discovery that some clones were immune to wart disease (*Synchytrium endobioticum*), and that this was a heritable trait, provided added impetus. In Scotland, for example, in the 1920s a group of farmers and merchants founded the Scottish Society for Research in Plant Breeding (SSRPB). In cooperation with the Department of Agriculture and Fisheries for Scotland (DAFS), the SSRPB set up what became the Scottish Plant Breeding Station and is now part of the Scottish Crop Research Institute (SCRI)(Gallie 1955; Simmonds 1968). Over the next half century, state support for this and other similar ventures increased and private sector involvement declined in proportion. Lack of legal protection for the breeders of cultivars contributed to a situation where plant breeding in general, and potato breeding in particular, became almost entirely a state funded activity. The situation is now changing and private sector involvement in plant breeding has increased substantially since the introduction of plant variety protection legislation. However, in the case of potatoes, with a single recent exception (see Dunnett, this volume), breeding remains a state funded activity. There are three centres of state potato breeding; the SCRI, Scotland; the Plant Breeding Institute (PBI), England and Loughgall, Northern Ireland (DANI). From this it might be deduced that outlining of the UK strategy would be fairly straightforward; but for historical and geo-political reasons, the organization and structure of Agricultural Research in the UK is complicated (Anon. 1983a).

ORGANIZATION AND STRUCTURE

Each of the three state supported programmes are independently funded; SCRI by direct grant from DAFS; PBI by the Agriculture and Food Research Council of England and Wales (AFRC) (which is further complicated by AFRC's dual funding by the Department of Education and Science (DES) and the Ministry of Agriculture, Fisheries and Food (MAFF)). Breeders at Loughgall are employed by the Department of Agriculture of Northern Ireland (DANI). The organization and funding of agricultural research and development in the UK is under review and the role of the state in the breeding of finished cultivars is one area under scrutiny (Anon. 1982, 1983b, 1985a,b). However, most of this bureaucratic complexity is concerned with policy; strategy being the responsibility of the individuals employed in the institutes. It is therefore difficult for the employee of one institute to summarize a general strategy encompassing all three. Nevertheless, several recent reviews; the institutes' annual reports and discussions with colleagues suggest that the objectives of SCRI, PBI and DANI programmes share a great deal in common and their strategies, designed to meet these objectives, differ more by degrees of emphasis than in absolute terms (Simmonds 1969; Holden 1977; Thomson 1980; Mackay 1982).

OBJECTIVES

The primary objectives of state supported potato breeding are to meet the needs of the UK potato industry, by breeding new improved cultivars for fresh use and processing within the UK. Recently, this brief has been extended to include the production of cultivars suitable for export. "Export potential" has been a particular objective of the DANI breeding programme since 1979. Northern Ireland is a seed producing area and of a total seed production of approximately 130 000 tonnes per annum, more than 10 000 tonnes are "exported" to the British mainland and over 50 000 tonnes to various foreign markets in the Mediterranean basin or elsewhere (Anon. 1984a, Howard Lee pers. comm.). In 1982, the SCRI was also commissioned by DAFS to extend its brief to include "export potential", in recognition of this important component of the Scottish Seed Potato Industry (c. 60 000 tonnes per annum).

The UK ware potato areas are dominated by rather few cultivars (Figs. 1, 2), all of which possess recognized faults, in particular susceptibility to many of the major pests and diseases. Consequently, a great deal of effort is devoted to breeding for resistance to these diseases and pests, principally to late blight (Phytophthora infestans), potato cyst nematodes (Globodera spp.) and the common viruses (Mackay 1982). However, in terms of priorities, the breeders' efforts are often constrained by the disease screening techniques available. There is broad agreement, for example, that resistance to the Erwinia complex, soft rot and blackleg, should be a major objective of breeding programmes, but suitable means of testing are still under development. It is also acknowledged that, whilst improved disease and pest resistances are highly desirable objectives, yield and quality are the prerequisites of a successful cultivar. It is important to stress that as state funded organizations, SCRI, PBI and DANI breeders are also heavily committed to strategic research of a more fundamental nature with longer term objectives, which it is not possible to encompass in this brief review.

CURRENT STRATEGIES

Most, if not all, currently available and recently bred cultivars are the product of "traditional" breeding at the tetraploid level. Traditional breeding relies almost entirely on phenotypic recurrent selection, whereby parental clones, often cultivars with complementary features, are crossed and the breeders attempt to identify recombinants amongst their progenies which combine the best features of both parents (Howard et al. 1978). The cultivar breeding programmes of SCRI, PBI, DANI and other similar organizations are therefore broadly similar (French 1980). They will probably continue to be the main source of new cultivars in the foreseeable future. There is, however, considerable room for improvement, by the application of scientific method. Research into the efficiency of early generation selection for agronomic traits (Brown et al. 1984); the purposive breeding of parental material multiplex for major gene disease and pest resistance loci (Dale & Mackay 1982; Mackay 1982); and the development and use of progeny tests for polygenically inherited resistances (Phillips & Dale 1982; Caligari et al. 1984) offer much promise for substantial improvements in the near future.

The use of novel germplasm(s) as part of, or in parallel with, a traditional breeding scheme is common to all three programmes. The incorporation of several resistances to diseases and pests from primitive cultivars (S. tuberosum ssp. andigena) and wild species (S. demissum, S.

simplicifolium, *S. vernei*, *S. stoloniferum*, etc.) has been successfully achieved, are available in modern cultivars and now form part of the Tuberosum (*s. lato*) gene pool. Nevertheless an alternative strategy to traditional methods of introgression by hybridization and backcrossing has been a feature of the SCRI programme. In this case locally adapted populations have been bred by cyclical recurrent selection from primitive 4*x* (Andigena) and 2*x* (*S. phureja*) cultivars. This strategy might enable such material to be immediately exploited as parental clones, without the time consuming need for extensive backcrossing (Simmonds 1976; Glendinning 1979; Carroll 1982). The fact that the first Neotuberosum x Tuberosum hybrid

Figure 1. Changes in percentage area of early cultivars planted in Great Britain, 1973-1984.

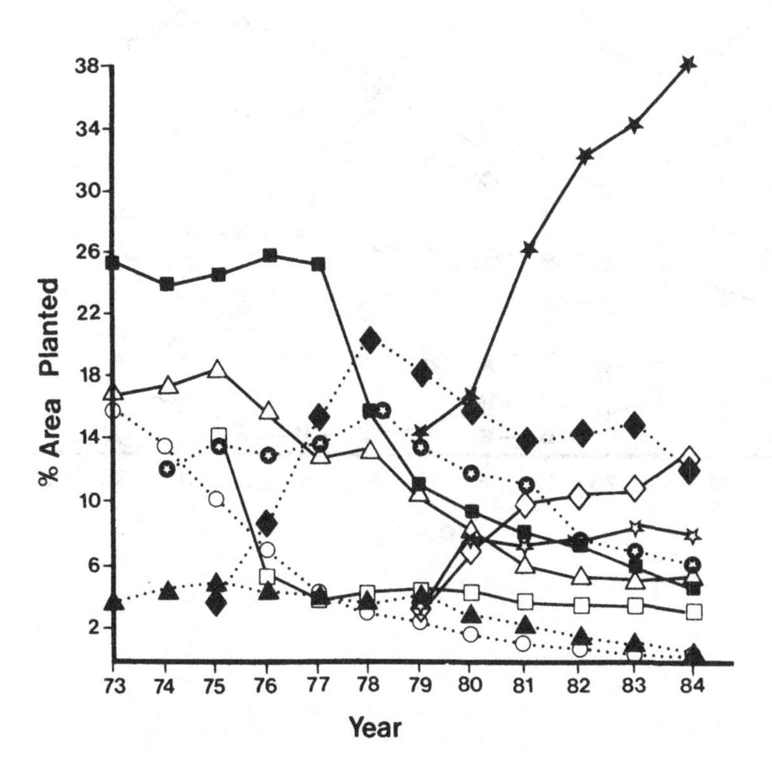

clone to enter UK National List Trials has recently been named (Shelagh) lends hope to this possibility (Anon. 1984b).

The potato's vegetative means of reproduction and the ease with which it lends itself to propagation by various tissue culture techniques, has led to its rapid use as a vehicle for a great deal of research into the use of novel techniques. For example, induced somaclonal variation may offer

Figure 2. Changes in percentage area of maincrop cultivars planted in Great Britain, 1973-1984.

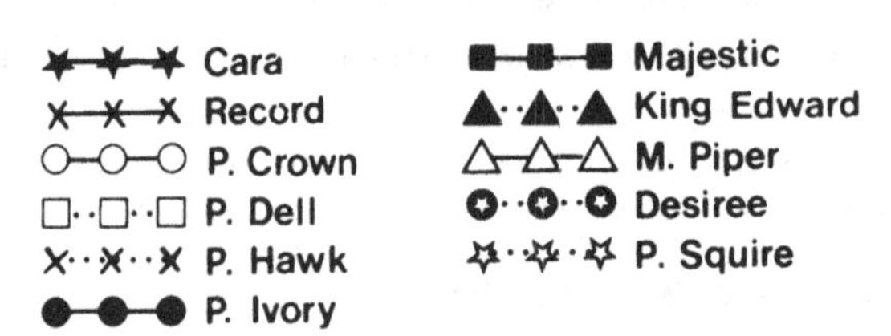

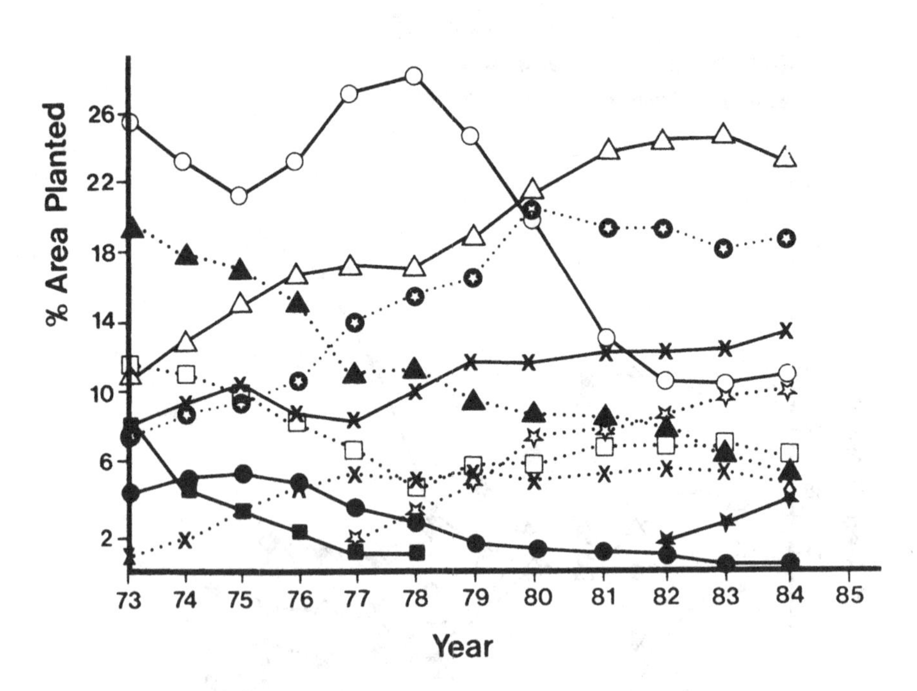

the means to improve, by degrees, existing successful cultivars with recognized weaknesses. This has been a particular strategy of the PBI since 1980 (Anon. 1981; Thomson, this volume); although both DANI and SCRI have carried out some investigative studies, their commitment is relatively minor (Howard Lee, pers. comm.). Opinions vary as to the potential of this technique, and some authorities regard its value mainly merely as a means of increasing the frequency of variants which can occur spontaneously (Sanford et al. 1984). The problems such an approach pose for those responsible for the statutory implementation of Plant Variety Rights (Distinctness, Uniformity, Stability tests) are also obvious. SCRI is investigating the use of irradiated pollen as a means of limited gene transfer (Jinks et al. 1981; Borrino et al. 1985) and has a longstanding involvement in the production and use of dihaploids, as a breeding/research tool (De Maine 1984). More recently the latter work has been extended to include studies on the production of monoploids. By combining various emergent techniques such as dihaploids and monoploid induction, tissue culture and protoplast fusion, it may be possible to more rapidly achieve the production of improved parental clones, multiplex at several disease resistant loci, and/or produce material for research purposes, which will provide a greater insight into the genetic architecture of the species. These are all elements of SCRI's longer term strategic breeding programmes.

The production of cultivars from true seed (TPS) has been mooted as a novel strategy (Anon. 1985c) but seems more likely, if successful, to be of interest to the third world; it is, however, not entirely neglected in the UK, for PBI are currently engaged on studies on TPS in association with the University of Birmingham, supported by funds from the UK Overseas Development Agency (Anon. 1985d; Jackson, this volume).

In conclusion, it seems that UK breeding strategy can be summarized as retaining and improving upon traditional methods for cultivar breeding; but insofar as finance and Government policy permits, researching and developing on as broad a front as is possible, thus keeping all options open.

REFERENCES

Anon. (1981). An. Rep. Pl. Breed. Inst., 1980, pp. 39-40.

Anon. (1982). Towards a policy for R and D in Agriculture and Food London: Her Majesty's Stationary Office.

Anon. (1983a). Organisation of R and D for the Agriculture and Food Industries, MAFF, HMSO. Dd8247890. 9377.

Anon. (1983b). Agriculture and Food Research Council Corporate Plan 1984-1988. December 1983.

Anon. (1984a). Seed Potatoes from Northern Ireland. Seed Potato Promotions (NI) Ltd. 1984.

Anon. (1984b). An. Rep. Scott. Crop Res. Inst., 1983, p. 69.

Anon. (1985a). A Long Term View. Agriculture and Food Research Council discussion paper, May 1985.

Anon. (1985b). A Strategy for Agricultural Research and Development. Department of Agriculture and Fisheries for Scotland Consultation paper, May 1985.

Anon. (1985c). True Potato Seed. In Potatoes for the Developing World, pp. 90-4. Lima: International Potato Center.

Anon. (1985d). An. Rep. Pl. Breed. Inst., 1984, pp. 48-9.

Borrino, E.M., Caligari, P.D.S., Powell, W., McNaughton, I.H. & Hayter, A.M. (1985). Cytological observations on the effects of pollen irradiation in diploid and polyploid crops. Heredity, 54, 165-70.

Brown, J., Caligari, P.D.S., Mackay, G.R. & Swan, G.E.L. (1984). The efficiency of seedling selection by visual preference in a potato breeding programme. J. Agric. Sci., Cambs., 103, 339-46.

Caligari, P.D.S., Mackay, G.R., Stewart, H.E. & Wastie, R.L. (1984). A seedling progeny test for resistance to potato foliage blight (Phytophthora infestans (Mont.) de Barry). Potato Res., 27, 43-50.

Carroll, C.P. (1982). A mass selection method for the acclimatisation and improvement of edible diploid potatoes in the United Kingdom. J. Agric. Sci., Cambs., 99, 631-40.

Dale, M.F.B. & Mackay, G.R. (1982). Breeding for PCN resistance. Nematologia, 28, 143 (Abstract).

De, Maine, M.J. (1984). Patterns of variation in potato dihaploid families. Potato Res., 27, 1-11.

French, W.M. (1980). Potato varieties and variety testing in Denmark, France, West Germany and the Netherlands. Potato Marketing Board James E. Rennie Award Report No. 9.

Gallie, R.J.L. (1955). The Scottish Plant Breeding Station, An Historical Review. An. Rep. Scott. Pl. Breed. Inst., 1954 (reprinted as an Occasional Paper).

Glendinning, D.R. (1979). Enriching the potato gene-pool using primitive cultivars. Proc. Conf. Broadening Genet. Base Crops, Wageningen, pp. 39-45. Wageningen: Pudoc.

Holden, J.H.W. (1977). Potato breeding at Pentlandfield. An. Rep. Scott. Pl. Breed. Stat., 1976-77, pp. 66-97.

Howard, H.W., Cole, C.S., Fuller, J.M., Jellis, G.J. & Thompson, A.J. (1978). Potato breeding problems with special reference to selecting progeny of the cross of Pentland Crown x Maris Piper. Rep. Pl. Breed. Inst., 1977, pp. 22-50.

Jinks, J.L., Caligari, P.D.S. & Ingram, N.R. (1981). Gene transfer in Nicotiana rustica using irradiated pollen. Nature, 291, 586-8.

Mackay, G.R. (1982). Breeding for resistance to diseases and pests. In: Producing Quality Seed Potatoes in Scotland. Bull. No. 1., Proc. Scott. Soc. Crop Res., pp. 27-36.

Phillips, M.S. & Dale, M.F.B. (1982). Assessing potato seedling progenies for resistance to the white potato cyst nematode. J. Agric. Sci., Cambs., 99, 67-70.

Robb, W. (1921). Breeding, selection and development work in Britain. In: Report of the International Potato Conference, London, ed. W.R. Dykes, pp. 27-34. London: Royal Horticultural Society.

Robb, W. (1948). Research and the Farmer. VI. Plant Breeding. Scott. Agric., 27, no. 3, 2-8.
Sanford, J.C., Weeden, N.F. & Chyi, Y.S. (1984). Regarding the novelty and breeding value of protoplast-derived variants of Russet Burbank (Solanum tuberosum L.). Euphytica, 33, 709-15.
Simmonds, N.W. (1968). The Scottish Plant Breeding Station. Scott. Agric., Spring 1968.
Simmonds, N.W. (1969). Prospects of potato improvement. An. Rep. Scott. Pl. Breed. Stat., 1968-9, pp 18-38.
Simmonds, N.W. (1976). Neotuberosum and the genetic base in potato breeding. A.R.C. Research Review, 2, no. 1, 9-11.
Thomson, A.J. (1980). Potatoes 3: Breeding for temperate agriculture. Span 23, no. 2, 70-2.

PRIVATE POTATO BREEDING IN THE UNITED KINGDOM

J.M. Dunnett

The Caithness breeding programme was initiated in 1976 by an experienced potato breeder turned Scottish seed grower (the Writer) who first linked up with a London-based exporter and then with three of the larger Scottish grower/dealers. Thus the Caithness Group, which aspires to breed new varieties and sell seed of them worldwide, could be described as a vertically integrated seed potato business on the Dutch model, the only one in the United Kingdom.

The first breeding objective was to improve upon Desiree, the Dutch red-tubered variety which is a leading maincrop in the UK, as well as in important seed potato export markets. Desiree lacks resistance to cyst nematode (*Globodera* spp.) and common scab (*Streptomyces* spp.).

Maris Piper, a leading British variety from the Plant Breeding Institute, Cambridge, has the gene H1 for resistance to *Globodera rostochiensis*, so the first cross was Desiree X Maris Piper, or DXMP in the short notation which will be developed in a self-evident way. Pentland Crown, a leading variety bred at the Scottish Plant Breeding Station near Edinburgh was used as the source of resistance to common scab in the second generation cross D.MPXPC. Up to this point, therefore, the breeding had the advantage of relying solely on three leading varieties of English, Dutch and Scottish origin.

No leading variety or even remotely commercial variety possessed a high level of resistance to *Globodera pallida*, a character which was only available in sub-commercial material ex *Solanum vernei* (denoted by V). The third generation comprised D.MP..PCXV and PCXV and the fourth and final generation is intended to be D.MP..PC...VXPC.V. Some additional material in the form of D.MPXV and PC.VXPC.V was generated and the class V parents comprised subclasses such as Vrt, Vs and Vsrt, i.e. selected primarily for resistance to and tolerance of *Globodera pallida*, or for characters such as tuber size and shape.

In general, the breeding plan could be described as a sequence of polycrosses, usually with a limited number of highly selected seedlings on one side and a larger sample of less highly selected seedlings on the other. Seed was always bulked.

The total turnover is 5000 seedlings per annum, every generation is expected to yield up to four varieties worth entering for National List Trials, and from start to finish the programme is expected to require about 20 years.

SELECTION AND SCREENING METHODS

THE EFFICIENCY OF EARLY GENERATION SELECTION

J. Brown

The most intense selection in many potato breeding programmes is carried out in the early generations. The conditions under which these generations are grown are often not typical of normal agronomic practice. The large number of genotypes that are assessed usually dictates that each genotype is grown in small plots, in many cases at a single location without replication and the "better" clones are identified by visual appraisal. In this study progenies from eight potato crosses were examined. From each cross 70 clones were grown from true potato seed in a glasshouse (GH), in single plant plots in the first clonal year (FCY) and 3-plant plots in the second clonal year (SCY). A random subsample of 25 clones per cross was also grown in 5-plant plots in the third clonal year (TCY). Subject to the availability of tubers, the FCY, SCY and TCY plots were replicated twice at two locations (BB and MURR). After harvest in each year all the plots were visually assessed independently by four potato breeders on a 1-9 scale of increasing desirability. The data presented in this report are the mean scores of the four breeders.

Correlations of mean breeders' preference between the different years'assessment were all found to be significantly greater than zero (Table 1). Scores in the GH correlated most highly with those recorded in the FCY and lowest with those recorded in the TCY. Similarly, the correlation between the FCY and SCY was larger than that between the FCY and TCY. This would suggest that the association between successive years is likely to be due,

Table 1. Correlation of mean breeders' preference, on a clonal basis, between different years. Correlations involving the TCY are based on 200 clones while all others are based on 560 clones.

	GH	FCY	SCY
FCY	0.48		
SCY	0.36	0.64	
TCY	0.30	0.43	0.76

in part, to a carry over effect of the mother tuber. If the trials in the TCY are taken as being the closest to normal agricultural practice and therefore the most accurate assessment of each clone's worth, then the GH scores showed a lower repeatability than the scores in the FCY and hence suggest that selection in the FCY would be more effective than in the GH.

In parallel with this experiment, 1000 seedlings of one of the crosses examined (C1) were grown and evaluated through our normal breeding scheme. From the original 1000 seedlings, 337 clones were selected in the GH, 59 were reselected in the FCY, 18 were reselected in the SCY and finally five clones were selected in the TCY for a further year's assessment. In the TCY the 25 clones from this cross, taken at random at the SCY stage with no selection exerted in the GH, FCY or SCY, were grown alongside the 18 clones that had survived the "commercial selection". From these 25 random clones, eight were worth consideration in a fourth year. Seedlings from the other seven crosses examined were also grown and assessed commercially in larger numbers. From amongst the clones in the experiment, 45 were considered suitable for trialling in the fourth clonal year while none of the "commercially" selected clones survived selection in the early generations. Therefore, despite the positive correlations found in Table 1 it would appear empirically that individual clonal selection in the early generations was having an effect that was worse than random.

Examination of the progeny rankings of breeders' preference (Table 2) showed that the progeny assessment was repeatable, irrespective of the environment in which the progeny was grown. It was also found that the proportion of clones that was considered to have merit at the SCY stage was directly proportional to the progeny rank. Even when progeny assessment

Table 2. Progeny rankings of breeders' preference for eight progenies grown in the glasshouse and at BB and MURR in the FCY and SCY. Ranks are based on the assessment of 70 clones per cross.

	GH	FCY BB	FCY MURR	SCY BB	SCY MURR	% Selected in SCY
C1	1	3	2	2	1	48.6 (1)
C2	4	4	6	4	6	11.4 (5=)
C3	6	7	5	7	4	11.4 (5=)
C4	3	5	3	6	5	24.3 (4)
C5	8	6	7	5	7	10.0 (7)
C6	5	2	4	3	2=	41.4 (3)
C7	2	1	1	1	2=	42.8 (2)
C8	7	8	8	8	8	7.1 (8)

was carried out with a random sample of as few as 25 clones per cross, a good indication was obtained as to the worth of each cross. If progeny assessment, rather than individual clonal assessment, was carried out in the early generations large numbers of crosses could be evaluated by examination of only a few individuals from each cross. Larger numbers of clones from the "better" crosses could then be multiplied to a level that would allow a more accurate assessment in larger plots.

PROBLEMS ASSOCIATED WITH EARLY GENERATION SELECTION OF POTATO CLONES IN WEST SIBERIA

B. Dorozhkin and B. Cadychegov

Although potato breeding strategy must take account of new developments such as *in vitro* tissue culture, somatic fusion, dihaploids and true potato seed, further improvements still need to be made in the classical techniques, including the selection of parents, the choice of crosses, and hybridization and selection of resulting progeny. This report discusses selection in early generations.

It is well known that the identification of desirable genotypes is complicated by variability in yield and other characters due to such factors as year, location, agronomy and experimental method. These environmental factors affect all stages in the breeding programme but are a particular problem in the early stages when there is no replication and only 1-10 plants per clone; nevertheless 99% of the initial population is discarded at this stage in the Omsk programme.

The influence of year, agronomy, variety and plant spacing on yield, starch content and protein content of individual tubers has been studied. Overall, genotype had a greater influence on starch content than on yield and protein content. There was a close correlation between protein content and fertilizer treatment. The correlations between first tuber generation, or A clones, and second tuber generation, or B clones, varied between r = -0.014 and r = +0.301 depending on the character (Table 1). This

Table 1. Correlation between clones selected at single-plant stage (A clones) and 10-plant stage (B clones)

Character	Correlation coefficient	
	minimum	maximum
Yield	+0.006*	+0.248
Starch content	-0.010*	+0.301
Protein content	-0.014*	+0.229

* nonsignificant, *p* = 0.05

was because of the strong influence of the environment, particularly of the year, on the various genotypes. In practice, B clones are subjected to even greater environmental variation, as virus infected plants are rogued from plots three times during the season. This affects plant spacing and hence varietal evaluation.

After the B clone stage, breeding material is sent to the sub-taiga zone, because of the high virus risk in the southern forest steppe zone of West Siberia. Here the clones are multiplied in virus-free conditions and, at the same time, evaluated under these environmental conditions.

The importance of evaluating breeding material in different environments has been stressed by a number of workers. Results from trials of three varieties of differing maturity, grown at four sites in the Omsk region for 10 years (Figure 1) showed that 64% of the variation in yield

Figure 1. Contribution of various factors to potato yield (A) and starch content (B).

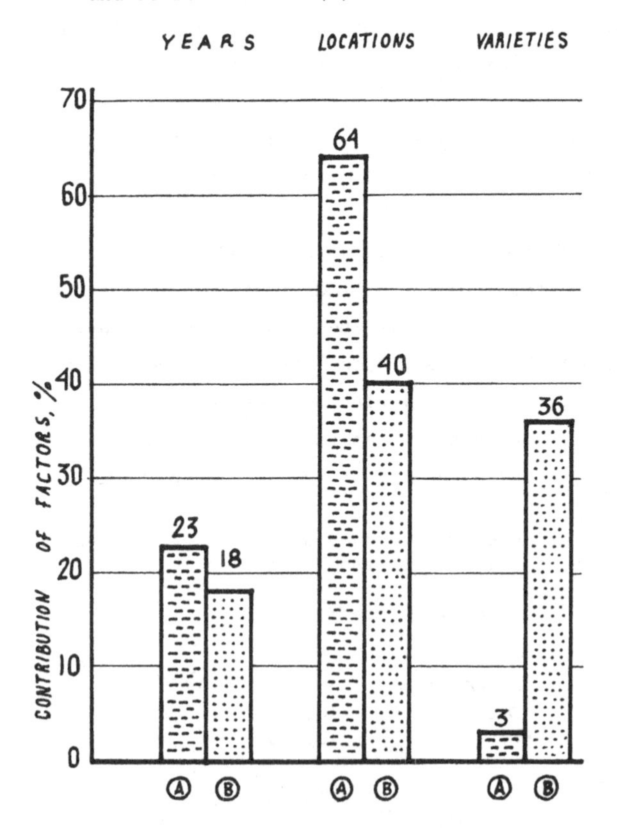

was due to site, 23% was due to year and only 3% to variety. Starch content was more influenced by variety (36%) but site was still the major source of variation (40%).

CONCLUSIONS

These results confirm that it is difficult to select for yield, starch and protein content in the early generations. Tai & Young (1984), recognizing the important contribution of year to variation in quantitative characters, proposed an intermediate 4-plant plot stage between the A (1 plant) and B (10 plant) stages. An alternative, however, is to reduce selection pressure for yield, starch and protein content in the early generations, although it is still necessary to discard selections with very low yields or starch content. This can be done by comparison with the variability of the standard control varieties.

Priority at the early stages should be given to more stable characters which can be evaluated more efficiently. These include tuber shape, skin colour, flesh colour, eye depth, stolon length, tuber uniformity and appearance and tuber abnormalities. At present, resistance to pests and diseases is often not considered until after the B clonal year. Many of these screens should be applied earlier, however, using artificial inoculation methods as described, for example, by Swiezynski (1984), Plaisted et al. (1984), Lacey et al. (this volume) and Martin (this volume).

REFERENCES

Plaisted, R.L., Thurston, M.D., Brodie, B.B. & Hoopes, R.W. (1984). Selecting for resistance to diseases in early generations. Am. Potato J., 61, 395-403.

Swiezynski, K.M. (1984). Early generation selection methods used in Polish potato breeding. Am. Potato J., 61, 385-94.

Tai, G.C.C. & Young, D.A. (1984). Early generation selection for important agronomic characteristics in a potato breeding population. Am. Potato J., 61, 419-34.

INFLUENCE OF WEIGHT OF SEED TUBERS ON SELECTION OF FIRST YEAR CLONES: PRELIMINARY RESULTS

K.M. Louwes and A.E.F. Neele

INTRODUCTION

In the Netherlands most potato breeders grow seedlings in the glasshouse in pots. The heaviest tuber per plant is harvested as seed for the first clonal generation. The seed tubers are small and moreover weight varies enormously within populations. As seed tuber weight affects the phenotype of the first year clones, the variation in weight has a negative effect on selection efficiency. This has been confirmed by the results of Blomquist and Lauer (1962), Swiezynski (1968), Brown *et al.* (1984) and Maris (pers. comm.). These workers all found that plants obtained from heavier seed tubers had better chances in plant selection but none could specify a weight interval at which tuber weight plays a less decisive role in the selection of first year clones.

Research on this selection problem will be presented in this paper.

MATERIALS AND METHODS

The influence of weight of seed tubers on the phenotype of first year clones is being studied in two similar 2-year experiments. Experiment 1 started in 1984 and ended in 1985 whilst experiment 2 started in 1985 and will be concluded in 1986. During the first experimental year, first year clones were grown in the field. Each genotype was represented by six plants (experiment 1) and eight plants (experiment 2), obtained from tubers differing in weight. For experiment 1 tuber weight ranged between 1.6-80.3 g and for experiment 2 between 1.5-77.5 g.

In both experiments a second year of testing is needed to obtain a more precise assessment of the genotypic value of each clone. Second year clones have common seed tuber sizes ranging between 30-40 mm and each genotype is represented by four plots of two plants each. Two plots are harvested in July and two in September.

For experiment 1, 216 clones out of nine populations were tested. Four varieties were included and grown in replications to assess the environmental variation. For experiment 2, 173 clones including one variety, are being used. Except for one extra population the genotypes were derived from the same crosses as for experiment 1.

RESULTS AND DISCUSSION

Regression analysis revealed a significant positive influence of weight of seed tubers of first year clones on total tuber yield, yield of tubers over 30 mm and biomass production.

Figures 1 and 2 show that there was an increase in the proportion of plants selected on appearance at harvest with increasing weight of seed tubers. Only plants with a rating from 7 to 9, on a 1-9 scale were selected. This resulted in a selection intensity of 12.6% (experiment 1) and 13.7% (experiment 2). These percentages correspond with those obtained in the first clonal generation by practical breeders in the Netherlands. In Figure 2 the selection intensity of 13.7% is indicated in order to mark the transition between tuber weights with a worse than and a better than average chance of being selected. Results from first year clones in both experiments showed that plants from tubers weighing less than 8 g have very little chance of being selected with the present selection procedure, irrespective of the genotypic value of the clone.

Regression analyses showed that 25% of the differences in tuber yield between first and second year clones are due to the weight of seed tubers in the first year. Coefficients of determination between first and

Figure 1. Distribution of weight of seed tubers of 1st year clones planted in 1985. The dark parts indicate the plants selected in each weight class.

Figure 2. Percentage of plants selected in the various classes of seed tuber weight in 1985.

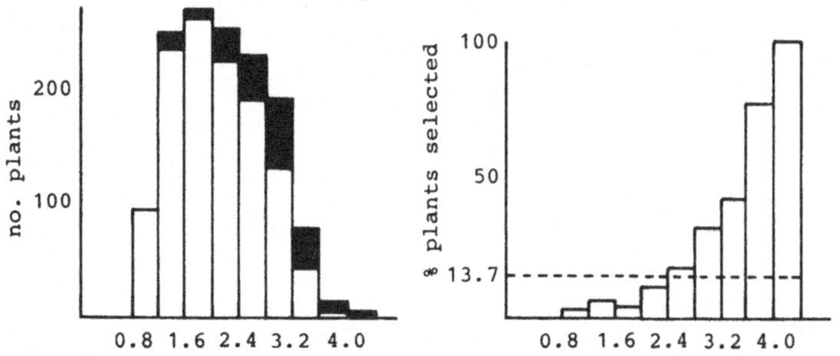

second year clones for mean tuber yield and plant appearance were calculated separately for first year clones with a tuber weight below and over 8 g. The respective values for the former group were 47 and 26% and for the latter 58 and 35%.

From the results it seems advisable to group the seedling tubers according to their weights and select the first year clones within these groups. Varieties with corresponding tuber weights should be included. To reduce differences in seed tuber weight seedlings should be grown in larger pots when fewer small tubers will be produced.

REFERENCES

Blomquist, A.W. & Lauer, F.I. (1962). First clonal generation potato progeny performance at two Minnesota locations. Am. Potato J., 39, 460-3.

Brown, J., Caligari, P.D.S., Mackay, G.R. & Swan, G.E.L. (1984). The efficiency of seedling selection by visual preference in a potato breeding programme. J. Agric. Sci. Camb., 103, 339-46.

Swiezynski, K.M. (1968). Field production of first-year potato seedlings in the breeding of early varieties. Eur. Potato J., 11, 141-49.

A JOINT CYST NEMATODE/LATE BLIGHT TEST FOR EARLY GENERATION SCREENING OF POTATO CLONES

C.N.D. Lacey, G.J. Jellis, N.C. Starling and S.B. Currell

This paper describes a technique for screening breeding material in the first clonal year for resistance to *Globodera rostochiensis*, *G. pallida* and *Phytophthora infestans*. Seedlings were grown in pots in an aphid-proof glasshouse and after harvest, when tubers with poor colours, extremely long stolons or very poor shape were rejected, tubers were stored at 5°C until dormancy had broken. Tubers were removed from store at weekly intervals and kept at room temperature to induce sprouting. One small tuber was planted on the surface of 100g of compost at 30% water content in each closed transparent plastic container, as described by Phillips *et al.* (1980). The inoculum used was a suspension of equal numbers of eggs/larvae of both *G. rostochiensis* Ro1 and *G. pallida* Pa2-3 collected from beds where they had been maintained on the appropriate differential host variety. The containers were kept in a dark store room (15 to 20°C) for 6 to 7 weeks while the cysts were forming. Recording commenced when the majority of the cysts of *G. rostochiensis* were yellow and *G. pallida* were white. A few mature cysts were brown and hence not assignable to species, while some *G. rostochiensis* cysts were still white and wrongly attributed to *G. pallida*, but as only clones with dual resistance were selected this was not of practical importance.

Recording was done by counting cysts visible through the walls of the plastic container. Resistance was determined by reference to control clones. Normally only those seedlings which produced no brown or yellow cysts and only two or three white cysts were selected. Results were checked by comparison with predicted genetic ratios and clones that passed further screens for characters such as quality and yield will be retested at a later date using conventional single pathotype washing-out tests (Goody 1963).

The plants recorded as resistant were transferred from their containers to 15cm pots and grown in a heated glasshouse. When they were about 45cm tall they were transferred to a polythene tunnel within a glasshouse. Here they were kept constantly damp with automatic intermittent overhead irrigation and sprayed with a suspension of zoospores of a complex isolate of P. infestans. Susceptibility to late blight was assessed by leaf area destroyed at set times in comparison with control clones, which behaved as predicted from their known field reactions. Examples of the results as the percentage of apparently resistant clones are given for three typical families in Table 1.

For G. pallida the figures were as expected, with a considerable increase in resistant progeny when both parents carried resistance. G. rostochiensis resistance, being based on a single dominant gene, would be expected to give about 46% resistance for the cross nulliplex x simplex and 71% for the cross simplex x simplex.

Those tubers with apparent dual resistance were passed on to the blight screen where they were scored on a 1 to 9 scale. The control clones included Cara (7) and Stormont Enterprise (8) as resisters with King Edward (4) as the most susceptible control. All clones scoring 5 or less were rejected. Overall results for 1985 are given in Table 2.

The overestimate of G. rostochiensis resistance shown in Table 1 could either be due to defects in the testing procedure, in which case susceptibles erroneously selected as resistant will be found in later re-tests, or due to reciprocal effects of the two resistances (Phillips & Trudgill 1983). The use of the container test for screening unselected progenies makes the identification of duplex clones routine and could lead to smaller population sizes without the loss of potential varieties; it is

Table 1. Progeny results for dual nematode container tests

Parent resistances	n	Percentage resistant to: G. rostochiensis	G. pallida	both
Pa x RoPa	121	66	87	51
Pa x Ro	64	68	48	26
Ro x RoPa	104	87	40	37

Table 2. Summary of nematode/late blight screen

Progenies tested	42	
Individuals tested	4967	
Dual nematode resistant	1581	(31.8%)
Successful blight tests	1369	
Blight 'resistant' (>5)	627	(45.8%)
Overall selection rate		(13.2%)

the number of genotypes screened effectively that is important, not the number grown.

REFERENCES

Goody, J.B. (1963). Laboratory methods for work with plant and soil nematodes. Tech. Bull. No. 2, Ministry of Agriculture, Fisheries and Food. London: HMSO.

Phillips, M.S., Forrest, J.M.S. & Wilson, L.A. (1980). Screening for resistance to potato cyst nematode using closed containers. Ann. Appl. Biol., 96, 317-22.

Phillips, M.S. & Trudgill, D.L. (1983). Variations in the ability of *Globodera pallida* to produce females on potato clones bred from *Solanum vernei* or *S. tuberosum* ssp. *andigena* CPC 2802. Nematologica, 29, 217-26.

SCREENING FOR RESISTANCE TO DISEASES IN A POTATO BREEDING PROGRAMME

G.J. Jellis, R.E. Boulton, N.C. Starling and A.M. Squire

Disease resistance is an important component of the breeding programme at the Plant Breeding Institute (PBI). Early generation screening is carried out on material bred for a specific character or when clones used as parents are known to be susceptible to a particular disease. Families segregating for gene *Ry* which confers extreme resistance to potato virus Y, for example, are sprayed with the virus at the seedling stage (Wiersema 1972). Susceptibility to blight (*Phytophthora infestans*) is inherent in sources of resistance to the potato cyst nematodes *Globodera rostochiensis* and *G. pallida*, so a combined early generation screen for nematodes and blight has been developed (Lacey *et al.*, this volume).

Table 1. Disease screening at the PBI

Disease (Pathogen)	References to test methods (sometimes modified)	Year testing commences
mild mosaic (potato virus X)	Howard *et al.* (1977)	5
severe mosaic (potato virus Y)	Davidson (1973)	5
leafroll (potato leafroll virus)	Davidson (1973)	5
gangrene (*Phoma exigua* var. *foveata*)	Jellis (1982)	6
dry rot (*Fusarium solani* var. *coeruleum*)	Jellis & Starling (1983)	7
common scab (*Streptomyces* spp.)	Jellis (1975)	6
blight (*Phytophthora infestans*)	Stewart *et al.* (1983) Howard *et al.* (1976)	4 (tubers) 7 (foliage + tubers)
blackleg (*Erwinia carotovora* ssp. *atroseptica*)	Lapwood & Gans (1984)	7
wart (*Synchytrium endobioticum*)	Noble & Glynne (1970)	6*
powdery dry rot (*Fusarium sulphureum*)	Jellis & Starling (1983)	7
powdery scab (*Spongospora subterranea*)	Jellis *et al.* (1984)	8
spraing (tobacco rattle virus)	Richardson (1970)	8*

*Tests done by Agricultural Development and Advisory Service

All breeding material is assessed from the fourth year onwards for resistance to diseases caused by a range of fungal, bacterial and viral pathogens (Table 1). Most of the tests are now well established but the powdery scab (*Spongospora subterranea*) glasshouse test is still at the development stage and procedures for assessing resistance to blackleg (*Erwinia carotovora* ssp. *atroseptica*) are at present being evaluated at a number of centres in the UK.

Except in early generation screening, susceptibility to any one specific disease is not necessarily considered a sufficient reason for rejecting a particular clone. Decisions are based on all available data, including yield, quality, appearance and physiological defects as well as resistance to pests and diseases. In practice, however, we have never submitted a wart susceptible clone for National List trials.

REFERENCES

Davidson, T.M.W. (1973). Assessing resistance to leafroll in potato seedlings. Potato Res., *16*, 99-108.

Howard, H.W., Langton, F.A. & Jellis, G.J. (1976). Testing for field susceptibility of potato tubers to blight (*Phytophthora infestans*). Pl. Path., *25*, 13-4.

Howard, H.W., Cole, C.S., Fuller, J.M., Jellis, G.J. & Thomson, A.J. (1977). Potato breeding problems with special reference to selecting progeny of the cross of Pentland Crown x Maris Piper. Ann. Rep. Pl. Breed. Inst., 1976, pp. 22-50.

Jellis, G.J. (1975). The use of polythene tunnels in screening potatoes for resistance to common scab (*Streptomyces scabies*). Pl. Path., *24*, 241-4.

Jellis, G.J. (1982). Laboratory assessments of the susceptibility of potato tubers to gangrene (*Phoma exigua* var. *foveata*). Pl. Path., *31*, 171-7.

Jellis, G.J. & Starling, N.C. (1983). Resistance to powdery dry rot (*Fusarium sulphureum*) in potato tubers. Potato Res., *26*, 295-301.

Jellis, G.J., Aubury, R.G., Boulton, R.E., Squire, A.M. & Starling, N.C. (1984). Ann. Rep. Pl. Breed. Inst., 1983, pp. 42-3.

Lapwood, D.H. & Gans, P.T. (1984). A method for assessing the field susceptibility of potato cultivars to blackleg (*Erwinia carotovora* subsp. *atroseptica*). Ann. Appl. Biol., *104*, 315-20.

Noble, M. & Glynne, M.D. (1970). Wart disease of potatoes. FAO Pl. Prot. Bull., *18*, 125-35.

Richardson, D.E. (1970). The assessment of varietal reactions to spraing caused by tobacco rattle virus. J. Natn. Inst. Agric. Bot., *12*, 112-8.

Stewart, H.E., McCalmont, D.C. & Wastie, R.L. (1983). The effect of harvest date and the interval between harvest and inoculation on the assessment of the resistance of potato tubers to late blight. Potato Res., *26*, 101-7.

Wiersema, H.T. (1972). Breeding for resistance. *In* Viruses of Potatoes and Seed-Potato Production, pp. 174-87. Wageningen: Pudoc.

BREEDING FOR RESISTANCE TO POTATO VIRUSES WITH SPECIAL REFERENCE TO cDNA PROBES

R.E. Boulton, G.J. Jellis and A.M. Squire

The virus programme, which was initiated in 1978, has given priority to breeding for resistance to the economically important potato viruses Y (PVY) and leafroll (PLRV). Sources of resistance to PVY were initially obtained from the Scottish Crop Research Institute, the Netherlands and Germany. Many of the accessions possess gene *Ny* from *Solanum demissum* which provides comprehensive hypersensitive resistance. Some have excellent PLRV resistance, and also field immunity to potato wart disease (*Synchytrium endobioticum*). More recently the *Solanum stoloniferum* gene *Ry* for extreme resistance, from sources such as the varieties Corine and Pirola, has been introduced into the programme, and we have also acquired accessions identified as PVY resistant by the International Potato Center (CIP). The development of lines duplex for *Ny* or *Ry* is in progress.

Breeding for resistance to infection with PLRV has utilized those PVY-resistant parents which also have PLRV resistance, and more recently accessions from CIP. Since PLRV resistance is less effective under severe infection pressure we are investigating the possibility that the gene for lethal hypersensitivity to PLRV present in varieties such as Monza, when incorporated into clones already possessing resistance to infection, will reduce incidence of the virus in the field. Tetraploid hybrids derived from *S. brevidens* will also be investigated as sources of PLRV resistance.

Many useful parental lines have been identified and potentially high-yielding clones possessing high resistance to PVY and PLRV are at all stages of selection. Breeding requires the accurate assessment of large numbers of genotypes. We have developed cDNA probes for the detection of PVY, PLRV and potato virus X, and these are routinely applied in the breeding programmes (Baulcombe *et al*. 1984; Boulton *et al*. 1984). The technique, which we call SASH (sap spot hybridization) is at least as sensitive as DAS-ELISA and has the distinct advantage of allowing more samples to be processed accurately, saving both time and labour available.

Seedlings specifically bred for PVY resistance conferred by genes Ny or Ry are inoculated with PVY by spraygun. Tubers from the plants selected as resistant are grown the following year and indexed by SASH for the presence of PVY before entering the next stages of multiplication. PVY-resistant clones derived from parents possessing resistance to potato cyst nematodes (Globodera rostochiensis, G. pallida) are also screened at this stage for resistance to cyst nematode and to late blight (Phytophthora infestans) in an early-generation test (see Lacey et al., this volume). Large numbers of individual clones cannot be selected easily for PLRV resistance in early-generation virus tests, since infection cannot be induced by sap inoculation. Clones in PBI's breeding programmes are therefore assessed for resistance to PVY and PLRV from their fifth year by a recently developed procedure (Boulton et al. 1984). Clones are exposed to the viruses in large-scale field trials; eye-pieces are cut from the harvested tubers in the following February and planted in a glasshouse with supplementary lighting. Sap from these plants is tested by SASH for presence of PVY and PLRV, and virus-susceptible clones discarded from the programmes. This procedure is very accurate and saves considerable time and land-use over previous methods.

REFERENCES

Baulcombe, D.C., Boulton, R.E., Flavell, R.B. & Jellis, G.J. (1984) Recombinant DNA probes for detection of viruses in plants. Proc. 1984 Brit. Crop Prot. Conf., Brighton, pp. 207-13.

Boulton, R.E., Jellis, G.J., Baulcombe, D.C. & Squire, A.M. (1984) The practical application of complementary DNA probes to virus detection in a potato breeding programme. Proc. 1984 Brit. Crop Prot. Conf., Brighton, pp. 177-80.

SCREENING FOR RESISTANCE TO DISEASES AND PESTS

G.R. Mackay

In potato breeding programmes much of the testing for disease and pest resistance consists of laboratory or field trials of clonal material, which are costly of time and effort. The selection pressure that has therefore been imposed for these characters has been weak. The Scottish Crop Research Institute (SCRI) has been investigating ways of increasing the efficiency of selection by identifying parental clones which pass on their resistance with an high probability.

Several economically important resistance traits are governed by single dominant effective factors inherited in a disomic manner, e.g. *H1* gene conferring resistance to *Globodera rostochiensis* (Ro1); *Rx* genes giving immunity to potato virus X (and B) and *Ry* genes providing extreme resistance to potato virus Y. Selective breeding combined with test crossing of resistant clones with susceptibles and assessing the progenies can produce parental material with increased "copies" of such genes. Clones which have a single copy of the gene (simplex) are intercrossed and by test crosses with susceptibles, duplex clones are identified amongst their progeny (Figure 1). The duplex clones are then intercrossed or selfed and the resulting clones again crossed to susceptibles to identify those with three or four copies (triplex or quadruplex) of the gene (Dale & Mackay 1982; Solomon & Mackay person. comm.).

The number of clones that need to be tested to distinguish between the various ratios is readily calculated (Mather 1938). As shown in Table 1, the use of triplex or quadruplex parents ensures resistance in the progenies. The eventual aim is to produce parental clones with all such resistance genes present in at least triplex form.

Progeny testing is being extended to include quantitatively inherited disease and pest resistance such as to: potato leafroll virus (PLRV), potato cyst nematodes (PCN), late blight (*Phytophthora infestans*), gangrene (*Phoma exigua*) and common scab (*Streptomyces* spp.). Clones which show resistance of this type are, as for the simply inherited characters, crossed to a range of susceptibles and their progenies are then tested for resistance. Clones which are empirically shown in this way to produce high levels of resistance in their progenies are used as parents. Such progeny tests can also be used to investigate the genetic control of these characters, thus allowing their more efficient exploitation.

One of the most powerful forms of progeny test utilizes plants grown directly from true seed. Such tests have been developed at SCRI and are currently used for PCN (Phillips & Dale 1982) and foliage resistance to late blight (Caligari et al. 1983). Tests for other resistances such as to gangrene, PLRV, tuber late blight and common scab need to be assessed on tuber grown material; these tests are at an advanced stage of development.

Table 1. The expected result of crossing clones with different 'copies' of a dominant resistance gene with a susceptible, ignoring the possibility of double reduction

Cross			R:S	% Resistant in progeny
Simplex	(Aaaa)	x Nulliplex (aaaa)	1:1	50%
Duplex	(AAaa)	"	5:1	83%
Triplex	(AAAa)	"	1:0	100%
Quadruplex	(AAAA) x	"	1:0	100%

Figure 1. Scheme for producing triplex/quadruplex parents

Simplex (Aaaa) x Simplex (Aaaa)

Duplex : Simplex : Nulliplex

test cross with susceptible to identify duplex

intercross/self duplex

Quadruplex : Triplex : Duplex : Simplex : Nulliplex

test cross with susceptible to identify quadruplex and triplex

REFERENCES

Caligari, P.D.S., Mackay, G.R., Stewart, H.E. & Wastie, R.L. (1983). A seedling progeny test for resistance to potato late blight (Phytophthora infestans (Mont.) de Bary). Potato Res., 27, 43-50.

Dale, M.F.B. & Mackay, G.R. (1982). Recent developments in breeding for resistance to potato cyst-nematode. Abst. XVIth. Int. Symp. Eur. Nem. (St. Andrews). Nemat., 28, 142-3.

Mather, K. (1938). The measurement of linkage in heredity. London: Methuen.

Phillips, M. & Dale, M.F.B. (1982). Assessing potato seedling progenies to the white potato cyst nematode. J. Agric. Sci., 99, 67-70.

BREEDING FOR RESISTANCE TO AND TOLERANCE OF POTATO CYST NEMATODE

M.F.B. Dale

Progress towards breeding potato varieties resistant to *Globodera rostochiensis* (Woll.) and *G. pallida* (Stone) has advanced considerably over the past decade. Recent developments at the Scottish Crop Research Institute (SCRI) in testing progeny with quantitatively inherited resistance, i.e. those derived from *Solanum vernei* (Bitt *et* Wittm.) and *S. tuberosum* ssp. *andigena* (Juz *et* Buk.) CPC 2802, have allowed genotypic parental values for resistance to *G. pallida* to be estimated (Phillips & Dale 1982). Data from such seedling progeny tests also allow the plant breeder to identify the most resistant progenies for further assessment and subsequent selection.

In parallel with improved progeny and parental selection for quantitative resistance to *G. pallida* has been the selective breeding and test crossing of parental material to produce parents which are triplex or quadruplex for the major gene *H1*, which confers resistance to *G. rostochiensis* (Ro1 and Ro4). The use of such parents guarantees that all members of derived progenies are resistant, eliminating the need for routine clonal testing for the *H1* gene and thus releasing resources (Mackay, this volume).

After sowing the selected progenies, clones produced are tested for resistance to both *G. rostochiensis* and *G. pallida* in the third clonal year using closed containers (see also Lacey *et al.*, this volume). Within these tests resistant, partially resistant and susceptible standard controls are used, allowing easy reliable comparisons both within and between tests and also between laboratory, glasshouse and field tests.

After further assessment and selection for a wide range of agronomic, disease and pest resistance and quality characteristics, material identified as resistant to both *G. rostochiensis* and *G. pallida* is further tested in field tolerance trials in potato cyst nematode (PCN) infested soils with the collaboration and assistance of ADAS. While resistant varieties will control and reduce PCN levels in the soil it is equally important, if not more so, that they produce acceptable yields in the absence of nematicides i.e. that they can tolerate a degree of root damage caused by the nematode. Such tolerance trials, as well as providing valuable information on yield potential, give important information on how different levels of quantitative resistance perform in the field and what level of control is exerted on PCN populations. Table 1 gives an example of some of the data collected from such trials.

From such data it is evident that the initial population density in the soil has a significant effect on subsequent multiplication levels, perhaps in part as a result of competition effects. The canister test gives a reliable estimate of resistance, with rankings between the three sites and the canister results correlating well. These trials also give an indication of what resistance levels should prove of value to agriculture when placed in the context of a rotation. Further work is in progress at SCRI to develop a reliable glasshouse test to assess tolerance characteristics which could allow larger numbers of clones to be tested earlier in the programme. Results of trials indicate that tolerance and resistance are independent of each other. However there appear to be a number of environmental, and as yet unknown, factors involved, making progress in this area limited. The result is that breeders are still reliant on field tests.

A measure of the success of the breeding programme for PCN resistance is the recent addition to the UK National List of one SCRI variety, Morag, which has good resistance to both *G. rostochiensis* and *G.*

Table 1. Comparison of % resistance to *G. pallida* (canister test) with field data (Pf/Pi) from 3 sites in 1983 (ADAS/SCRI field trials)

Cultivar	% res. (*G. pall.*)	rank	Leeds (Pi=89)	rank	Ramsey (Pi=70)	rank	Nocton (Pi=31)	rank
Cara	0	10=	1.60	10	5.56	10	56.7	10
P. Dell	0	10=	0.89	8	4.66	8	27.6	9
12243	70	8	1.32	9	5.13	9	14.1	8
12382	71	7	0.89	6	2.01	7	6.8	6
12290/1	80	6	0.42	7	0.65	4	5.3	5
11305	83	5	0.58	5	0.59	2	8.7	7
12290/2	84	4	0.50	4	0.70	5	6.6	3
12288/2	86	3	0.42	3	1.00	6	2.7	1
12288/1	92	1=	0.31	2	0.64	3	5.0	4
11233	92	1=	0.31	1	0.50	1	2.6	2

(Pi = initial population as eggs gm^{-1} soil)

pallida as well as a high level of tolerance. Another three potential varieties are currently undergoing National List Trialling.

ACKNOWLEDGEMENTS

SCRI wishes to acknowledge the co-operation and assistance of the Agricultural Department and Advisory Service (Cambridge, Leeds and Shardlow, England) in carrying out these trials.

REFERENCE

Phillips, M.S. & Dale, M.F.B. (1982). Assessing potato seedling progenies for resistance to the white potato cyst-nematode. J. Agric. Sci., 99, 67-70.

BREEDING MULTI-RESISTANT POTATO GERMPLASM

M.W. Martin

INTRODUCTION

Recurrent selection and mass intercrossing are used to enhance resistance in domestic potatoes to potato virus Y, potato leafroll virus, *Verticillium* wilt, Columbia root knot nematode, and deep-pitted scab. Lines resistant to one of these diseases or pests are intercrossed to enhance that single attribute. Resultant true potato seed (TPS) is bulked and sown in a field nursery where severe selection pressure will identify increased resistance to that disease or pest. Clones from selected TPS plants are re-tested in the same nursery and also screened in other nurseries and performance trials to select for other attributes. Surplus pollen from inter-crossing within each parental group is combined with pollen from the other four groups and used to intercross all five. It is also used to pollinate a parental group of breeding lines with superior horticultural attributes. TPS from these between-group mass intercrossings is sown in a field exposed to many diseases, pests and stresses to identify clones with multi-resistance. A number of multi-resistant breeding lines have been developed and are maintained in a disease and pest-free state by meristem culturing.

CHOOSING PARENTS FOR CROSSING GROUPS

To determine the best parents, we have tested known resistant germplasm from throughout the USA, Canada and elsewhere. In addition, several hundred advanced and early-generation selections are being obtained annually from US and Canadian breeding programmes. Over an 8-year period these have been screened, tested and retested in disease and pest nurseries, along with several thousand selections from our recurrent selection programme. Much disease and pest resistance has been found. Several parents have been identified for each intercrossing group and better ones keep emerging. As higher levels of resistance are identified or developed by recurrent selection, combining horticultural acceptability with resistance becomes easier.

MASS INTERCROSSING WITHIN AND BETWEEN PARENTAL GROUPS

Mass intercrossing is accomplished by collecting pollen from all flowering plants within a parental group. When possible an equal quantity is collected from each line in the group. This pollen mixture is used to pollinate all plants within that group. Surplus pollen from within-group pollination is combined into one capsule and used to intercross a second set of cuttings from all groups combined. It is also used to pollinate breeding lines with few resistances but superior horticultural attributes. Mass intercrossing is done on potted plants or on cuttings in painted fruit jars of aerated water. Air, delivered by emitters from a modified drip irrigation system, keeps water fresh and reduces stem rot. Bulked TPS from each within-group crossing is screened through the appropriate field nursery to find improved resistance. Bulked TPS from between-group intercrossing is sown in a field providing as many selection pressures as possible. Survivors of either within or between-group screening trials are retested as clones several times to verify resistance. Resistant clones are tested in other nurseries and performance trials to develop breeding profiles that will determine their use in future crossing groups. Over 200 multi-resistant lines have been selected as useful parental lines. These are maintained in a disease and pest-free state by meristem culture. Nine of these lines have combined resistance to six or more serious diseases or pests; 15 are resistant to five and the rest are resistant to up to four diseases or pests.

REFERENCES (for further reading)

Martin, M.W. (1978). Use of mass selection in early stages of potato breeding. Amer. Potato J., 55, 386. (Abstr.)

Martin, M.W. (1979). Breeding for potato disease resistance. Proc. Washington State Potato Conf., 18, 95-100.

Martin, M.W. (1981). Simultaneous mass-intercrossing within and between groups of parents solves some problems encountered in breeding potatoes. Amer. Potato J., 58, 510. (Abstr.)

Martin, M.W. & Thomas, P.E. (1982). Parental line development and maintenance for breeders. Amer. Potato J., 59, 477-8. (Abstr.)

Martin, M.W., Thomas, P.E., Santo, G.S. & Pavek, J.J. (1983). Resistance to Columbia root knot nematodes and other Northwest potato maladies. Proc. Washington State Potato Conf., 22, 83-91.

Martin, M.W. (1984). Development and maintenance of multi-disease resistant potato breeding parents. Amer. Potato J., 61, 529-30. (Abstr.)

RESISTANCE TO STORAGE DISEASES IN BREEDING STOCKS

A. Pawlak, J.J. Pavek and D.L. Corsini

INTRODUCTION

Soft rot (Erwinia carotovora ssp. atroseptica) and dry rot (Fusarium sulphureum) are serious storage diseases which must be considered in potato breeding. Resistance of potato cultivars and breeders' selections to both pathogens has been investigated. So far, resistant cultivars have not been developed. Differences within Solanum tuberosum are slight and are restricted primarily to grades of susceptibility.

MATERIALS AND METHODS

Soft rot: Tubers of 269 cultivars and breeders' selection were evaluated during 1983. These were divided into five groups according to origin. There were 42 cultivars, 116 S. tuberosum (tbr) breeders'selections, 66 hybrids of tbr x S. tuberosum ssp. andigena (adg), 12 hybrids of cv. Butte x S. microdontum (mcd) and 33 hybrids of tbr haploids x S. phureja (phu) or S. stenotomum (stn). A second test in 1983 and two tests in 1984 were conducted with a limited number of clones, including some which were very resistant.

Erwinia carotovora ssp. atroseptica (Eca) was isolated on Stewart's pectate medium from potato stems with typical blackleg symptoms. Dilution of Eca was made with sterile distilled water to obtain 5×10^6 colony forming units/ml. Ten ml of inoculum were injected at a depth of 2cm into each wound. The amount of tuber rot was calculated as the volume of decayed tissue (cm^3) using the formula of Sorensen & Sparks (1980) for determination of the volume of bruises. An arbitrary scale based on mean tuber rot was used to group cultivars and breeding clones into five categories: resistant (R) rot volume 0.5 cm^3; moderately resistant (MR) 0.51-2.50 cm^3; intermediate (I) 2.51-7.00 cm^3; moderately susceptible (MS) 7.01-15.00 cm^3 and susceptible (S) > 15.01 cm^3.

Dry rot: Tubers of 117 tbr breeders' selections were inoculated with F.

sulphureum at Zamarte Breeding Station during 1981-1983. The dilution of the inoculum was adjusted to 8 x 10^5 macroconidia/ml. An arbitrary scale based on tuber rot size was used to group clones into five categories.

RESULTS

Classification of clones within potato germplasm according to tuber resistance to inoculation with Eca is shown in Figure 1. Results of

Figure 1. Classification of clones within various groups of potato germplasm based on tuber response to inoculation with Erwinia carotovora ssp. atroseptica.

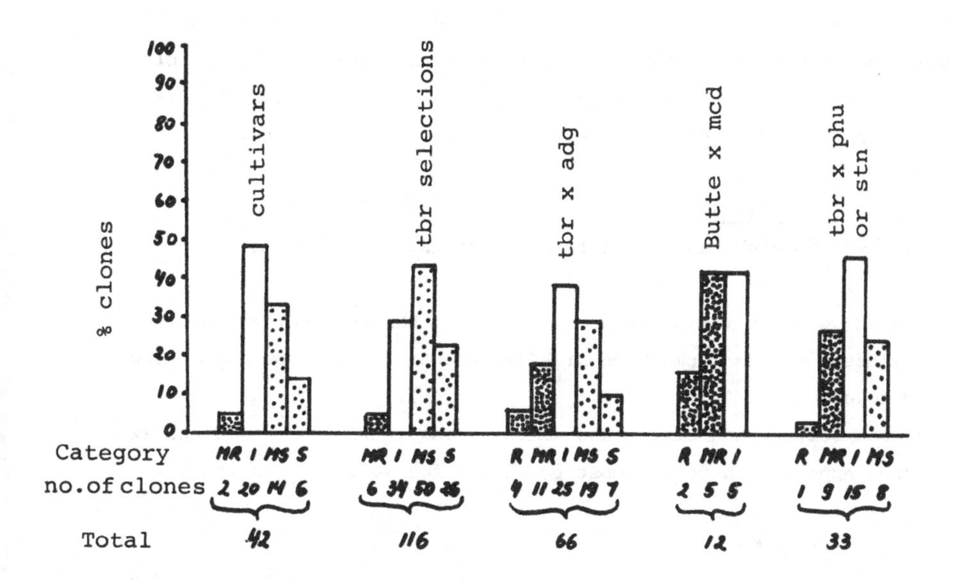

these tests indicate that there is little chance of finding high resistance in cultivars and their progenies, possibly because of the relatively narrow genetic base prevailing in tbr.

Greater variability in soft rot response was demonstrated in adg hybrid progenies. Four clones were classified as resistant: these included JB1-3, an adj selection, and two hybrids derived from JB1-3 (A78141-4, A78191-4).

A few triploid progenies of cv. Butte x mcd and diploid progenies of tbr x phu were represented in this evaluation. These progenies also showed a relatively high number of resistant and moderately resistant clones in comparison with tbr cultivars and breeders' selections (Figure 1). ADX513-1, ADX248-16 (both phu/stn x tbr hybrid progenies), and ATD63-2 and ATD63-7 (mcd x tbr) were most resistant. More than 80% of breeding clones tested for dry rot were moderately susceptible or susceptible. The reaction of 117 clones to _Fusarium_ was highly consistent between years.

CONCLUSIONS

1. There was no high resistance to soft rot and dry rot in evaluated material of tbr.
2. Existing genetic variability in hybrids of tbr and another species provides a source from which high resistance to soft rot may be bred and selected.
3. Crosses between adg or phu and 4_x_ tbr clones, and stn or mcd and 2_x_ tbr clones appear to have the most potential for soft rot resistance.

REFERENCE

Sorensen, L.H. & Sparks, W.C. (1980). A method for determination of the bruise resistance of potatoes. Am. Potato J., _57_, 494 (abstract).

PHYTOPHTHORA RESEARCH AT THE FOUNDATION FOR AGRICULTURAL PLANT BREEDING (SVP), THE NETHERLANDS

A.G.B. Beekman

Research at the SVP on resistance to *Phytophthora infestans* focuses on:

1) the development of efficient methods to evaluate the level of race non-specific haulm resistance to *P. infestans* and to distinguish between different components of resistance;
2) the detection of genotypes with race nonspecific resistance to *P. infestans*;
3) the study of resistance mechanisms and the inheritance of the resistance.

In this paper these research topics are highlighted and future research plans are mentioned.

DEVELOPMENT OF TEST METHODS

A field test has been developed at the SVP to assess leaf resistance to *P. infestans*. Clones to be tested are planted between spreader rows of a susceptible variety, in replicated microplots of six plants each. Standard varieties are included for comparison. All plants are inoculated artificially with a zoospore suspension. The percentage of diseased leaf area is assessed seven times per clone at weekly intervals, starting one week after inoculation. These seven scores are transformed to a weighted mean (WM) value using a formula (Tazelaar 1981). The WM values of 30 varieties lacking major (R) genes corresponded well with the known scores for resistance to *P. infestans* in the Dutch variety list ($r = -0.85$). Moreover the results of this test appeared to be highly reproducible.

However, correlation analyses showed that there are high positive coefficients of correlation between:

1) the WM values, based on seven observations, and the sum of two of these scores, the first of these two observations being made when 50% of the leaf area of the variety Bintje is blight infected. ($r > 0.95$, $N > 50$).

2) the WM values based on two or three replications ($r > 0.92$, $N > 50$).

Therefore it is concluded that the testing can be simplified by basing it on two replications and on two scores for the percentage of diseased leaf area, assessed with an interval of 1 week.

DETECTION OF RESISTANCE TO P. INFESTANS AND DEVELOPMENT OF PROGENITORS WITH RESISTANCE

At the SVP much effort is directed towards the investigation of wild *Solanum* species from the German-Netherlands Potato Department of the Gene Bank in Braunschweig (BGRC), for leaf resistance to *P. infestans*. Results of these evaluations have been published (Van Soest *et al.* 1984; Tazelaar 1981) or are made available on request from the BGRC in Braunschweig. Apart from these wild species, clones from programmes of the SVP, the plant breeding department of the Agricultural University (IVP) and the Dutch breeders have been evaluated for resistance to *P. infestans* at the SVP.

A number of *Solanum* species has been selected for further study of resistance to *P. infestans*. The resistance sources are used at the SVP to produce clones with early tuberization and a high level of race non-specific resistance to *P. infestans*. Very resistant, middle-early to middle-late clones have been produced. Further research has to be carried out to see if and to what extent this high level of resistance can be incorporated in earlier maturing genotypes.

STUDIES OF LEAF RESISTANCE MECHANISMS

More effort has been put into this aspect of *Phytophthora* research at the SVP recently. The research focuses on the analyses of resistance mechanisms in different wild and primitive *Solanum* species. Different components of resistance are studied and the inheritance is analysed. Data on the mechanism of resistance in *Solanum* species could serve as a basis for an informed choice of parents with the aim of accumulating resistance genes.

In 1985 the resistance of 80 genotypes from 12 species was studied in more detail using the leaf disc method described by Hodgson (1961). Analyses of variance and covariance showed large genotypic differences in the number of leaf discs which showed penetration symptoms, the number of discs with mycelium, the number of sporangia produced per leaf area or per disc with mycelium, and the latent period. Correlations between these characters varied from $r = +0.37$ to $r = +0.95$. Resistant genotypes were identified in _S. berthaultii_, _S. microdontum_, _S. verrucosum_, _S. sucrense_ and _S. venturii_.

REFERENCES

Hodgson, W.A. (1961). Laboratory testing of the potato for partial resistance to _Phytophthora infestans_. Am. Potato J., _38_, 259-64.

Soest, L.J.M. van, Schöber, B. & Tazelaar, M.F. (1984). Resistance to _Phytophthora infestans_ in tuber-bearing species of Solanum and its geographical distribution. Potato Res., _27_, 393-411.

Tazelaar, M.F. (1981). The screening of _Solanum_ species for horizontal resistance against late blight (_Phytophthora infestans_) and its use for breeding programmes. Abstracts, 8th Trienn. Conf. EAPR, Munich, pp. 34-6.

EFFECT OF TIME INTERVAL BETWEEN INOCULATION AND ASSESSMENT ON RELATIVE CONTENT OF POTATO VIRUS Y^N IN LEAVES OF POTATO PLANTS

Xu Pei Wen

INTRODUCTION

Screening for resistance to virus diseases based on both qualitative and quantitative measurements of virus in plants grown under controlled conditions may increase the efficiency of resistance breeding. Moreover, a measurement of changes in virus content in host plants could be of great help in studying the nature of resistance. It has been suggested that the enzyme-linked immunosorbent assay (ELISA) is a suitable technique for the assessment of virus content and a negative correlation between the extinction values in ELISA and the level of resistance to PVY^N in some potato varieties has been found. However the assessment may be influenced by a number of factors. This paper describes the results of an investigation on the effect of time interval between inoculation and assessment on the relative content of virus in leaves of plants infected with PVY^N.

MATERIALS AND METHODS

Three experiments were carried out in 1984 and 1985 at the Foundation for Agricultural Plant Breeding (SVP) in Wageningen. Plants grown in a glasshouse at 22 to 24°C and a daily photoperiod of 16h were inoculated mechanically at different plant ages and assessed for relative virus content by ELISA at different time intervals after inoculation. Leaves of inoculated plants were sampled at different positions according to an appropriate experimental design. ELISA was done according to the method of Clark and Adams (1977) with the modification of De Bokx and Maat (1979). The varieties used in the experiments differed in resistance to PVY^N (Anon. 1985). Healthy potato plants were used as controls.

RESULTS

In Experiment 1, significant differences in extinction values were found between time intervals of 9, 15 and 22 days after inoculation for sap from the top leaves and from the uninoculated lateral leaflets. Similar results were found in Experiments 2 and 3 using more varieties. There was a tendency to higher extinction values with longer time intervals.

There were significant interactions between varieties and sampling times and these affected the correlation between relative virus content and varietal resistance.

The potato varieties differed in the content of virus in the inoculated terminal leaflets relative to that in the lateral leaflets. This might point to genetic differences between varieties in reducing multiplication of the virus. Differences were smallest at the longest time interval after inoculation. Further assessments should be made to explain this phenomenon. Significantly higher extinction values for sap from the uninoculated lateral leaflets of the susceptible varieties suggest a higher rate of virus multiplication or a more rapid translocation of the virus in the more susceptible varieties. Differences in the increase in the relative content of the virus in the uninoculated lateral leaflets tested at different time intervals may point to a difference in resistance to virus translocation. Apart from the interaction between variety and time interval, significant interactions between time interval and leaf position were also found.

In conclusion, the relative content of PVY^N in leaves of primary infected plants is influenced by the time interval between inoculation and assessment. The time interval interacted with both plant genotype and leaf position. Therefore, in order to obtain a reliable estimate of varietal differences in virus content, as measured by ELISA, it is important to establish the optimum time between inoculation and assessment and whether to test the inoculated leaflet or adjacent ones. In our experiments, the inoculated terminal leaflets and the uninoculated lateral leaflets tested about 13 or 16 days after inoculation gave the best results.

ACKNOWLEDGEMENT

The author thanks Prof. Sun, Dr De Bokx and Dr Beemster for valuable suggestions and discussions, and Dr Dellaert, Ir Beekman and Dr Dolstra for valuable advice on the manuscript.

REFERENCES

Anon. (1985). 60e Beschrijvende rassenlijst voor landbouwgewassen, pp. 252-8. Wageningen: RIVRO.

Bokx, J.A. de & Maat, D.Z. (1979). Detection of virus Y^N in tubers with the enzyme-linked immunosorbent assay (ELISA). Meded. Fac. Landbouw. Rijksuniv. Gent, 44, 635-44.

Clark, M.F. & Adams, A.N. (1977). Characteristics of the microplate method of enzyme-linked immunosorbent assay for the detection of plant viruses. J. Gen. Virol., 34, 475-83.

SELECTION AND EVALUATION OF POTATOES FOR IMPROVED TOLERANCE OF ENVIRONMENTAL STRESSES

D. Levy

SELECTION OF HEAT TOLERANT CLONES

The adverse effect of high temperatures on tuber yield and quality is a major obstacle to potato production in hot regions (Ewing 1981). In the Mediterranean region, as well as in subtropical parts of Asia and Africa, potatoes are exposed to high day and night temperatures and a comparatively dry atmosphere. Almost all *Solanum tuberosum* cultivars from Europe or North America respond to such conditions with a significant loss in tuber yield and quality (Levy 1985). Local breeding of "heat tolerant" cultivars has been necessary to improve potato production in hot regions.

The selection of heat tolerant clones is carried out in the field or in environmentally controlled glasshouses. Should controlled conditions not be available, selections can be made in aphid-proof screen-houses. Avoidance of virus infections is a major obstacle because of the high aphid populations for most of the year.

The inhibition of tuberization by high temperatures is illustrated in Figure 1. Seedlings from open pollinated seeds of Desiree were grown in the field in Israel under 40-mesh screens, seeds were sown on the first day of each month and seedlings examined for tubers 90 days later. Each point on the graphs represents results obtained from 111 to 264 seed-lings (Levy 1984).

FIELD EVALUATION OF HEAT AND DROUGHT TOLERANCE

Although potatoes grown in hot, semi-arid conditions are irrigated throughout most of the growing season, they are nevertheless commonly exposed to water stress resulting from high ambient temperatures and low humidity during the spring and summer. This may reduce both tuber yield and quality. There are cultivar differences in tolerance to water deficits (Levy 1983a,b) that may help in the selection of genotypes suitable for growing in arid and semi-arid environments. The effects of high temperatures and water deficit were investigated under field conditions. Potatoes were grown in the relatively cool spring and at high temperatures in the summer season. In both seasons the potatoes were grown under three water regimes: adequate water supply, moderate water deficit, and severe water deficit. These regimes were achieved by a modification of the single-line source sprinkler irrigation system (Hanks et al. 1976). Severe drought reduced tuber yields in both seasons. Some tolerance of a moderate water deficit in the spring was exhibited by Draga, Desiree and Monalisa. Late and intermediate cultivars produced high tuber yields in the spring season, and early cultivars had relatively smaller yield losses in the summer. Early maturation was closely correlated with lower yield losses in the summer, pointing to the possible involvement of an escape mechanism in the tolerance of heat stress. The extent of sprouting, rotting and malformation of tubers varied considerably. High temperatures enhanced all of these and drought increased sprouting and malformation.

The differential response of various genotypes to environmental stresses indicates that it should be possible to improve stress tolerance in the potato.

Figure 1. Effect of temperature and day length on tuberization

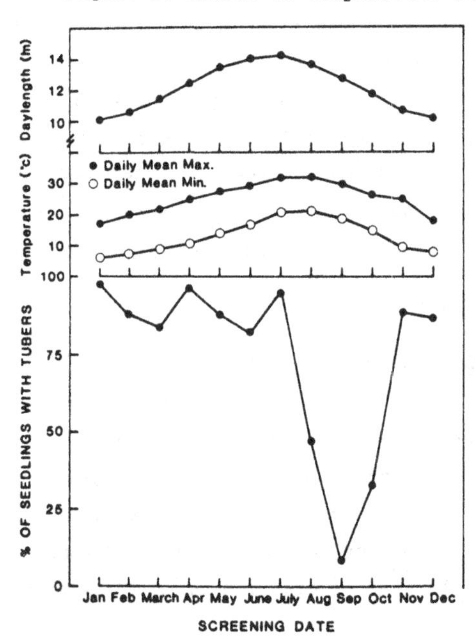

REFERENCES

Ewing, E.E. (1981). Heat stress and tuberization stimulus. Am. Potato J. 58, 31-49.

Hanks, R.J., Keller, J., Rasmussen, V.P. & Wilson, G.D. (1976). Line source sprinkler for continuous variable irrigation crop production studies. Soil Sci. Soc. Amer. J., 40, 426-9.

Levy, D. (1983a). Varietal differences in the response of potatoes to repeated short periods of water stress in hot climates. 1. Turgor maintenace and stomatal behaviour. Potato Res. 26, 303-13.

Levy, D. (1983b). Varietal differences in thr esponse of potatoes to repeated short periods of water stress in hot climates. 2. Tuber yield and dry matter accumulation and other tuber properties. Potato Res. 26, 315-21.

Levy, D. (1984). Cultivated *Solanum tuberosum* L. as a source for the selection of cultivars adapted to hot climates. Tropic. Agric. (Trinidad), 61, 167-70.

Levy, D. (1985). Tuber yield and quality of several potato cultivars as affected by seasonal high temperatures and by water deficit in semi-arid environment. Potato Res. (in press).

VARIETY ASSESSMENT

POTATO VARIETY ASSESSMENT IN THE FEDERAL REPUBLIC OF GERMANY

Walter Bätz

INTRODUCTION

In our country potato breeding is done by private breeders, with the exception of the Bayerische Landesanstalt für Bodenkultur und Pflanzenbau, Freising. Institutes like Max-Planck-Institut für Zuchtungsforschung in Cologne evaluate species from South America with resistance to diseases or other special qualities, and release their adapted clones to the breeders. The assessment of potato varieties is done by the Bundessortenamt at Hannover in cooperation with many other institutions. Like a patent office, the Bundessortenamt grants protection rights to the breeder for a new variety - Plant Breeders' Rights. This is based on the Variety Protection Act which was revised in 1985.

This Seed Act confirms that seed of agricultural species cannot be offered for sale unless the variety concerned is on the registered list of the Bundessortenamt. In order to be registered the variety must be distinct from other varieties, sufficiently uniform and stable, of value for cultivation and use and suitably named for registration. A variety is considered to be of value for cultivation and use if, in comparison with similar registered varieties, there is a clear improvement in characters concerned with crop cultivation, the utilization of harvested crops or any product obtained from such crops.

DISTINCTNESS, UNIFORMITY AND STABILITY

Assessment of DUS is based on sprout morphology under diffuse light and a field trial at one site with three replications for 2 to 3 years. At the completion of tests successful varieties are granted Plant Breeders' Rights. Protection has been extended from 25 to 30 years by the recent 1985 Act.

VALUE FOR CULTIVATION AND USE

For assessing value for cultivation the new variety is grown at several sites in the Federal Republic of Germany. The test is conducted by different institutions, such as chambers of agriculture and university institutes, on behalf of the Bundessortenamt in accordance with the guide lines established by the Bundessortenamt and the institutions. The normal duration of the test is 3 years. The varieties are cultivated according to their maturity group, as shown in Table 1. There are standard control varieties in every group.

The seed of the new varieties is supplied by the breeders. The standard control varieties are multiplied at one site by the Bundessortenamt and sent to the different test stations. It is not considered practicable to multiply the new varieties at a single site.

The Bundessortenamt provides test stations with booklets for recording observations. These are returned after harvest. The figures are computed by means of electronic data processing equipment and are compiled in detailed performance reports which are given to the breeders. The Bundessortenamt informs people who are in charge of the tests by circulars, training courses and visits to the test stations during the summer. The trials consist of four replications, and each plot has 60 plants.

The observations in the field include the number of days from planting to senescence, the number of gaps, dwarf plants, maturity, stem canker (*Rhizoctonia solani*), blackleg (*Erwinia carotovora*), early blight (*Alternaria solani*), late blight (*Phytophthora infestans*) and virus diseases. After lifting, yield, grading, content of starch, percentage of tubers infected by late blight, common scab (*Streptomyces* spp.), rattle virus, hollow heart and cracking are determined. Records are also made of the incidence of green and deformed tubers, the depth of the eyes, shape of tubers, colour of flesh and skin conditions. After winter storage, sprouting characters are observed.

The performance tests in the field are not sufficient for a

Table 1. Number of sites at which different maturity groups are tested each year

	Very early			Early			Mid early			Mid Late		
Year of testing	1	2	3	1	2	2	1	2	3	1	2	3
Number of trials	9	9	9	5	12	12	5	12	12	5	12	12

comprehensive assessment. Special tests are necessary for determining the resistance to wart disease (Synchytrium endobioticum), cyst nematode (Globodera spp.), virus, late blight in tubers, common scab and rattle virus. For wart disease the Biologische Bundesanstalt and two other institutes use the methods of Glynne & Lemmerzahl and Kotthoff & Spieckermann (see Ross 1986). Cyst nematode resistance is measured using the container biotest (Behringer 1967) and the pot test. The tests for rattle virus are done in the field and the tests for common scab in a special bed containing infected sand.

Resistance to mechanical damage is of great importance. The new varieties are assessed during the third year of testing and in the two years following registration. For this purpose they are grown in the field and lifted by a potato harvester. After this they pass over a shaker and grading machine, and the damage is assessed 6 weeks later.

The cooking qualities are described after a cooking test according to the scheme developed by the EAPR, using a 1 to 9 scale. Varieties for processing are tested for the special qualities required.

After 3 years the breeder gets a report which is the basis for a decision on registration. A committee of three members of the Bundessortenamt makes this decision. The breeder is invited to discuss the details of the variety with the committee. If the variety is not registered the breeder has the option of applying for a second discussion with another committee. This comprises two members of the Bundessortenamt and five members of other institutions carrying out the tests. If the breeder does not agree with their decision he can apply to a Court of Justice.

The variety is registered for 10 years, but the breeder can withdraw it at any time. An extension is possible, if the breeder demands it and if the variety is still important for cultivation and use. The number of potato varieties for which breeders have made application for registration during the last 11 years, and the number of those varieties which have been registered are shown in Table 2.

Some of the problems of the system which should be considered are:

1. It may happen that during the 3 years of testing there are none which are extremely dry, wet, hot or cold.
2. Whilst trials are carried out in well cultivated fields, conditions are not always so favourable on commercial farms.

3. For many reasons there are no special tests for stem canker, blackleg and dry rot (Fusarium spp.).
4. The test for rattle virus should be done at several sites.
5. Tests for mechanical damage should be done after different storage conditions.
6. During the winter physiological stability of the ware potato varieties should be tested. This means determining the levels of sucrose, ascorbic acid and citric acid in the tubers.

Table 2. Comparison of varieties submitted for registration and those registered by the Bundessortenamt

Year	Number of varieties submitted for registration	Number of registered varieties
1975	46	9
1976	33	8
1977	43	12
1978	36	12
1979	48	11
1980	48	8
1981	54	16
1982	42	14
1983	44	9
1984	54	10
1985	39	7

REFERENCES

Behringer, P. (1967). Der Biotest - Nematodennachweis der Zukunft. Kartoffelbau, 18, 170-1.

Ross, H. (1986). Potato Breeding - Problems and Perspectives. Adv. in Pl. Breeding, 13. Berlin & Hamburg: Paul Parey.

POTATO VARIETY ASSESSMENT IN FRANCE

P. Perennec

INTRODUCTION

In 1960 a "Catalogue of Cultivated Species and Varieties" was officially introduced and this provided a restricted list of varieties of certain species, the seed of which can be sold in France.

A variety is registered by the Ministry of Agriculture according to the advice of the "Comité Technique Permanent de la Sélection des Plantes Cultivées" (Permanent Technical Committee for Cultivated Plant Selection (CTPS)). This advice is based on the results of trials made by the "Groupement d'Etude et de Contrôle des Variétés et Semences" (Group for Study and Control of Varieties and Seeds (GEVES)) of INRA. Registration is for 10 years. It can be renewed for successive 5-year periods at the breeder's request and with the agreement of CTPS.

A breeder or his representative who wants a variety to be registered makes an application to CTPS. He has to pay registration fees for expenses for experimentation and has to provide seed equivalent in health status to basic seed, class E.

EXPERIMENTAL CONDITIONS

Official technical regulations include details of experimental procedures. The experimentation lasts at least 2 years. A variety can only be registered if it appears distinct from other varieties, stable and uniform after examination of its different characters. This examination is at present made by RIVRO, the Netherlands (see van der Woude, this volume), and was part of an agreement for cooperation between the two countries. The agronomic value and quality of the variety is established by considering a range of characters: yield, tuber appearance (shape, regularity and depth of eyes), cooking quality, storage and processing ability, susceptibility to diseases, pests and physiological defects. In the trials, varieties are classified into several groups according to their utilization and maturity.

There are control varieties specified by CTPS in every group.

At present, registered varieties can be classified into three categories:

1) Varieties for human consumption with firm flesh - tubers mainly middle grade; good eating quality with no disintegration on cooking. Most of these varieties are of French origin.

2) Varieties for human consumption - these do not have such good freedom from disintegration but they can be cooked in a larger number of ways. Early varieties are cultivated for the production of 'Primeurs', for which tubers are harvested before foliage maturity. Mid-early and mid-late varieties are harvested when they are fully mature and can be stored for consumption during winter and spring. Varieties for processing are in this group.

3) Starch varieties - mainly for starch production.

Table 1 shows the number of potato varieties entered for registration during the last few years and the numbers of varieties registered.

EVALUATION FOR AGRONOMIC CHARACTERISTICS, DISEASE RESISTANCE AND QUALITY

The methods which are used have been described by Perennec (1968) and presented in "Bulletin des Variétés, Pomme de terre (1983)." Yield of tubers, for varieties for human consumption, and of starch, for starch varieties, is assessed using a network of trials in which new varieties are compared with standard controls. Trials are done at 15 to 20 sites distributed all over France. Each trial comprises five replications, each plot containing 60 to 100 plants.

Observations during the growing season include speed of emergence, vigour of plants, earliness of tuber set, foliage maturity, frequency of late blight (*Phytophthora infestans*) and of virus diseases.

At harvest, tubers are weighed and graded, and the percentage of tubers with late blight is estimated. Samples of tubers are taken from some trials in order to make other observations, such as shape and regularity of tubers, colour of the flesh and skin, dry matter content and the

Table 1. Number of varieties registered in recent years

Year	1981	1982	1983	1984
Tested varieties	12	11	15	19
Registered varieties	6	5	8	9

percentage of misshapen tubers, cracked tubers or other defects including hollow heart and internal rust spot.

Susceptibility or resistance to the main diseases is evaluated using specific trials either in the laboratory or in the field. These include:- 1) Resistance to wart disease (Synchytrium endobioticum). Four tests are done, two using artificial inoculation and two relying on natural infection. 2) Susceptibility to potato virus Y (PVY) and potato leafroll virus (PLRV). Systematic trials are done in fields where conditions for virus infection are favourable and the frequency of virus infected plants is determined in clonal progenies. 3) Hypersusceptibility or extreme resistance to potato viruses X, Y and A. The virus is introduced by grafting or by sap inoculation. 4) Resistance to cyst nematode (Globodera rostochiensis, pathotype Ro1). The test is done by artificial inoculation in petri dishes on agar medium using the method of Mugiery (1982). 5) Susceptibility to late blight on foliage and tubers. Plots of each variety are grown with controls in the field without any protective treatment. 6) Susceptibility to common scab (Streptomyces scabies). Trials are done in infested fields where potatoes have been grown frequently.

Cooking quality (texture, disintegration and after cooking blackening) is evaluated on samples from different trials using the methods proposed by the EAPR. Taste is determined at the same time by a tasting panel. Varieties for processing are examined for special qualities such as dry matter content, reducing sugar content, colour of fried product and storability.

EVALUATION OF VARIETIES FOR REGISTRATION

After the completion of trials the variety is evaluated, using rules specified in the official technical regulations. The main characters are first scored using a 1-9 scale by comparison with standard control varieties. These scores are weighted according to the agronomic or economic importance of the character.

Bonuses or penalties are added depending on other varietal characteristics; for example, resistance to a particular disease will result in a bonus whereas high susceptibility will result in a penalty. The final score is established by adding together all the scores multiplied by their weighting factor and adding or subtracting bonuses or penalties. An example is given in Table 2.

CTPS recommends registration of varieties for which the total sum is equal or superior to a minimum value which is fixed in advance. At present this is 60 points for varieties for human consumption but if a variety has a total I (see Table 2) equal to or higher than 46 points and a grand total of 57 points, it is also recommended for registration. The minimum value for starch varieties is 57 points.

GENERAL OBSERVATIONS

This method of assessment eliminates varieties in which there will be no interest, but it cannot predict the success of registered varieties, as the crop is very dependent on cultural conditions. Some defects may not be revealed during the two experimental years.

It is necessary to perfect specific tests for evaluation of susceptibility to some diseases (late blight, virus, *Fusarium* disease, *Phoma*,

Table 2. Evaluation of a variety: an example

I. Category: Human Consumption Skin: Yellow Flesh: Yellow Shape: Oval

II. Earliness: Mid-early

III. Yield and value for use	1-9 Score	Weighting Factor	Evaluation
Yield	7	4	28
Shape (Regularity)	7	1	7
Eye depth	7	0.5	3.7
Cooking quality	4	2	8
Bonus or penalty (+ or -)			
Storability			0
Grading			0
Disintegration and texture			+2
After cooking blackening			0
Discoloration after cooking			0
Quality for processing			
Defects (susceptibility to cracking damage, blue discoloration, hollow heart etc)			-2
		Total I:	46.5 pts

IV Susceptibility to diseases	1-9 Score	Weighting Factor	Evaluation
Foliage blight	6	1.5	9
Common scab	4	0.5	2
Wart (Race 1)	9	0.5	4.5
Bonus or penalty (+ or -)			
Tuber blight			0
Other diseases (nematode, virus)			+1
Ageing			-1
		Total II:	15.5 pts
		General Total:	62.0 pts

blackleg, (Erwinia carotovora), Rhizoctonia, etc.) in order to evaluate these characters better and to avoid the problems associated with current field trials. A simple and reliable test to measure susceptibility to mechanical damage is not yet available.

Adaptation of the varieties to diverse French cultural conditions cannot always be determined because of the limited number of experimental sites.

REFERENCES

Anon. (1983). Bulletin des variétiés, Pommes de terre. INRA, Station d' Amélioration de la Pomme de terre et des Plantes à bulbes, 29207 Landerneau et GEVES, La Miniere 78280 Guyancourt.

Mugniery, D. (1982). A method for screening potatoes for resistance to *Globodera* spp. under laboratory conditions. *In* Research for the Potato in the year 2000, ed W.J. Hooker, p. 137. Lima: International Potato Center.

Perennec, P. (1963). Les modalites de l'inscription au Catalogue Francais des Variétés de Pommes de terre. La pomme de terre francaise, **327**, 8-15.

VARIETY ASSESSMENT IN THE NETHERLANDS

K. van der Woude

INTRODUCTION

The potato is the most important field crop for arable farming in the Netherlands. In the past 10 years the area under potato has varied from 160 000 to 173 000 ha. Well over 40% is used for the cultivation of ware potatoes, nearly 40% for starch potatoes and about 20% for seed potatoes. In each case a large part of the produce must be exported, either fresh or in processed form. The cost price therefore has to be low and the quality requirements high.

Breeding has done a great deal for farmers trying to meet these requirements. Furthermore, a comprehensive system of assessment of the value for cultivation and use has made it possible for new, improved, products of breeding to be identified rapidly, while on the basis of the same objective data the introduction of less promising entries has been prevented.

The importance of potato varieties in the Netherlands is demonstrated by the fact that 96 of the 262 recommended varieties of field crops which are registered in the Descriptive List of Varieties of Field Crops are potato varieties. A few of these 96 are grown on a very large area, and as many as 48 are grown on more than 500 ha and/or cover 100 ha for the production of seed potatoes (Anon. 1985; Daemen 1985).

VARIETY RESEARCH: STATUTORY REGULATIONS AND TRADE

In this context variety research means the comparison and investigation of varieties by an independent authority as a basis for statutory regulations concerning the approval of varieties and their admission in the variety lists. The aim of these statutory regulations is, among other things, to stimulate breeding work (breeders' rights) and to inform the farmer as to which varieties will give the best results under Dutch cultural practices. Variety research in the Netherlands is entrusted to the Government Institute for Research on Varieties of Cultivated Plants

(RIVRO). In a number of cases RIVRO asks the assistance of specialized institutes (Figure 1).

The results of this research are reported to two independent committees, one being responsible for the Netherlands Register of Varieties (NRV) and the other for the Descriptive List of Varieties of Field Crops. The breeders do not serve on these committees; however, they can submit their opinion on varieties to the committees.

There are two distinct fields of investigation: (i) registration research and (ii) research on the value for cultivation and use (VCU). Registration research is necessary for a variety to be entered onto the NRV, while the investigation of the VCU is concerned with entry onto the Descriptive List. Entry onto the NRV is a prerequisite for admission onto the Descriptive List and for obtaining breeders' rights. Admission to the Descriptive List of Varieties is followed by inclusion in the Common Catalogue of EEC. After this, commercialization of seed material within the Common Market is not subject to regulations other than those of a phyto-sanitary nature. Except for restricted quantities that are needed for experimental purposes, seed potatoes of registered varieties which are not on the National List can only be exported to countries outside the Common Market.

Registration research

During registration research the applications are assessed for their official recognition (distinctness, uniformity and stability) and, in the case of breeders' rights, also for their "newness". This work takes two growing seasons and is done at two locations in trials with two replications of 50 plants each. In the field a great number of morphological characteristics is determined. Laboratory tests are also carried out, using the light sprout method. This technique is very efficient and discriminating. In certain instances electrophoresis is successfully used for characterizing varieties (Houwing *et al*. 1985). The reports are drawn up in accordance with the UPOV directives, and by virtue of bilateral agreements the Netherlands perform registration research for France, Denmark and Belgium. Switzerland and Ireland have already adopted many reports.

Research on the value for cultivation and use (VCU)

Both the cultural value and the value for the user determine whether a variety can be profitably grown; the cultural value defines the

Figure 1. Programme for research on value for cultivation and use (VCU)

Year	Stage and programme
1	Preliminary Stage (PS)
1	PS 1 — For consumption 140 clones; For starch 60 clones
	4-5 field performance tests; plot size 16 plants Y^N and leafroll virus susceptibility Susceptibility to external damage Susceptibility to blue discoloration+ Consumer quality Processing (French fries and crisping) quality+
2	PS 2 — For consumption 35 clones; For starch 15 clones
	6-7 field performance tests; plot size: 16 plants Programme as PS1 *Phytophthora* - foliage inoculation field trial + - tuber inoculation field trial+ *Dormancy period and sprout growth vigour Central multiplication of seed
	Official Stage (OS) — For consumption (mainly Netherlands); For consumption (mainly abroad); For starch
3	OS 1
	25-40 observation trial fields; plot size 36 plants Programme as PS 2 X and A virus susceptibility Spraing and scab resistance Processing quality as mashed potatoes+ * Storability of starch varieties+
4-6	OS 2-4
	25-40 observation trial fields 2-7 interprovincial trial fields; plot size 50-70 plants Programme as OS 1 Resistance to cyst nematodes (+) and to wart disease (+) Viruses: extreme resistance or immunity (+) Seed potato performance Starch quality (+) * Susceptibility to ageing * *Phoma* and *Fusarium* tuber susceptibility * *Verticillium* susceptibility (+)

+ Tested by or in cooperation with specialized institutes.
* This research is in the process of development and is not yet applied routinely.

minimum supply price (cost price) and the users' value the maximum demand price, the market price of the respective variety. Only when this market price is higher than the cost price, can the variety be grown profitably.

Besides being the basis for approval, the assessment of VCU is also the starting point for advice on varieties. Advice is only useful when past results predict something about the future performance of a variety with a certain degree of reliability. From an advisory point of view the prediction must be such that the individual farmer can apply it to his own situation. The predictability is dependent on trial error, variety x location interaction and variety x year interaction. The standards that are imposed on the reliability of the predictions and on their relevance to the individual farmer partly determine the size and the design of the varietal investigation. For potato the value of the results can be enhanced by pooling yields from the same type of soil.

Testing programme. As described by van Loon (this volume) there are many small breeding programmes and some larger ones in the Netherlands. A comprehensive system of comparison is required to select from the numerous domestic and foreign varieties and clones those that are the most promising for Dutch growers. This is illustrated in Figure 1.

The comparison is done in the first place on the trial fields of the breeders or association of breeders.

The intermediate stage between the private and official trials is called the Preliminary Stage. It lasts 3 years and starts with 200 clones in the first year which are reduced to 50 clones in the second. Fifth-year and older clones can be admitted to the Preliminary Stage; this means that official varietal trials begin at a very early stage. Admission to the Preliminary Stage is granted on the strength of the varietal characteristics of the clones in question. These are determined in private trials, and on the basis of the extent of the breeding programme of the individual breeder. In the second year of the Preliminary Stage, central multiplication of seed begins for the following Official Stage trials.

During the next 3-4 years the promising varieties are tested in the Official Stage trials. Varieties are tested in 25-40 observation trial fields annually. These are unreplicated tests conducted by interested farmers, breeders and research institutes. The seed is obtained from the central multiplication source. The observations are done by those conducting the trials using special report forms. Each plot consists of 36 plants.

During the final 3 years of the Official Stage, performance is also tested in the interprovincial trials. These trials consist of larger plots (50-70 plants per plot), usually with three replications. They are often used by the advisory service to draw the attention of the farmers to the new varieties.

At the end of every season, including those of Official Stage, the experimental results are summarized and a decision is made about the continuation of each variety being trialled. Great weight is given to the results from the tests conducted by the breeder in that respective year. Investigations may temporarily be stopped because, for instance, one wishes to get more information on the potential of a certain variety in other countries. During the Preliminary Stage a distinction is made between varieties for starch and varieties for consumption purposes. The distinction between varieties mainly intended for use in the Netherlands and those mainly for export is made in the course of the tests, consumption and processing quality being the dominant criteria.

In the field tests (Figure 1) for performance as ware and seed, attention is paid to various characteristics of the foliage, to drought resistance, to yield, grade, underwater-weight and to the many tuber characteristics.

Disease tests are done for 3-5 years to establish the susceptibility to virus infection (neighbour infection test, four plants per plot, with four replications). From each plant three tubers are taken. One sprout, 20 cm long, is tested for the presence of virus using the ELISA technique. Clones that do not show symptoms of potato viruses A, X or Y^N in the field trials, are grafted to infected potato plants to find out whether they have extreme resistance or immunity to these diseases and also to the potato virus Y strains Y^C and Y^O.

Phytophthora in the foliage. Susceptibility is determined in inoculated plots (three replications with three plants per plot). The inoculum is a mixture of races with all known virulence genes. The trial field is usually inspected three times during the build-up of an epidemic. In the final evaluation more weight is given to the first observation. Varieties and clones whose results on inoculated and uninoculated fields suggest the presence of R-genes are examined in separate trials.

Phytophthora in the tubers. Susceptibility is determined in one inoculated field trial per year. This is supplemented with data from the field performance tests, some of which are not, or only poorly, protected chemically in the latter half of the growing season.

Spraing and scab. Each year three trial fields with two or three replications per variety are laid out on soil infested with tobacco rattle virus (spraing) and its vector, the nematode *Trichodorus*, and *Streptomyces* spp. (common scab). The infection levels vary considerably from year to year and from location to location.

Phoma and Fusarium. Tubers are inoculated and the spread of infection in the tuber is measured. The incubation temperature with *Phoma exigua* var. *foveata* is *c*. 7°C, with *F. solani var coeruleum* and *F. sulphureum* it is *c*. 15°C.

Verticillium dahliae. Young cuttings are inoculated by submerging roots in a suspension of mycelium (van der Spek *et al*. 1985).

Resistance to cyst nematode (Globodera spp) and to wart disease (Synchytrium endobioticum). These tests are done in the laboratory using many replications. Investigations concern resistance to races 1 and 2 of wart disease and to pathotypes Ro1, 2, 3, 4 and Pa2 and 3 of cyst nematode.

Quality research with ware potatoes.

- Consumer quality: cooking tests, with attention to taste, discoloration, texture, protein and glycoalkaloid content.
- Quality for French fries: frying tests at different times during the storage season. Attention is paid to discoloration in the par-fried product and to flavour and texture in the reconstituted product.
- Crisping quality: frying tests at different times during the storage season. Attention is paid to crisp colour and consistency.
- Quality as mashed potatoes: quality tests during the storage season. Assessment in accordance with the Karlsruher Bewertungs Schema.

Quality research with starch potatoes.

- Starch quality: in **samples from the field trials**, assessments are made for the size of the starch granules, the viscosity (phosphate determination), the dry matter, sugar and total protein and the coagulative protein content.

Susceptibility to external damage. Immediately after harvest, samples from field trials are bumped about in a flat-bottomed shaking machine in which a digger chain has been mounted (Figure 2). The number of severely, moderately or slightly damaged tubers is a measure of susceptibility (Meyers 1981).

Susceptibility to blue discoloration. After some weeks storage at 10°C samples from field trials are shaken with the same shaking machine as mentioned above, but without the digger chain. Three weeks later they are peeled. The number of severely, moderately or slightly discoloured tubers is a measure of susceptibility.

Figure 2. Shaking machine, used for the determination and susceptibility to external damage and to blue discoloration. In the first case a digger-chain is mounted (as shown here); in the latter case the machine is used without this chain.

Storability of starch varieties. Samples from several field trials are placed in clamps, after which susceptibility to rot and tendency to sprouting are determined. In further investigations the varieties are assessed for loss of dry matter and of starch during storage caused by conversion to sugars and by respiration.

Physiology,

- Dormancy period during storage in conditioned rooms at 12°C.
- Sprout growth vigour during storage in conditioned rooms at 12°C.
- Ageing. Storage till mid-January at *c.* 6°C, afterwards at *c.* 16°C, removing the sprouts several times. After planting in the field this treatment gives an indication of susceptibility to little potato.

Tests should only be done if good techniques are available. Unfortunately, for a number of important characteristics this is not the case, e.g. for resistance to bacterial diseases. The scheme (Figure 1) indicates test methods which are being developed or modified. In a number of cases a method may be too complex or not feasible, or may be too unreliable.

Some mathematical aspects of the research procedure

Pooling of variety tests. One of the most striking aspects of varietal research is the investigation of incomplete series. Every year new varieties are included in the tests while the testing of others is finished. Because of this the variety means over the years and the trials for a given character cannot be compared directly. In order to make such comparisons possible a special mathematical method is used. Taking into account the effects of location and varieties this technique provides the comparable variety means (Corsten 1957). In many cases the direct results are pooled. In some cases the figures are first transformed, to reduce the year effects as much as possible. This holds, for instance, for the figures for susceptibility to virus diseases. Figure 3 shows that from year to year, not only the level of infection, but also the range between slightly susceptible varieties and highly susceptible ones can vary considerably. Therefore the infection percentages themselves are not pooled but are first transformed to valuation figures for each individual field trial. The pooling over years is then done on the basis of these individual valuation figures.

The coefficients of variation for various characteristics

Table 1 shows the coefficients of variation (CV) for the varietal means per field trial, derived from poolings over years and trials for tuber yields, underwater weights and values of a number of characteristics. The magnitude of the CVs depends on the experimental error ($\sigma^2 e$), the number of replications (r) per experiment, the variety x location interaction ($\sigma^2 v \times p$), the variety x year interaction ($\sigma^2 v \times j$) and the general mean ($\bar{x}$) derived from the pooling, according to the formula

$$CV = \frac{1}{\bar{x}} \left(\frac{\sigma^2 e}{r} + \sigma^2 vxp + \sigma^2 vxj\right)^{\frac{1}{2}} \qquad \text{(Formula 1)}$$

The greater the CV the more tests must be conducted to obtain an accurate assessment of a variety. However, one will always have to accept a certain

Figure 3. Transformation of data from PVY^N susceptibility trials.

As a result of the differences in number and in activity of aphids, the infection percentages (IP) in tests measuring susceptibility to PVY^N fluctuate considerably between years and between varieties differing in susceptibility. The graph shows these fluctuations in IP (a) for three varieties differing in susceptibility. Figure (b) is derived from a transformation per trial of IP of the same three varieties. In the latter case the year-to-year and the variety-to-variety fluctuations are much smaller.

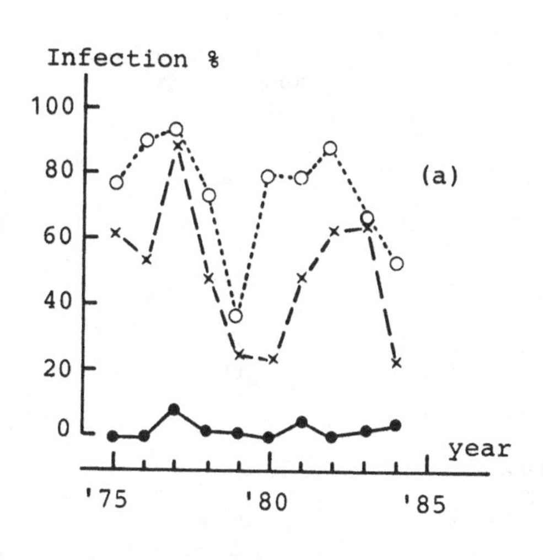

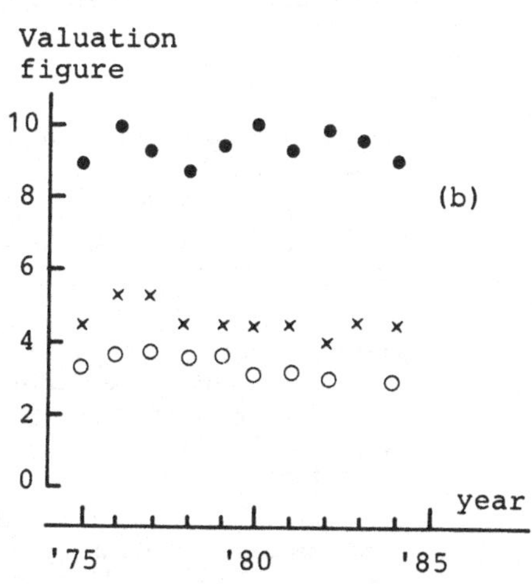

amount of inaccuracy in decisions on varieties. The CVs differ considerably for the various characteristics; the CV for under-water weight is much lower than that for tuber yield and the CV for leafroll virus is slightly higher than that for PVY^N. Among the characteristics for quality, the CV for mashing quality is relatively low, but that for discoloration of par-fried French fries is high. This is very disappointing since it is an important character for the trade.

The CVs for the two methods of measuring susceptibility to external damage are about 12.5%. The shaking test (Figure 2) is a simulation technique, which can be conducted using a relatively small sample (about 40 tubers, size 40-60 mm). The lifting test is a technique that closely resembles every-day practice. In addition to the similar CVs, there is reasonable agreement in the ranking order of varieties for the two methods. The CV for blue discoloration is clearly greater than that for external damage.

Size and shape of the experimental design

As described in the section on the experimental programme, the field performance tests are mainly conducted as unreplicated trials. Apart from the fact that small experiments appeal more readily to private farmers, there are considerable variety x year and variety x location interactions

Table 1. Coefficients of variation for varietal means per experiment (with r replications) for tuber yield, under-water weight and some disease and quality characteristics derived from poolings over years and trials.

Characteristics	Coefficient of variation (Formula 1) (%)
Ware varieties on clay-soil (r = 3)	
Tuber yield	9.5
Under water weight	4.8
Susceptibility to virus (r = 4)	
PVY^N	13.7
Leafroll virus	15.8
Quality research (r = 1)	
Discoloration of par-fried French fries	20.8
Quality of reconstituted French fries	11.3
Colour of crisps	12.5
Mashing quality	8.3
Susceptibility to external damage and blue discoloration (r = 1)	
External damage - lifting machine	12.6
- shaking test	12.1
Blue discoloration	19.0

in potato trials so that the reliability of the results is optimized by using a great number of trial sites.

This is clear from Table 2, which shows the reliability of the estimates of varietal differences using different experimental designs, i.e. 15 trials with three replications or 45 unreplicated trials, for two crops that are quite divergent for variety x year variety x location interactions. In the table a distinction has been made between decisions on the basis of the average trial results within one year and those that are taken on the strength of results after several years (e.g. 3 years). In the latter case the variety x year interactions play a dominant role.

To compute the values for Table 2, the following formulas were applied:

(Formula 2) σ^2 var.diff. between trials within years $= \frac{2}{p}\left(\frac{\sigma^2 e}{r} + \sigma^2 vxp\right)$

(Formula 2)

(Formula 3) σ^2 var. diff. between trials and years $= \frac{2}{j}\left(\frac{\sigma^2 e}{rxp} + \frac{\sigma^2 vxp}{p} + \sigma^2 vxj\right)$

(Formula 3)

Table 2. The effect of the experimental design used (i.e. many experimental fields versus few fields with many replications) on the reliability of the estimates of varietal differences for two crops that vary for the magnitude of environmental interaction.

	magnitude of interaction components*			σ^2 varietal differences between trials within 1 year (Formula 2)		σ^2 varietal differences between trials and years (Formula 3)	
	$\sigma^2 e$	$\sigma^2 vxp$	$\sigma^2 vxj$				
	g/m^2	g/m^2	g/m^2	g/m^2	%	g/m^2	%
Sugar beet							
	250	26	35				
r = 3 p = 15				15	100	28	100
r = 1 p = 45				12	80	27	96
Ware potato							
	11910	11000	4090				
r = 3 p = 15				1996	100	3390	100
r = 1 p = 45				1018	51	3066	90

*see text for explanation of symbols

The results in Table 2 demonstrate that for potato an increase in the number of trial fields at the expense of the number of replications greatly diminishes the variance for varietal differences, whereas for sugar-beet the effect is relatively small.

From Formula 3 it appears that with both field crops the effect of the variety x year interaction is the most important component in the variance for varietal differences over several years. This implies that an extended trial period is needed to obtain reliable results.

CONCLUSIONS

1. Both the registration research and the investigation of cultural value have an effect on the market potential of varieties.
2. The test procedure for VCU is characterized by an early start (from the fifth-year clonal generation onwards) and a long duration.
3. The investigation is started with many clones; less promising clones are discarded early. It is also possible to discontinue the investigations at a later date.
4. In the experimental programme early attention is given to characteristics that are the most important for VCU of the crop in the Netherlands, whereas less important characteristics are dealt with at a later period.
5. The magnitude and the nature of the genotype x environment interaction receive full consideration. Many field performance tests are conducted as unreplicated trials.
6. The possibility of assessing VCU correctly is restricted while test methods for some important characteristics are rather unreliable or lacking.

REFERENCES

Anon. (1985). Beschrijvende Rassenlijst voor Landbouwgewassen 1985. Maastricht: Leiter-Nypels.

Corsten, L.C.A. (1957). Vectors, a tool in statistical regression theory. Wageningen: Veenman & Zonen.

Daemen, M.V.P. (1985). De belangrijkste aardappelrassen in 1984. De Pootaardappelwereld 38, 22-4.

Houwing, A., Suk, R. & Ros, B. (1985). Generation of light sprouts suitable for potato variety identification by means of artificial light. Acta Hort. (in press).

Meijers, C.P. (1981). Diseases and defects liable to affect potatoes during storage. In: Storage of Potatoes, ed. A. Rastovski & A. van Es, pp. 138-66. Wageningen: Pudoc.

Spek, J. van der, Mulder, A., van der Woude, K. 1985. Verwelkingsziekt van aardappelen, een zorg erbij. Informa 6, 8-10.

POTATO VARIETY ASSESSMENT IN POLAND

Julia Borys

The potato crop is of great economic importance in Poland. The area of land under potatoes is 2.2 million ha, which is about 15% of total arable land. This is due to the predominance of light soils (nearly 60% of total land) and suitable natural conditions for potato growing. Total annual potato production is 40 to 50 million tonnes. About 55% of total production is for fodder, mainly for pigs, and nearly 12% is used for domestic consumption. Potato is almost the only raw material available for alcohol and starch production. Considerable efforts are being made to increase exports and usage for industrial processing, i.e. alcohol, starch and especially dried food production.

During the next few years the area under potato is expected to decrease, but the average potato yield is expected to increase. The introduction of new, more valuable varieties is one of the main factors leading to the increase of yield.

The breeding of new potato varieties in Poland is coordinated by the Institute for Potato Research at Bonin, which has four Experimental Stations. Another five Plant Breeding Stations belong to the Association of Plant Breeding and Seed Production Enterprises. Preliminary trials are carried out at five or six centres over 3 years before official trials.

Official variety assessment has a long tradition in Poland. Since 1969, it has been the responsibility of the Research Centre for Cultivar Testing (COBORU) at Slupia Wielka. The management of the Register of Plant Varieties and decisions concerning the release and regionalization of varieties has been the responsibility of the Ministry of Agriculture, Forestry and Food Economy but this will pass to COBORU in the near future.

Variety assessment is a necessary and statutory requirement. The main tasks of COBORU are:

1) The preparation of the information concerning the introduction of varieties to, or cancellation from, the Register of Plant Varieties;

registration is necessary before a variety can be grown and sold for seed;

2) the comprehensive determination of the characteristics of new varieties;
3) the recommendation of the regional suitability of varieties;
4) the formulation of essential advice on individual cultivation requirements;
5) the publication and dissemination of information about registered varieties.

The potato trials and tests are carried out at 66 Experimental Stations, which represent almost all climatic and soil conditions in Poland.

Every year new varieties become available as a result of plant breeding activity. During the past 10 years (1975-1984) 122 new potato varieties have been tested in the official trials. In that time 38 new varieties were registered. This means that every year about 30% of 10 to 20 new varieties are registered. A new variety must be considered as a compromise between what is desirable and what is achievable (Werner 1978).

Whether or not a new variety is registered depends on the result of:

1. the tests on botanical and morphological characters;
2. the evaluation of the economic value;
3. the quantity and the quality of seed material.

The tests on botanical and morphological characters which are used to confirm the originality, distinctness, uniformity and stability are done by the Department of Variety Identification in COBORU. The economic value of new varieties is determined on the basis of numerous field trials and the quality analyses.

Super elite seed of all varieties is provided by the breeders and multiplied at one centre situated in the North in good conditions for potato seed production. These uniform conditions for growing and storage of healthy seed material of all potato varieties improve comparisons between varieties and the determination of relative yields.

Main trials are carried out in four maturity groups: very early varieties at 28 centres, 36 plants per plot, three lifts (after 40 days, 50 days and at maturity); early varieties at 28 centres, 48 plants per plot, two lifts (after 65 days and at maturity); mid-early varieties at 38 centres, 60 plants per plot; mid-late and late varieties at 36 centres, 60 plants per plot; mid-late and late varieties at 36 centres, 60 plants per plot. Registration trials of foreign varieties are carried out at five centres in three maturity groups.

Each variety after registration is continued in main trials until it is withdrawn from the Register of Plant Varieties and from the planned seed production.

The trials of very early and early varieties are laid out in a split-block design. All trials with later varieties are laid out in incomplete block designs or alpha designs (Patterson & Williams 1976) in four replications. Use of incomplete block designs instead of complete block designs in potato field trials has been found to decrease mean squares for error in about 90% of trials (Pilarczyk 1985). Analyses of variance and co-variance are used to examine main effects in single trials. Analytical techniques are used to interpret genotype x environment interactions (Calinski *et al.* 1979, 1980, 1983).

Degeneration trials of registered varieties are carried out at 16 centres. They provide information on yield reduction of particular varieties multiplied three times in each region and also serve to check their resistance to virus diseases. These results are very useful for indicating regional suitability of varieties.

Maturity trials are done at six centres. The results of these trials give information on the length of the growing period of particular varieties.

Cooking tests are done at ten centres, using a 1-9 scale. A synthetic index of quality for table potato varieties has been devised by COBORU. Varieties for processing are tested in the laboratories of the Institute for Potato Research and by the Potato Industry.

Resistance to wart disease (*Synchytrium endobioticum*), cyst nematode (*Globodera rostochiensis*, pathotype RO1) and viruses is determined by the laboratories of the Institute for Potato Research.

Susceptibility of varieties to bacterial and fungal diseases is tested by COBORU. On the basis of numerous observations and laboratory tests, the Institute for Potato Research evaluates varietal resistance to each pathogen on a 1-9 scale by comparison with standard control varieties.

Mechanical damage tests of tubers are very important, but the methods of testing are not very satisfactory. In future, tests will be made by harvester at several centres.

The improvement of methods of variety assessment is undertaken by the Institute for Potato Research in close cooperation with COBORU.

Caliński, T., Czajka, S. & Kaczmaret, Z. (1979). Methods for studying the genotype - environment interaction. Quaderni di Epidemiologia Supplemento al., No. 1, pp. 11-29.

Caliński, T., Czajka, S. & Kaczmarek, Z. (1980). Anliza jednorocznej serii ortogonalnej doswizdczen odmianowych ze szczegolnym uwzglednieniem interakcji odmianowo-srodowiskowej. I. Analiza ogolna. Biul. Oc. Odm., 2(12), 67-81.

Caliński, T., Czajka, S., Kaczmarek, Z. (1983). Analiza jednorocznej serii ortogonalnej doswiadczen odmianowych, ze szczegolnym uwzglednieniem interakcji odmianowo-srodowiskowej. II. Analiza szczegolowa. Biul. Oc. Odm., 1(15), 39-60.

Patterson, H.D., Williams, E.R. (1976). A new class of resolvable incomplete block designs. Biometrika, 63, 83-92.

Pilarczyk, W. (1985). Use of a measure of accuracy of field trials for detecting untypical trials. 1st Eurpoean Biometrical Conference, Budapest, p.122.

Werner, E. (1978). Breeding new potato varieties in Poland. 7th Triennial Conference of the European Association for Potato Research, pp. 177-8.

POTATO VARIETY ASSESSMENT IN THE UK

D.E. Richardson

In order to comply for entry to a National List a variety must be distinct, uniform and stable (DUS), and be of value for cultivation and use (VCU). The DUS requirement is also necessary to establish eligibility for Plant Variety Rights. In the UK there are both statutory and advisory VCU systems. The statutory National List Trials are largely used to eliminate inferior varieties, and the advisory Recommended List Trials aim to select the superior varieties.

NATIONAL LIST

National List testing is conducted on an interdepartmental basis by the agricultural departments for England and Wales, Scotland and Northern Ireland. The Scottish Department coordinates the work, collates the results and prepares the reports. The National Institute of Agricultural Botany acts as agent for the Ministry in England and Wales. A National List DUS/VCU Group makes recommendations on the distinctness and merits of new potato varieties to the UK National List Committee.

Both DUS and VCU tests take place over 2 years. Distinctness tests are based on morphological characters, which closely follow the guidelines of the International Union for the Protection of New Varieties of Plants (UPOV). Uniformity and stability rarely present problems in potatoes.

VCU tests include yield trials and tests for disease resistance and quality. Yield trials consist of two replicates of 100 plants at three centres - Cambridge, Edinburgh and Belfast. Seed is provided by breeders. The different seed origins and the small number of trials limit the value of the yield data. The use of micropropagation for the multiplication of seed for trials might overcome the seed supply problem (Wooster & Dixon 1985). In addition to information on yield these trials also provide agronomic data and material for quality tests. Special tests are used to assess external and internal damage to tubers.

Disease tests are done at one or more of the three test centres on the following: wart (Synchytrium endobioticum), blight (Phytophthora infestans), potato cyst nematode (Globodera rostochiensis and G. pallida), gangrene (Phoma exigua var. foveata), dry rot (Fusarium solani var. coeruleum), skin spot (Polyscytalum pustulans), common scab (Streptomyces spp.), and viruses PVA, PVX, PVY and leaf roll (PLRV).

Quality tests are done on the produce from the field trials for the following characters: dry matter, enzymic browning, after cooking blackening, colour boiled, colour crisped, consistency, dryness, fibrous tissue and off flavours. The EAPR guidelines on cooking assessments are used. Glycoalkaloid contents are also monitored.

The EEC directive with regard to acceptance of varieties onto National Lists is that they should show a clear improvement over existing varieties on the list. Hitherto the interpretation of this directive has been arbitrary, and because of the limitations of the testing system, few potato varieties have been rejected. However, with the steady rise in numbers - from five in 1975 to 20 in 1985 - setting more precise standards is being attempted (Talbot, this volume).

Breeders may appeal against National List decisions to the Controller of the Plant Variety Rights Office. An informal hearing is arranged with members of the Potato DUS/VCU Group and the National List Committee. If differences of opinion cannot be settled at this stage, the appeal is considered by a Tribunal of members appointed by Ministers.

National List trial results are published in the UK Plant Varieties and Seeds Gazette. The NIAB Classified List provides summary information on varieties on the UK National List, together with some of the Common Catalogue varieties which have been tested in advisory trials in the UK.

RECOMMENDED LIST

NIAB Recommended List trials are advisory and results are mainly applicable to the ware growing areas of England and Wales. There is close collaboration with the Agricultural Development & Advisory Service (ADAS), Potato Marketing Board (PMB) and the advisory bodies in Scotland and Northern Ireland. NIAB coordinate these advisory trials through its Potato Trials Advisory Committee which consists of 15 experts drawn from all sectors of the potato industry and the three Agricultural Departments. Recommendations are submitted for agreement to Council, which is the Institute's governing body.

Yield trials are for 3 years at five to 20 centres for maincrop varieties, five to 10 centres for first early and second early varieties. Trials consist of three replicates of 100 plants, with split plots if rate of bulking data is required as in the case of early varieties and also a limited number of maincrop trials. Seed is all of common origin, which is regarded as an important prerequisite for making valid yield comparisons (Flack 1983). Special tests are carried out to assess resistance to both external damage and internal bruising (blackspot) on samples from three centres. Disease and quality tests are similar to those for National List purposes. Additional diseases are included, such as spraing (tobacco rattle virus - TRV), powdery scab (Spongospora subterranea) and blackleg Erwinia carotovora ssp. atroseptica). Observations are made on other diseases and pests, especially stem canker/black scurf (Rhizoctonia solani), silver scurf (Helminthosporium solani) and slugs (Mollusca spp.). Cooking tests are done on samples from at least six different soil types both before and after storage, and there are additional tests for chipping and baking.

Provisional recommendations ('P') are made after 3 years in Recommended List trials. Developmental trials (e.g. spacing, sprout manipulation, irrigation treatments) are carried out on 'P' varieties by ADAS for a further 2 or 3 years. Also in this period PMB carry out surveys on storage quality, marketability and consumer acceptability of the varieties produced in bulk (Storey & Hampson, this volume).

If early promise is maintained, 'P' varieties are given full recommendation for general purpose - 'G' or for special purpose - 'S'. 'S' varieties are those which have special attributes such as suitability for crisping, very early bulking, long storage, or for situations where resistance to a particular disease or pest is essential. Certain poor characters make them unsuitable for general recommendation.

Breeders may appeal against a Recommended List decision to NIAB Council. This is rare, since there is close liaison with breeders both before and after decisions are made, and the Advisory Committee consider the breeders own data. Varieties are always considered alongside other candidate varieties and those already on the list, and never in isolation.

Results are available each year, but publication is usually delayed until after the third year of trials; they are included in the NIAB Bulletin of Crop Varieties and Seeds. The Recommended List is reviewed and published each year in Farmers Leaflet No 3, with ratings for most important characters on the 1-9 scale. The PMB also publish annually its Commercial Assessment of Recently Introduced Potato Varieties.

FURTHER CONSIDERATIONS

The increasing numbers of varieties being produced and offered for testing coincides with a reduction in resources. It is therefore important to make the fullest use of all available information by cooperation between organisations both at home and abroad.

The Recommended List policy of selecting superior varieties - 'as good as or better than the best' is restrictive and is sometimes disappointing for breeders. However, this limited choice is generally regarded as an advantage for growers and for the development of centralized grading, marketing and processing. 90% of varieties grown in the main ware areas are NIAB Recommended.

There is a need for better standardization of disease testing methods (Gans & Wooster, this volume). Minimum disease standards are usually applied voluntarily by breeders and testing authorities, but there are no statutory minimum levels. Nevertheless care is needed to prevent the introduction of varieties with very high susceptibilities. The classic control of wart disease - which was once rampant in the UK - was achieved by growing immune varieties.

Minimum quality standards might be appropriate for certain characters like the tendency to cracking, internal rust spot, after cooking blackening or excessive disintegration on boiling. Glycoalkaloid content is a special consideration (White, this volume); it would seem to come under the directive that a variety can be refused entry or removed from a National List if it is likely to adversely affect the health of persons, animals or plants; setting standards, however, is more appropriate for the public health authority.

No system of testing is perfect, and it has been suggested that promising varieties may be lost because they needed special treatments for optimum performance. Since it is difficult to predict and control certain growing conditions, a variety is generally of little value unless its average performance is good over a range of conditions. However, since it is predicted that up to 50% of the potato acreage will soon be irrigated, it is necessary to include irrigation in the routine testing procedures (Flack, this volume).

More special quality tests may be necessary, as for crisping, if there is a continued trend towards "added value" products. On the other hand, there is an increasing emphasis on human health which could slow down the trend towards some processed products with potentially harmful additives.

Balancing the good and bad characters is likely to continue to cause assessment problems. High quality can sometimes compensate for moderately low yield or disease resistance, but high disease resistance may not compensate for low yield or quality, nor high yield for low disease resistance and quality.

REFERENCES

Flack, S.J. (1983). Effect of seed origin on potato yields. J. Natn. Inst. Agric. Bot., 16, 267-71.

Wooster, P. & Dixon, T.J. (1985). Application of micropropagation to potato seed production. J. Natn. Inst. Agri. Bot., 17, 99-106.

USE OF COMMON ORIGIN SEED FOR POTATO TRIALS

S.P. Kerr

There is a need for common origin seed for potato trials because both physiological and pathological factors can affect earliness, rate of bulking, grading and final yield as well as reactions to disease. The effect of seed origin on subsequent performance is well documented (Wurr 1978). Rainfall, temperature, time of planting, defoliation and harvest all influence the physiological condition of the seed and subsequent crop performance. It is for these reasons that seed for NIAB Recommended List trials is grown at a seed multiplication site in Cumbria under healthy and uniform conditions.

Effects of seed origin have been found to produce highly significant differences in marketable yields of Pentland Crown from two seed sources as shown in Table 1. These differences occurred despite the proximity of the two sites and similar management regimes.

Comparison of yields of the first early variety Arran Comet from more widely separated seed sites showed the differences normally associated with 'once grown' seed. (Table 2). Seed was from Cumbria and

Table 1. Comparison of yield (t/ha) of Pentland Crown seed stocks from two seed sources (Cumbria & Durham) at 65 trial sites (from Flack 1983).

	Mean diff. t/ha	t value
Total Yield	1.37	2.96 ***
80 x 40mm Yield	2.60	5.19 ***

*** indicates statistical significance at $p = 0.001$

Table 2. Marketable yields in t/ha > 25mm. Mean of three trials. Lifts at weekly intervals.

	Lift 1	Lift 2	Lift 3	Lift 4
Arran Comet (Cumbria)	9.06	15.65	22.98	29.18
Arran Comet (once grown)	12.83	21.70	27.00	31.09
LSD ($p < 0.05$)	2.361	3.707	3.368	5.571

"once grown" at the three trial sites in Devon and Kent. The relative differences are greater at the earlier lifts.

For trials with first early varieties where early bulking is of prime importance, the use of once grown seed has been considered. A seed multiplication site near to the early growing areas would be required. This would supply seed which was physiologically aged more than Cumbrian seed, but there could be problems associated with virus health.

The virus health and fungal and bacterial inoculum levels present on the seed also affect variety performance, and are influenced by seed source. A high standard of virus health can be achieved with breeders' seed but differences in fungal health (Phoma, Fusarium, Rhizoctonia, Helminthosporium etc) and blackleg (Erwinia) will vary with inoculum levels and growing conditions at seed sites. Examples of high incidence of these diseases have been found in seed stocks which bear no relationship to varietal susceptibility. Common origin seed does not necessarily eliminate these diseases but differences between varieties are likely to be due to variety rather than seed site.

REFERENCES

Flack, S.J. (1983). Effect of seed origin on potato yields. J. Natn. Inst. Agric. Bot., 16, 267-71.

Wurr, D.C.E. (1978). 'Seed' tuber production and management. In The Potato Crop, The Scientific Basis for Improvement, ed. P.M. Harris, pp. 327-54. London: Chapman & Hall.

MICROPROPAGATION - AN AID IN THE PRODUCTION OF NEW VARIETIES

P. Wooster and T.J. Dixon

INTRODUCTION

Potato breeding involves selection over several years from a large number of clones, and the breeder has many varieties to maintain. Seed numbers are inevitably low for both practical and economic reasons. This is often a limiting factor in the progress of a variety through advisory and developmental trials. It is therefore advantageous to accelerate seed production at this stage by means of micropropagation.

The Potato Section of the National Institute of Agricultural Botany (NIAB) produces its own seed of new varieties for Recommended List trials and other advisory trials. Prior to 1982 primary propagation was carried out by virus tested stem cuttings (VTSC); this gave improved multiplication rates over tubers and the plants were free from pathogenic fungi (Hirst & Hide 1966). In 1982 NIAB replaced stem cuttings with *in vitro* propagation (micropropagation); this gave better control of multiplication, improved hygiene and easier transportation of material. Micropropagation reduced seed production time for most varieties by one year.

METHOD

The method consists of a phase of micropropagation followed by transfer to soil under glass and finally transplantation into the field (Wooster & Dixon 1985). Polythene tunnels or screenhouses may be used when field conditions are not suitable for high value seed.

Laboratory stage

The potato is relatively easy to propagate *in vitro* using axillary buds as the basic propagule. Initial explants can be taken from plant stems or tuber sprouts. The time from initial explanation to the first subculture is generally longer than the time between subsequent subcultures. Multiplication rates of up to tenfold per month can be achieved (Hussey &

Stacey 1981). There are differences in multiplication rates between varieties (Table 1).

Multiplication rates can be altered by the addition of growth regulators to the media, e.g. nodal production can be enhanced by the addition of gibberellic acid (GA_3), but care must be taken as excess GA_3 may depress nodal production and also encourage adventitious meristems (Firmin 1984). The addition of cytokinin and auxins can be used to increase shoot or root growth (Hussey 1980). Tuberization _in vitro_ can be encouraged by the use of benzyl-amino purine in the media (Hussey & Stacey 1984).

Storage

Cold storage ($3^{o}C$) of cultures is used to prevent difficult labour peaks during subculturing. When required for transfer to soil medium or further subculturing cold-stored cultures are placed at room temperature for 24 h to avoid tissue damage. Cultures can be stored for up to 100 days at $3^{o}C$ with no deleterious effects on subsequent growth under glass. If the cultures are to remain _in vitro_ (e.g. for germplasm storage) the time between subcultures can be extended to 6 months or more. The production of tubers _in vitro_ can reduce the frequency of subculture even further.

Field stage

Micropropagated plantlets are transferred from _in vitro_ to peat blocks and grown for 3 to 4 weeks in a glasshouse until large enough (8-10 cm high) for transplanting.

Wind damage and drought stress are reduced by planting into furrow bottoms on pre-ridged land. The small plants are progressively earthed up until well established. In the first instance this needs to be done carefully by hand in order to avoid covering plants completely.

Where wild grazing animals like rabbits are prevalent, precautions should be taken to exclude them. Young transplanted micropropagated plants

Table 1. Mean number of nodes produced per plant per month (NIAB 1983).

Cultivar	Maturity	Nodes/month
Maris Bard	early	3.1
Ulster Sceptre		5.9
Fronika	mid early	7.5
Wilja	mid early	4.0
Desiree	late	4.8
Pentland Crown	late	2.8

appear to be much more palatable than plants grown from tubers, which are usually rejected. For control of rabbits, electric netting for the first 3 weeks after transplanting is very effective.

In order to avoid frost damage transplanting must be delayed until June (in Cumbria). This makes maturity 2 to 3 weeks later than plants grown from tubers. Therefore there may be a slightly greater risk of virus spread resulting from late summer aphid migrations. Normal precautions are taken to control both viruses and late blight (*Phytophthora infestans*).

Transplanted micropropagated plants outyield transplanted stem cuttings, but yields are usually lower than from tuber-grown plants. Tuber numbers per plant are also higher in micropropagated plants than stem cutting plants (Wooster & Dixon 1985). In 1983 the mean tuber number per plant ranged from 6.4 to 19.2.

Uniformity and stability

It is important in the propagation of potato seed stocks that genetic variation is minimised. In the method described here the basic propagule is the axillary bud and propagation does not involve the production of adventitious meristems. Studies have shown that cultures propagated in this manner have very good genetic stability and chimaeras are unlikely to occur any more frequently than in propagation by tubers (Denton *et al.* 1977, Westcott 1981). In some varieties micropropagated plants produce elongated or irregular tubers, but in subsequent generations these irregularities do not recur suggesting that this is an environmental effect (Wooster & Dixon 1985).

CONCLUSIONS

Micropropagation enables a large number of plants to be produced in a small space over a short period of time and also gives greater flexibility to seed production programmes. Micropropagation has been adopted by Agricultural Departments for the multiplication of definitive stocks prior to classification (certification) and by breeders for the rapid multiplication of new varieties. *In vitro* cultures are easily transported and may even be sent by post. The cultures are sterile and, if virus tested, provide an ideal method of moving plant material from one country to another for seed production and plant breeding purposes.

REFERENCES

Denton, I.R., Westcott, R.J. & Ford-Lloyd, B.V. (1977). Phenotypic variation of Solanum tuberosum L. cv Dr. McIntosh regenerated directly from shoot tip culture. Potato Res., 20, 131-6.

Firmin, D.M. (1984). Gibberellic Acid as a media additive for in vitro propagation of Potato (Solanum tuberosum L). J. Agric. Sci. Camb., 103, 703-4.

Hirst, J.M. & Hide, G.A. (1966). Attempts to produce pathogen free stocks. Rothamsted Exp. Sta. Rep. 1966, p. 129.

Hussey, G. (1980). In vitro propagation. In Tissue Culture Methods for Plant Pathologists, eds. Ingram, D.S. & Helgeson, J.P. pp. 51-61. Oxford: Blackwell Scientific Publications.

Hussey, G. & Stacey, N.J. (1981). In vitro propagation of Potato (Solanum tuberosum L.) Ann. Bot., 48, 787-96.

Hussey, G. & Stacey, N.J. (1984). Factors affecting the formation of in vitro tubers of potato (Solanum tuberosum L.). Ann. Bot., 53, 565-78.

Westcott, R.J. (1981). Tissue culture storage of potato germplasm I. Minimal Growth Storage. Potato Res., 24, 331-42.

Wooster, P. & Dixon, T.J. (1985). Application of micropropagation to potato seed production. J. Natn. Inst. Agric. Bot., 17, 99-106.

TESTING POTATO VARIETIES FOR RESPONSE TO DROUGHT AND IRRIGATION

S.J. Flack

Since water availability is a major factor in the production of potatoes, it must also be a major factor in the testing of new varieties. This becomes increasingly important as the potato area under irrigation increases, and the trial system is based on local practice to ensure that results are relevant to current cultural methods. The percentages of NIAB Recommended List trials at present irrigated are: 56% first early, 38% second early and 32% maincrop trials. These closely reflect the national situation in the UK.

The usefulness of drought resistant varieties is obvious in dry situations. There is also a need to know the response of varieties to irrigation. Since irrigation is costly and sometimes limited, it is important to know the likely return of marketable yield from different varieties when planning irrigation schedules. It should be possible to consider irrigation water in the same way as other inputs like fertilizers, fungicides or aphicides.

Hitherto, the assessment of drought resistance has been based on yield results at centres affected by drought, so information depended on the seasons in which the variety was in trials. However, it is difficult to determine differential variety effects of drought and irrigation from the data. Therefore, in order to reduce dependence on standard field trial data, a two-bed system was used to develop a standard test (Flack & Richardson 1984).

The two bed system consisted of a variable depth bed lined with an impervious membrane to give a linear decrease in available soil water, and a variable irrigation bed with a sprinkler line designed to give a linear decrease in applied water.

The deep end of the variable depth bed produced similar yields to those in the unirrigated part of the variable irrigation bed. Taking these as the base level yields, it was possible to quantify decrease in yield due to drought and increase in yield due to irrigation. Drought produced a yield decrease of 46% in Desiree and 91% in Romano. Irrigation produced a yield increase of 17% in Romano and 36% in Maris Piper.

Calculation of 1 to 9 ratings in Table 1 were derived from the data according to the following formulae. Because of the limited range of varieties included in the developmental test beds a 2 to 8 scale was used:-

$$\text{Drought Resistance} = 8 - 6\,\frac{v - d}{D - d}$$

where v = % yield loss in test variety
D = % yield loss in most severely affected variety
d = % yield loss in least severely affected variety

$$\text{Irrigation Response} = 2 + 6\,\frac{v - i}{I - i}$$

where v = % yield gain in test variety
I = % yield gain in most responsive variety
i = % yield gain in least responsive variety

There was good agreement with Recommended List drought resistance ratings derived from standard field trials. There was no direct linkage between drought resistance and irrigation response.

Table 1. Variety ratings for drought resistance and irrigation response (figures in brackets are Recommended List ratings)

Drought resistance		Irrigation response	
Desiree	8.00 (7)	Maris Piper	8.00
Pentland Crown	7.83 (7)	Record	4.83
Cara	7.64 (7)	Pentland Crown	4.68
Kingston	6.81 (6)	Kingston	4.60
Maris Piper	5.40 (4)	Foxton	4.42
Record	4.83 (4)	Desiree	2.63
Foxton	3.75 (4)	Romano	2.00
Romano	2.00 (4)	(Cara not included)	

REFERENCE

Flack, S.J. & Richardson, D.E. (1984). Testing potato varieties for response to drought and irrigation. Abstracts, 9th Trienn. Conf. EAPR, Interlaken, p. 148.

TESTING VARIETIES FOR RESISTANCE TO AND TOLERANCE OF *GLOBODERA PALLIDA*

G.M. Gurr and S.P. Kerr

RESISTANCE

Resistance to potato cyst nematodes is defined as the ability of a variety to prevent nematode multiplication. Until recently, tests for resistance have been done in pots containing predetermined nematode populations. Resistance has been expressed as Pf : Pi ratios, where Pi = initial population at planting and Pf = final population at harvest. Resistance to *Globodera rostochiensis* is almost complete, when derived from *Solanum tuberosum* ssp. *andigena*, and testing then represents few difficulties. Resistance to *G. pallida*, however, is partial, making assessment less straightforward.

A *G. pallida* resistant variety has usually been regarded as one which reduces the nematode population, i.e. Pf : Pi < 1. Varieties with Pf : Pi ratios of approximately 2 have been considered to have partial resistance. However, nematode multiplication is readily affected by initial populations and other environmental factors (Phillips 1984) and varieties permitting higher rates of multiplication may still be useful in the field. It is also necessary to be able to advise growers on rotations and nematicide usage for varieties with different levels of resistance.

In order to standardize variety assessment, a 1 to 9 scale would be valuable. It is suggested that this could be based on the difference between initial and final populations (Pf minus Pi) expressed as a percentage of a fully susceptible control variety as follows:-

	Very resistant							Very susceptible	
Scale value :	9	8	7	6	5	4	3	2	1
(Pf-Pi)% control :	<0	<5	<10	<15	<20	<25	<30	<35	<40

The linear scale could be replaced by a curvilinear transformation if data showed that this would be statistically more satisfactory.

The rating of 9 would be given to a variety with high resistance similar to that of current varieties with the H1 gene for resistance to G. rostochiensis, i.e. a Pf : Pi ratio at < 1. Varieties with very low resistance would be given a rating of 1, which would be the same as that for fully susceptible varieties.

In order to illustrate the application of this proposed scale, ratings have been calculated for several varieties in recent pot and field tests (Table 1). These confirm the view that varieties with quantitative resistance control nematode multiplication in the field better than in traditional pot tests using inoculated sterile soil.

The supplementary effect of nematicides in controlling nematodes on resistant varieties is shown in results from a field trial (Table 2). This indicates that Pf : Pi ratios are reduced much more by variety than by nematicides.

TOLERANCE

Tolerance is the ability of a variety to grow and yield well despite being attacked by nematodes. This is determined by comparing yields of field trials grown on nematode-infested land with and without nematicide treatment (Table 3). Varieties which are resistant to G. pallida (Sante, Cromwell & Morag) were less tolerant than the control varieties.

The role of resistant varieties is likely to increase in the light of the growing concern over the use of agrochemicals, particularly the highly toxic nematicides. Until highly resistant and tolerant varieties

Table 1. Provisional G. pallida resistance ratings calculated from pot and field data using proposed 1-9 scale (mean of eight pot and three field tests).

Variety:	Sante	Cromwell	Morag	Maris Piper (C)	Pentland Crown (C)
Pot tests:	6	4	2	-	1
Field tests:	7	7	7	1	-

(C) = fully susceptible control

Table 2. Effect of nematicide treatment on G. pallida populations expressed as Pf : Pi ratios (mean of three trials)

Variety:	Sante	Cromwell	Morag	Maris Piper (C)	Cara (C)
Untreated	1.6	2.4	1.4	10.9	9.9
Treated (Temik)	0.6	1.1	1.3	4.2	5.3

(C) = fully susceptible control

are produced, there will be a need to integrate varietal control with nematicides and rotations.

Table 3. Marketable yields in t/ha of varieties grown in *G. pallida* infested soils with and without nematicide treatment (mean of three trials).

	Sante	Cromwell	Morag	Maris Piper (C)	Cara (C)
Untreated	23.74	22.30	23.57	28.73	22.77
Treated	35.16	32.38	34.14	35.45	27.57
% increase	48.1	45.2	44.90	23.40	21.10

REFERENCE

Phillips, M.S. (1984). The effect of initial population density on the reproduction of *Globodera pallida* on partially resistant potato clones derived from *Solanum vernei*. Nematologica, 30, 57-65.

TESTING FOR GLYCOALKALOIDS

J. White

INTRODUCTION

Glycoalkaloids are a group of toxic compounds found in potatoes. They are present at high concentrations in the leaves and shoots of potato plants and in tubers which have been exposed to light. Fortunately in the normal tubers of current popular varieties the level of glycoalkaloid is quite low (Parnell *et al*. 1984). It is thought that the function of glycoalkaloids might be related to disease and pest resistance.

Although the level of glycoalkaloids in popular varieties is low, there is a danger that modern breeding techniques may result in elevated levels in newly-bred varieties. The reason for this stems from the practice of introducing disease resistance characters from wild plants or primitive cultivars. For example, a variety of potato bred for resistance to Colorado beetle (*Leptinotarsa decemlineata*) showed unacceptably high levels of glycoalkaloid in the tubers (Sinden & Webb 1974).

Glycoalkaloid level is expressed as milligrams per 100g fresh tuber. Potatoes with less than 20mg/100g are considered entirely acceptable, potatoes with more than 20mg/100g often have a bitter, metallic taste and potatoes containing 50-100mg have caused severe illness.

Trials have consistently shown significant varietal influence on glycoalkaloid content. In view of this and the toxicity effects it is felt that glycoalkaloid screening is of considerable importance.

METHODOLOGY

At the National Institute of Agricultural Botany all National List candidate varieties of potato are screened for glycoalkaloid content. Samples from each of three sites are analysed in triplicate.

The method employed is that of Blincow *et al*. (1982) developed at the Food Research Institute in Norwich. The method involves the extraction of glycoalkaloids into acid solution from a known weight of tuber. The acid extract is heated to 90°C for 2 hours to effect hydrolysis of the glycoalkaloids into the oligosaccharide and alkaloid moieties. The hydrolysate is then made alkaline with ammonia and the alkaloids are extracted into dichloromethane. Three dichloromethane extracts are bulked and the dichloromethane is removed by rotary evaporation. The alkaloids are then redissolved in a known volume of dichloromethane and assayed spectrophotometrically using a bromothymol blue dye binding technique.

RESULTS

A summary of the results obtained from a number of years for some common varieties is presented in Table 1. This shows that the levels of glycoalkaloid in current varieties is extremely low and that there is no danger from eating them. However, varieties have been encountered with single samples exceeding 30 mg/100 g and some breeders report selections with levels in excess of 100 mg/100g, so there is a potential danger.

In the future, semi-automated techniques will be developed to replace the laborious manual method of glycoalkaloid testing used at present. Enzyme linked immunosorbent assay (ELISA) and gas liquid chromatography (GLC) are under consideration together with alternative colourimetric methods suitable for auto-analyser systems.

Table 1. Glycoalkaloid levels in Recommended List Potato Varieties at Cambridge (mg/100g fresh tuber)

Variety	Year				
	1980	1981	1982	1983	1984
Cara	4.3	7.2	-	-	-
Desiree	7.0	8.8	6.7	8.5	13.8
King Edward	5.8	7.0	-	-	-
Maris Piper	4.3	4.5	4.5	7.1	5.0
Pentland Crown	6.9	5.6	7.6	9.3	15.0
Pentland Dell	13.0	9.3	-	-	-
Pentland Squire	4.7	8.8	5.9	-	-
Kingston	4.2	6.0	4.3	-	-

REFERENCES

Blincow, P.J., Davies, A.M.C., Bintcliffe, E.J.B., Clydesdale, A., Draper, S.R. & Allison, M.J. (1982). A screening method for the determination of glycoalkaloids in tubers of potato varieties. J. Natn. Inst. Agric. Bot. 16, 92.

Parnell, A., Bhuva, U.S. & Bintcliffe, E.J.B. (1984). The glycoalkaloid content of potato varieties. J. Natn. Inst. Agric. Bot. 16, 535.

Sinden, S.L. & Webb, R.F. (1974). Effect on glycoalkaloid content of six potato varieties at 39 locations. Technical Bulletin 1472, US Department of Agriculture.

METHODS FOR CALCULATING 1-9 VALUES TO EXPRESS THE RESISTANCE OF POTATO VARIETIES TO DISEASES

P.T. Gans and P. Wooster

INTRODUCTION

Quantitative resistance to diseases of potato varieties is commonly expressed as a value on a 1-9 scale in publications by testing authorities, in advisory and technical literature and in information supplied by breeders prior to official trials. It is an effective means of summarizing a wide range of characteristics, whereby a high value indicates a desirable and a low value an undesirable quality. The values indicate relative susceptibility and may be interpreted by comparing values with those of known varieties.

Valid comparisons between 1-9 values obtained by different organisations can only be made with a knowledge of:-

(1) the experimental procedures used,

(2) the method of calculating 1-9 values from the experimental data and

(3) the control varieties used for these calculations.

EXPERIMENTAL PROCEDURES

A summary of the procedures used at the National Institute of Agricultural Botany (NIAB) is as follows:-

Late blight (Phytophthora infestans) in foliage: Glasshouse plants are sprayed with a suspension of zoospores and incubated at 15°C and 100% rh with artificial light (Saunders 1968). The proportion of the leaf surface affected is assessed after 7 days.

Late blight in tubers: Newly harvested tubers are sprayed with a suspension of zoospores and incubated at 15°C and 100% rh in the dark (Stewart *et al.* 1983). The proportion of the tuber surface affected is assessed after 10 days.

Common scab (Streptomyces spp.): A soil moisture deficit to induce common scab is created by erecting polythene tunnels to cover field plots immediately prior to tuber initiation (Jellis 1975). The proportion of the tuber surface affected is assessed in the autumn.

Gangrene (Phoma exigua var. foveata): Tubers are damaged by travelling twice over a commercial reciprocating-type grader at approximately 10°C (Gans, unpublished) and the freshly damaged tubers are rolled in a sand, vermiculite and maize-meal medium containing the pathogen (Gray & Paterson 1971). The proportion of the tuber surface affected is assessed after 8 weeks incubation at 10°C.

Blackleg (Erwinia carotovora ssp. atroseptica): Seed tubers are stab-inoculated with a darning needle prior to planting in the field. Soil moisture is maintained at field capacity by irrigation to encourage development of symptoms. The proportion of diseased plants is assessed (Lapwood & Gans 1984).

Powdery scab (Spongospora subterranea): Exposure trials are planted in fields known to be contaminated. In the autumn the proportion of the tuber surface affected by pustules and the severity of canker symptoms are assessed.

Potato leaf roll virus (PLRV): Two replicates of 25 plants per plot are grown in an area where aphid populations are normally high and are flanked by rows of plants known to be infected with PLRV. Infection of the progeny is assessed by visual inspection of post-harvest glasshouse-grown plants (Richardson 1976) and by enzyme linked immunosorbent absorbent assay (ELISA).

Potato virus Y (PVY): The method is similar to that used for PLRV, except that infector rows contain plants with PVY^O.

Spraing caused by tobacco rattle virus (TRV): Four replicates of five plants are planted in fields known to be infested with viruliferous nematodes. Each plot has an adjacent control because of the irregular distribution of nematodes. An index of incidence and severity is calculated (Richardson 1970).

Visual keys for the assessment of all these diseases, except blackleg, PVY and PLRV are published in 'Disease Assessment Manual for Crop

Variety Trials' (NIAB 1985).

CALCULATING 1-9 VALUES FROM EXPERIMENTAL DATA

The calculations follow three steps:

(1) Transformation of data prior to statistical analysis.

(2) Division of transformed variety means by the transformed mean of the set of control varieties.

(3) Linear transformation using a formula specific to each assessment method but which remains constant from year to year.

This procedure may be applied equally to yearly results as to means or adjusted means, derived by the fitting constants method for several years. An example is given in Table 1.

DISCUSSION

Transformation prior to analysis: Angular transformation is used on data expressed as a percentage to reduce statistical error. A curvilinear transformation may also redress the balance where it is clear that differences between varieties in the susceptible range are exaggerated when compared with differences in the resistant range. Empirically, angular transformation has been found convenient for both purposes.

Control varieties: It is useful to include varieties representing the full range of responses. It is also more convenient to use the same control varieties for all maturity classes and where possible to include controls used by other organisations.

Table 1. Derivation of 1-9 values for common scab susceptibility for some varieties in 1982 (Analysis for 52 varieties)

Variety	Surface area affected		
	%	Arcsin %	1-9*
Pentland Crown	3.5	10.80	8.3
Maris Peer	11.2	19.51	6.2
Estima	16.6	24.04	5.2
Home Guard	18.6	25.55	4.9
Maris Bard	22.3	28.19	4.3
Maris Piper	36.7	37.28	2.2
LSD (p=0.05)		5.30	1.2
CV%		16.7	
Mean of controls		24.19	

$$*\text{1-9 value} = 10.7 - 5.5\left(\frac{\text{Arcsin \% surface area affected}}{\text{mean of controls}}\right)$$

1-9 scale: The scale is relative and hence the linear formulae imply the amount of resistance which is considered to be beneficial, or the amount of susceptibility which is deleterious. In practice they have been chosen to give least divergence from existing values of known varieties. They may be altered in the light of changed objectives or demands.

Support from field data: Wherever possible results are supplemented with data from the field, for late blight from field inoculation trials of provisionally recommended varieties, for common scab and gangrene from observations in yield trials, and for blackleg from data supplied by the UK seed classification authorities. Occasionally these may influence the published 1-9 values.

REFERENCES

Gray, E.G. & Paterson, M.I. (1971). The effect of the temperature of potato tubers on the incidence of mechanical damage during grading and of gangrene (caused by Phoma exigua) during storage, Potato Res., 14, 251-62.

Jellis, G.J. (1975). The use of polythene tunnels in screening potatoes for resistance to Common Scab (Streptomyces scabies). Plant Path., 24, 241-4.

Lapwood, D.H. & Gans, P.T. (1984). A method for assessing the field susceptibility of potato cultivars to blackleg. Ann. Appl. Biol., 103, 63-70.

NIAB (1985). Disease Assessment Manual of Crop Variety Trials. Cambridge: National Institute of Agricultural Botany.

Richardson, D.E. (1970). Assessment of varietal reactions to spraing caused by tobacco rattle virus. J. Nat. Inst. Agric. Bot., 12, 112-8.

Richardson, D.E. (1976). Diagnosis of potato leaf roll virus in post-harvest glasshouse tests. Pl. Path. 25, 141-3.

Saunders, P.J.W. (1968). Testing procedures for the assessment of resistance of potato varieties to late blight (1965-1967). J. Nat. Inst. Agric. Bot., 11, 349-60.

Stewart, H.E., McCalmont, D.C. & Wastie, R.L. (1983). The effect of harvest date and the interval between harvest and inoculation on the assessment of the resistance of potato tubers to late blight. Potato Res., 26, 101-7..

ESTABLISHING STANDARDS IN VARIETY ASSESSMENT

M. Talbot

INTRODUCTION

Of all agricultural crops the potato is the most varied in its uses and markets. Potato breeders and agronomists face a difficult task in identifying improved varieties. This paper presents an approach to the problem of establishing standards for choosing varieties for inclusion on the UK National List.

NATIONAL LIST TRIALS

National List yield trials consist of two replicates of 100 plants at three centres (Belfast, Cambridge and Edinburgh) in each of two years (Richardson, this volume). Candidate varieties are planted with at least two control varieties.

A range of disease, damage and quality tests are also conducted at each of the three centres over two years. In all disease and damage tests there are included at least one susceptible and one resistant control variety (Gans & Wooster, this volume).

The results of yield performance trials are summarized as in Table 1.

Table 1. Summary of National List yield trial data

	Total marketable yield of tubers						
		1981			1982	1982	
Variety	Edin.	Belf.	Camb.	Edin.	Belf.	Camb.	Mean
				(t/ha)			
Control - Pentland Crown	60.4	32.1	64.6	66.0	26.7	40.7	48.4
				(% of control)			
Candidates - D	107	79	100	102	111	110	104
E	58	84	82	86	93	94	82
SE of difference	5.4	11.7	9.3	16.0	17.4	11.5	11.5

For disease and damage assessments (Table 2) observations are transformed to make the data more amenable to standard statistical techniques; percentage figures are converted to angles and lesion sizes to logarithms. Variety means over tests are calculated and using these means varieties are scored on a scale 1-9 (susceptible-resistant). The resistant control variety is scored 8 and the susceptible control is scored 2; the scores for other varieties are derived by linear interpolation from the variety means. Varieties significantly better than the resistant control, or significantly worse than the susceptible control are given scores of 9 or 1 respectively. Where the mean difference between susceptible and resistant controls is small relative to the precision of the tests then the resistant-susceptible score range is narrowed; for example, from 2-8 to 3-6 for foliage blight, dry rot and skin spot. The aim is to set the scale range so that as far as possible a difference of 2 score points represents a statistically significant difference ($\underline{p}$ = 0.05) between corresponding variety means.

ESTABLISHING STANDARDS

The NL regulations state with regard to value for cultivation and use: "The qualities of the plant variety shall, in comparison with other plant varieties in a National List constitute, either generally or as far as production in a specific area is concerned, a clear improvement either as regards crop farming or the use made of harvested crops or of products produced from those crops."

In interpreting the regulations it is necessary to establish for each character, what represents a clear improvement and what is the baseline from which the clear improvement is to be measured.

There are several possibilities for a baseline. The approach adopted has been to use the level of performance which separates the top

Table 2. Summary of National List blight data

Variety		Angular % number of tubers infected					
		1981		1982			Score
		Edin.	Belf.	Edin.	Belf.	Mean	(1-9)
Controls	- Home Guard	87	90	78	57	78	2
	King Edward	76	90	64	48	70	3
	Stormont Enterprise	6	27	16	9	14	8
Candidates	- D	39	27	30	0	24	7
	E	58	18	65	13	39	6
SE of difference		6.8	-	9.9	-	8.8	

two-thirds and the bottom one-third of established varieties when ranked by the character.

If one was able to measure a variety's performance precisely then a clear improvement could be determined solely by what is commercially significant. However the tests are imprecise and a clear improvement must sometimes be measured in statistical terms; in these circumstances a clear improvement may be seen as the difference between a candidate variety and a standard which will be exceeded in trials with low probability when there is no real difference. Using information from past trials it is possible to estimate for specified probability levels the size of such differences.

Table 3 outlines the calculation of acceptance standards for yield. For example, the control variety Home Guard has a mean performance relative to other first early varieties that is approximately 2% above the thirty-third percentile. Therefore the baseline is set at 2% below the yield of Home Guard - 98%. At the same time experience from past trials indicates that a difference of 12% between candidate and baseline is enough to make one 90% sure that the true performance of the candidate lies above the baseline; a clear improvement will be 98% + 12% = 110%. Varieties which produce

Table 3. Acceptance standards for marketable tuber yield

		% Number of listed varieties with score equal or less than*									Acceptance standard (as % marketable yield of control variety)		
Maturity	Control Variety	1	2	3	4	5	6	7	8	9	Base Line	Sig. Diff.+	Clear Improvement
First early	Home Guard				13	<u>30</u>	57	73	100		98	12	110
Second early	Maris Piper					<u>33</u>	33	67	89	100	100	12	112
Early maincrop	Pentland Crown		2	5	20	49	62	72	92	<u>100</u>	83	10	93
Late maincrop	Dunbar Standard		2	5	20	<u>49</u>	62	72	100	100	100	7	107

*Source - NIAB classified list of Potato Varieties 1985: Scored 1-9 where 9 identifies high yields.

+Least significant difference between a candidate and a control variety that will only be exceeded in 10% of cases when there is no real difference between the two varieties. Underlined percentages identify the score class in which the control variety lies.

at least this performance in trials can be said to show a clear improvement in yield.

A similar approach is used in deriving acceptance standards for disease and damage characters. For most characters a resistance score of 3 or 4 separates the top two-thirds from the bottom third of varieties and is used as a baseline. The precision of tests varies between characters but, in general, a score of 6 or 7 represents a clear improvement.

THE APPLICATION OF STANDARDS

An example of the application of acceptance standards to two early maincrop candidate varieties is given in Table 4. The average performance of the candidate varieties in 2 years of trials at three centres is

Table 4. Acceptance standards for early maincrop National List candidate varieties

	Standards*			Candidate performance	
	Base Line	Sig. Diff+	Clear Improvement	Variety D	Variety E
Total marketable yield	83	10	93	104+	82-
Unmarketable yield	22			11	10
Wart disease↓	FR			S-	FI
Gangrene	3	0.8	6	5	4
Tuber blight	3	1.6	6	7+	6+
Common scab	3	3.0	7	6	4
Virus Y	3	1.0	6	4	4
Leafroll virus	4	1.0	6	4	4
Cyst nematode (PA1 and PA2)			R	S	S
(RO1)			R	R+	R+
Foliage blight	3	1.9	6	3	7+
Dry rot	3	1.2	6	5	5
Skin spot	3	3.1	6	5	1-
Virus X			FI	S	FI+
A			FI	FI+	FI+
B			FI	S	FI+
C	FI			FI	FI
Tuber damage - external	3	3.0	6	4	5
- internal	3	2.3	6	3	2

*Yield is expressed as % of the control variety's marketable yield; all other measures are on a 1-9 scale where 9 is resistant; the letters represent, F-field, R-resistant, S-suceptible, I-immune; a plus sign marks a clear improvement and a minus sign indicates a disimprovement.

+Difference between a candidate and a control variety that is likely to be exceeded only in 10% of cases when there is no real difference between the two varieties.

↓For Latin binomials of pathogens and pests, see Gans & Wooster (this volume).

presented along with the corresponding baseline and clear improvement scores.

For some characters, only a baseline or clear improvement score is shown; in the case of cyst nematode resistance, since most established varieties are susceptible, only a clear improvement has meaning as a positive standard; also, since most varieties are wart resistant or produce acceptable levels of unmarketable tubers, the emphasis is placed on minimum standards for these characters.

In Table 4, clear positive or negative features of a variety's performance are highlighted by a plus or minus sign; scores less than the baseline are given a minus sign and scores as good or better than a clear improvement are identified by a plus sign.

Presentation of results as in Table 4, together with the results of assessments on cooking quality, processing quality, and tuber appearance, assists the potato experts in quickly identifying the good qualities and the defects. Attention can then be concentrated on assessing how far these characters in combination may represent a general improvement in performance, above that of the lower third of established varieties.

DISCUSSION

The use of an index to combine several characters into a single measure of value has been considered. As yet no satisfactory method of combination has been achieved, for several reasons;

1) complex relationships exist between the expression of many potato characters and the value of that expression to the user, and such a relationship would be difficult to incorporate in an index since, for example, the value of increments in disease resistance scores is not the same at all levels of resistance;
2) in an evolving industry it is important that criteria reflect future needs and conditions, and an index based on an analysis of decisions several years earlier may not be a satisfactory basis for predicting potential value several years on;
3) more than one value index would be required to reflect the different potato processing and consumer requirements;
4) in using an index circumstances could arise where a valuable improvement in one character might be outweighed by minor deficiencies in several other characters.

CONCLUSION

Successful criteria for assessing variety performance should support rather than supplant the experience and judgement of potato experts in evaluating the potential of new varieties. The criteria should reflect for each character both the precision of the tests and the distribution of the character amongst established varieties.

CONSUMER QUALITY REQUIREMENTS IN THE UNITED KINGDOM

R.M. Storey and C.P. Hampson

INTRODUCTION

The commercial success of a new variety depends upon a number of factors, including field and storage characteristics and its suitability for a particular market or use. Information on some of these aspects is obtained by National List trials and the more comprehensive Recommended List trials of the National Institute of Agricultural Botany (NIAB) (Richardson, this volume). However, detailed information on the bulk harvesting and storage properties and the consumer acceptability of a new variety is not usually available until it is grown on a commercial scale. This may be several years after initial acceptance on to the National List and to bridge this gap the Potato Marketing Board (PMB) carries out a series of commercial scale trials and surveys.

TRIALS

Trials are grown from seed of common origin as part of the UK collaborative trial system. New varieties and their controls are grown according to local practice on collaborating farms in large plots (0.2 ha) for 2 years. Three sites are used for first and second early varieties and five for maincrop varieties. After harvest, produce of second early and maincrop varieties is transported to Sutton Bridge Experimental Station, where samples are taken for damage and quality assessments. Approximately 6 tonnes per variety and site are stored in 1-tonne pallet boxes at 7°C for between 4 and 5 months. CIPC is used for sprout control.

Damage levels, disease incidence and ware out-turns are assessed before and after storage. Additionally dry matter, reducing sugars and cooking qualities, including fry colour of potential processing varieties, are measured.

SURVEYS

At harvest and store unloading, consumer acceptability surveys are undertaken for second early and maincrop varieties. One tonne of each variety from each site goes through the Experimental Station's commercial washing and grading lines. 1.4 kg samples are packed with a multiple choice questionnaire before distribution to consumers. The questionnaire attempts to assess consumer reaction to the appearance of the uncooked sample, freedom from after cooking blackening and disintegration, flavour, texture and suitability for boiling, baking, roasting and chipping.

With first early varieties, consumer surveys are carried out at harvest and the small quantities of produce required are packed and distributed directly from the trial centres. The results from the field and storage trials, consumer and producer surveys are collated and published annually in PMB publications such as "Commercial Assessment of Recently Introduced Potato Varieties", to give an overall view of the varieties' commercial acceptability.

The results of the 1981/82 survey emphasise the importance to the consumer of the appearance of a new variety. Freedom from damage, blemishes and disease are the most important factors affecting initial purchase of the potato and the subsequent satisfaction after cooking (Potato Marketing Board 1983). Skin and flesh colour and tuber shape are less important but an agreeable flavour and texture for the particular market, with good cooking qualities, are essential to maintain consumer interest in a particular variety.

REFERENCE

Potato Marketing Board (1983). Potatoes. Report on a Survey of Consumer Attitudes and Behaviour 1981/82.

THE EFFECT OF FERTILIZER TREATMENTS ON A RANGE OF OLD AND NEW EARLY-MATURING POTATO VARIETIES

A.J. Thomson

INTRODUCTION

Plant breeders have been criticized for selecting varieties which respond well to increasing levels of inputs, especially fertilizers and irrigation. It is implicit in this criticism that these varieties do not perform as well as their contemporaries at lower input levels and that profitability for the farmer depends on continuing use of high-input husbandry. A trial was grown in 1985 to start to examine these assertions. Only data from early-maturing varieties are reported here although more extensive trials of maincrop varieties were also grown.

MATERIALS AND METHODS

Seed tubers of the old varieties Puritan and Early Rose (both grown before 1850), British Queen (1884), Great Scot (1909) and King George (1911) were obtained from the Department of Agriculture and Fisheries for Scotland. In 1984 the seed of these varieties was multiplied using stem cuttings in an aphid-proof glasshouse at Cambridge along with seed of the newer varieties Pentland Javelin (1968), Wilja (1972), Maris Bard (1974), Marfona (1977), Ukama (1980), Provost (1981) and a new variety from the Plant Breeding Institute, Rocket (1986).

Four replicates of a split-plot field trial were grown in 1985. The whole plots comprised three fertilizer treatments: (1) the normal fertilizer rate used on potato trials at Cambridge (1.76 t/ha of a 10:10:15 compound fertilizer), (2) half the normal rate and (3) no fertilizer at all. Because of shortage of seed, only four tubers were planted per plot. Planting was on 17 April and harvest on 1 August. The data recorded were weight of foliage and weight of tubers and from these the total weight of foliage plus tubers and the harvest index (i.e. tuber yield as a percent of total foliage plus tuber yield) were calculated.

RESULTS

There were significant differences between varieties in foliage yield but they were not attributable to differences between the old and new groups (Table 1). However the significant differences between varieties in tuber yield were mostly due to the difference between the two groups, e.g. none of the old varieties had a bigger tuber yield than even the poorest of the new varieties. There were also significant differences between varieties in total yield of foliage and tubers and the higher total yields of the new varieties were due to their bigger tuber yields.

The harvest indices decreased as the fertilizer rate increased (Table 2) and this trend was significantly linear ($p<0.001$). The old

Table 1. Mean yields, kg per plot, of foliage, tubers and total biomass

	Foliage wt.	Tuber wt.	Total wt.
Old varieties:			
Great Scot	1.89	2.72	4.61
Puritan	1.19	1.68	2.87
British Queen	2.21	3.67	5.88
King George	2.23	3.16	5.39
Early Rose	1.56	1.97	3.53
New varieties:			
P. Javelin	1.33	3.73	5.06
Marfona	1.73	4.67	6.40
Wilja	1.59	4.63	6.21
Provost	2.12	5.91	8.04
Rocket	2.06	5.11	7.17
Maris Bard	1.38	4.25	5.64
Ukama	1.77	4.74	6.50
SED	0.16	0.37	0.49

Table 2. Mean harvest index (%)

	No fert.	Half fert.	Full fert.	Mean
Great Scot	67.5	60.8	48.9	49.1
Puritan	66.4	56.6	50.9	58.5
British Queen	70.1	64.8	56.0	62.5
King George	65.1	62.9	48.5	58.7
Early Rose	61.4	58.0	50.3	55.9
P. Javelin	77.7	73.1	71.4	73.7
Marfona	76.3	74.5	68.7	73.0
Wilja	76.6	74.8	72.8	74.5
Provost	76.9	75.0	70.3	73.6
Rocket	74.6	70.3	70.1	71.3
Maris Bard	81.7	75.4	70.0	75.5
Ukama	77.8	72.7	69.9	72.9
Mean	72.2	68.1	61.9	67.4

varieties had the lowest harvest indices at all fertilizer rates and the difference between the two groups of varieties widened with increasing fertilizer.

DISCUSSION

The results show that potato breeders have achieved increases in yield by increasing harvest index, as the new varieties have a bigger proportion of their total yield as tubers at all fertilizer rates. Harvest indices decrease with increasing fertilizer but the decrease seems to be steeper in the old than in the new varieties. Therefore the new varieties respond to increasing fertilizer better than the old ones. The magnitude of this trend is not great and more work on a larger scale is required to examine and quantify the response. However the preliminary evidence does not support the contention that new potato varieties need high fertilizer inputs to outyield older varieties as they were clearly superior in yield at all fertilizer rates.

VARIETY TRIALS IN EGYPT, WITH SPECIAL REFERENCE TO DORMANCY

M.K. Imam

INTRODUCTION

In Egypt and other countries with a similar climate, potatoes are grown as both a spring crop and an autumn crop. The spring crop is usually planted in January with imported seed tubers, while the autumn crop is planted in September using local seed tubers taken from the preceding spring crop. Attempts to grow the spring crop using local seed taken from the preceding autumn crop are usually hampered by the dormancy of the freshly harvested tubers. Potato varieties grown in Egypt show wide variation in length of dormancy and in their reaction to the different chemical agents that break dormancy.

METHOD

In the present investigation, seed tubers of five potato cultivars Alpha, Berolina, Domina, Granola and Hilta were purchased from the local market and grown on Assiut University Experimental Farm during the autumn season of 1984. After harvest in January 1985, tubers were exposed to the following treatments to break dormancy:

1. Cutting of tubers (cut, whole).
2. Pretreatment temperature (ambient temp. 5-22°C, high temp. 25-30°C).
3. Chemical treatment: a) soak treatment: thiourea, potassium thiocyanate.

   b) gas treatment; carbon disulphide (CS_2), ethylene chlorohydrin, rindite.

The tubers were planted on 19 February 1985 in a factorial experiment and data were recorded on rate of emergence, final plant stand, shoot length, number of stems per plant and total yield.

RESULTS

Alpha was the slowest cultivar to emerge and had the lowest number of stems per plant. Results for final plant stand (56 days after planting) are presented in Table 1. In most instances Alpha gave lower values in each treatment than the other cultivars. The highest value for final plant stand in Alpha was 17, while it was 18.0, 18.7, 18.5 and 18.3 in Berolina, Domina, Granola and Hilta respectively. In Alpha higher values were generally recorded in the thiocyanate and CS_2 treatment. However, in Berolina, all the chemical treatments under investigation gave good emergence. Most of the gas treatments gave lower values than the soak treatments in the other three cultivars. Alpha and Domina had the lowest total yields per unit area, while Berolina was the top yielding cultivar. Within

Table 1. Effect of different dormancy breaking treatments on final plant stand (max = 20) and yield (ton $fed.^{-1}$)+ for five potato cultivars.

Treatment*	Cultivar									
	Alpha		Berolina		Domina		Granola		Hilta	
	Stand	Yield	Stand	Yield	Stand	Yield	Stand	Yield	Stand	Yield
I Ia	9.7	3.4	14.3	6.9	16.0	3.1	15.0	1.5	16.7	5.1
b	10.0	1.8	15.0	4.1	18.3	4.9	16.0	3.7	17.3	4.3
c	14.8	3.2	17.3	3.6	17.3	4.7	17.3	3.3	17.3	3.7
d	6.7	2.5	8.3	4.3	5.7	3.0	7.0	2.2	12.0	0.8
e	13.0	2.5	16.3	7.7	12.7	3.1	15.0	1.3	16.3	5.5
f	4.3	0.9	14.0	9.7	3.7	1.6	4.5	1.1	7.7	6.3
IIa	3.0	0.1	13.7	3.0	14.3	3.7	13.7	2.9	13.0	2.6
b	7.7	1.5	14.0	2.9	15.7	4.0	15.3	3.2	18.3	3.5
c	12.3	2.6	18.0	8.5	16.3	6.6	16.5	4.6	17.7	3.6
d	8.7	1.0	14.7	8.3	12.7	2.8	8.5	1.4	16.3	3.3
e	11.0	1.6	17.3	7.6	17.0	2.6	4.0	1.3	16.3	4.3
f	9.3	0.9	17.7	8.4	13.0	2.8	7.5	0.4	13.0	4.7
IIIa	7.7	1.9	15.7	6.0	14.0	4.5	13.0	4.1	13.7	4.6
b	17.0	2.7	15.0	3.0	18.0	5.1	18.3	0.6	11.3	4.3
c	14.3	3.9	13.7	5.5	14.7	4.9	15.7	2.8	13.3	4.8
d	4.7	0.6	9.0	3.3	13.7	3.1	10.0	1.7	13.7	7.4
e	16.0	1.6	17.0	8.2	16.7	2.6	15.5	2.0	18.0	4.8
f	3.0	0.2	14.0	5.4	9.3	2.0	10.0	0.7	9.0	2.9
IVa	5.3	1.0	12.3	6.6	8.3	0.8	7.0	0.6	12.3	1.9
b	13.3	1.1	16.3	3.6	18.7	3.0	12.0	3.2	10.7	2.6
c	10.7	2.5	18.0	4.1	8.3	3.5	13.3	3.6	4.7	2.5
d	3.3	0.9	17.0	9.1	8.3	2.0	12.3	1.3	13.0	4.3
e	16.0	1.5	17.7	5.6	18.3	3.3	14.0	0.9	17.3	4.7
f	6.0	0.7	17.0	8.4	9.3	2.0	7.3	0.9	14.7	5.6
LSD										
(p=0.05)	5.08	NS	6.39	4.23	6.04	NS	7.25	2.48	4.78	2.66
(p=0.01)	6.78	NS	NS	NS	8.07	NS	NS	NS	6.39	NS

*I = Whole tuber, high temp; II = whole tuber, ambient temp;
III = cut tuber, high temp; IV = cut tuber, ambient temp;
a = control; b = thiourea; c = ethylene chlorohydrin;
d = carbon disulphide; e = rindite

+ 1 feddan = 0.42 ha

both Alpha and Granola, the rindite treatments generally gave lower tuber yields, while in Hilta and Berolina the gas treatments gave higher values for tuber yield per plot than the soak treatments.

The correlation between total yield and plant stand is shown in Table 2. A significant correlation between total yield and final plant stand (at 56 days) was found in Alpha and Domina but not in other cultivars. However, a significant correlation between total yield and plant stand was obtained in four cultivars (Alpha, Berolina, Domina and Hilta) when plants were counted 35 days after planting. This indicates the importance of early emergence of plants in reducing the failure of tuber formation.

CONCLUSION

It is recommended that further studies are carried out on early cultivars such as Berolina, which respond favourably to the chemicals tested. Such a combination of cultivars and chemical treatment may be promising for potato production in the spring planting in Egypt, using locally produced seed tubers taken from the preceding autumn crop. It would seem that late cultivars such as Alpha, which is very popular in Egypt, are not good candidates for local propagation by these techniques.

Table 2. Correlation between total yield and plant stand at two dates after planting

	Correlation coefficient (r) for	
Cultivar	Total yield and plant stand 35 days after planting	Total yield and plant stand 56 days after planting
Alpha	0.5595**	0.7252**
Berolina	0.6460**	0.3444
Domina	0.4923**	0.6901**
Granola	0.3920	0.3711
Hilta	0.4500*	0.3319

** Significant at $p = 0.01$
* Significant at $p = 0.05$

SEMI-CONVENTIONAL BREEDING METHODS

EFFICIENT UTILIZATION OF WILD AND PRIMITIVE SPECIES IN POTATO BREEDING

J.G.Th. Hermsen

INTRODUCTION

Wild species are products of natural evolution in centres of diversity. They are not manipulated on purpose or used by man. Their evolution is brought about by the interaction of abiotic and biotic factors with the genetically variable plant populations. This has resulted in the wealth of variation found among and within species. In the centres of diversity the species may coexist but remain largely separated by external and internal barriers developed in the course of evolution.

Domesticated or cultivated plants have derived from the wild species and are manipulated by man both agronomically and genetically in order to improve their adaptation to human needs. The potato has its main centre of diversity in the mountainous regions of Latin American countries. Here the potato was domesticated and has been grown for several millennia. These primitively cultivated potatoes comprise eight *Solanum* species: the diploids *S. phureja*, *S. stenotomum*, *S. goniocalyx* and *S. ajanhuiri*, the triploids *S. chaucha* and *S. juzepczukii*, the important tetraploid *S. tuberosum* ssp. *andigena* and the pentaploid *S. curtilobum*.

Not until the latter half of the 16th century was the potato introduced into countries outside Latin America, first into Europe, from there into North America and later on all over the world. The immediate ancestor of our present-day autotetraploid *S. tuberosum* ssp. *tuberosum* cultivars is the autotetraploid *S. tuberosum* ssp. *andigena*. Although initially only few genotypes of *S. tuberosum* ssp. *andigena* were introduced, a large number of different *S. tuberosum* ssp. *tuberosum* cultivars adapted to various conditions has been derived from that material. In the middle of the 19th century people became aware of the vulnerability of the potato crop and started thinking of introducing new material from Latin America. Systematic collection of wild potato species started only in the 1920s and has continued and increased since then. Initially diseases and pests stimulated collection of new genetic resources, both wild and primitive species. In the 1960s the desirability of broadening the genetic base of potato breeding (Simmonds 1966) was appreciated. This resulted in the introduction of a large variety of primitive forms, both diploid and tetraploid, and the initiation of adaptation programmes to long-day conditions, first in the UK and soon thereafter in the Netherlands and the USA (Plaisted, this volume). Large-scale collection of primitive forms has been performed by the International Potato Center (CIP) since its foundation in Peru in 1972. The great potential of wild species and primitive forms of potato has now been widely accepted and most new European and US cultivars have one or more wild or primitive species in their ancestry. The most extensive use of wild and cultivated *Solanum* species is being made by CIP, both directly and through research contracts with some Universities and Research Institutes in Europe and the USA.

Potato breeders have not always made efficient use of related species. In this chapter after a brief exposition of the valuable characters in tuber-bearing *Solanum* species, some methods will be described which allow more efficient use of these resources for potato breeding.

VALUABLE CHARACTERS IN TUBER-BEARING SOLANUM SPECIES

A survey of the valuable characters in tuber-bearing *Solanum* species must be incomplete, because these species have been studied only partially and only for a limited number of desirable traits. Furthermore, the aim of this section is not to give an exhaustive review. Therefore references will for the greater part be omitted.

Some general remarks may be useful before presenting relevant details about specific characters. *S. tuberosum* ssp. *tuberosum*, *S. tuberosum* ssp. *andigena*, *S. chaucha* and the triploid ecotypes of normally diploid species are autopolyploid. The other polyploid species are predominantly allopolyploid i.e. functional diploids. Approximately 70% of the tuber-bearing *Solanum* species are diploid. The total number of species according to Hawkes' (1963) revision amounts to 160. Since then several new species have been discovered and described, mainly by Ochoa (CIP, Peru). So the total number may now approach 200 although by no means all species have been investigated for breeding purposes. In addition (apart from a few exceptions) only few accessions per species have been evaluated and mostly for a low number of specific traits. Much work still needs to be done.

The use of wild species and especially primitive forms has

greatly increased since World War II. In some countries most new potato varieties have related species in their ancestry mainly because of the fight of breeders against late blight (Phytophthora infestans), virus diseases and cyst nematodes (Globodera spp.). The special position of CIP in this context is well known and does not need further elucidation. When working with potato species unusual results are not uncommon and may be ascribed to insufficient knowledge about valuable characters in the species. It is desirable or even essential to include potato species in breeding programmes when the characters required are

- largely lacking in S. tuberosum;
- present in S. tuberosum but at a low level;
- present in S. tuberosum, but are unstable or are inherited in a highly complex manner.

These general statements will be illustrated by relevant experimental data on some important characteristics.

Potato virus diseases

Monogenic dominant immunity or extreme resistance to potato virus X (PVX), potato virus Y (PVY) and potato virus A (PVA) has been detected in S. tuberosum ssp. andigena (PVX, PVY), S. acaule (PVX), S. stoloniferum (PVY, PVA), S. brevidens and S. etuberosum (PVY) and in S. chacoense (PVY, PVA). Immunity does not occur in pure S. tuberosum, but has been introduced into cultivars by breeding.

Valuable oligogenic resistance to potato virus M (PVM) has been found in S. gourlayi and S. spegazzinii and is being introduced into cultivars.

Polygenic partial resistance to infection by PLRV is known to occur in a group of related cultivars. However, very high levels of PLRV resistance have recently been discovered in S. brevidens, S. etuberosum and in S. acaule. The nature and inheritance of this valuable new resistance are being investigated and a high resistance to multiplication is known to be at least one of the components. At present this resistance is available in advanced material.

Bacterial wilt (Pseudomonas solanacearum)

Valuable bacterial wilt resistance, associated or not associated with resistance to root knot nematode (Meloidogyne incognita), has been detected in S. phureja (oligogenic and dominant), S. sparsipilum, S.

multidissectum, S. stenotomum, S. chacoense, S. raphanifolium and in S. pinnatisectum and S. jamesii. A broadly based breeding programme at CIP aims at combining bacterial wilt resistance with resistance to root knot nematode and heat tolerance. All these are needed - together with appropriate agronomic measures - to get bacterial wilt effectively under control in tropical lowlands.

Late blight (Phytophthora infestans)

Polygenic resistance to late blight is widespread among the tuber-bearing Solanum species and also among potato cultivars. Breeding for this type of resistance is hampered by its complicated inheritance and especially the association of polygenic resistance with lateness. Hypersensitivity ("R-gene resistance") appears to be concentrated in Mexican species. Little is known about the occurrence of R-genes in South American species. Because of geographic isolation it may be hypothesized that polygenic resistance in Mexican species is genetically complementary to polygenic resistance in South American species suggesting an obvious strategy for breeding for high-level and stable polygenic resistance. Nothing is known about a possible association, if any, between the expression of polygenes and R-genes. Knowledge is also lacking about the question as to why polygenic resistance is not present or not expressed in early varieties, at least not in the haulms. Although there is a wealth of sources of late blight resistance, it is hard to determine the best strategy for breeding for high and stable resistance in early varieties as long as the aforementioned problems have not been solved.

Potato cyst nematodes (Globodera rostochiensis and G. pallida)

More than 20 Solanum species, mainly from the Series Tuberosa, Megistacroloba and, more recently, Circaeifolia have been identified as sources of potato cyst nematode (PCN) resistance. Work in the Netherlands has shown that resistance to specific pathotypes is rather common in wild species. It has been widely used in breeding because it is very effective and it is simply inherited. However, this resistance regularly breaks down because of new pathotypes. A breeding strategy based on polygenic resistance, although complex, is the only way to a long-term solution. There are good sources of polygenes among wild species.

Resistance to insects

Insect resistance has been investigated most extensively in the United States and at CIP, Peru. Many accessions of a large range of Solanum species were assessed at the University of Minnesota for resistance to different aphids, leaf hopper (Empoasca fabae) and flea beetle (Epitrix cucumeris). Similar assessments were carried out in breeding material at Cornell University where research focussed on the mechanisms and inheritance of insect resistance especially in S. berthaultii and S. polyadenium. (Research on S. berthaultii was initiated in the UK by R. Gibson). The S. berthaultii resistance is based on the presence of glandular hairs. It is being included in potato breeding programmes in several countries. CIP has initiated research on resistance to tuber moth, Phthorimaea operculella, and detected resistance in S. sparsipilum, S. tarijense, S. phureja, S. stenotomum, S. tuberosum ssp. andigena and S. sucrense. Resistance to Colorado beetle (Leptinotarsa decemlineata) in S. chacoense and S. demissum was studied circa 1950 in Germany and The Netherlands, but research has not been continued because of effective means of chemical control.

Several wild species, including S. chacoense, S. demissum, S. berthaultii, S. polyadenium, S. stoloniferum and S. tarijense, have each shown resistance to several insects. It may therefore be efficient to breed for resistance to various pests in one programme, particularly when resistance is specific. Inheritance of resistance is as a rule complicated.

Frost tolerance

Frost tolerance has been demonstrated to occur in at least 35 Solanum species. Breeding for resistance has been carried out in several countries, using S. acaule as the main source. Complicated inheritance and assessment problems resulted in very slow progress. In the mean time basic research on the nature of cold tolerance has yielded valuable results. Useful laboratory methods have been developed for detecting cold tolerant genotypes.

Quality traits and yield

The primitive forms of potato are a rich source of various quality traits. Systematic studies in this field are rather scanty. There is much variability in taste and flavour, starch content, protein content and tuber texture in primitive forms. The low level of reducing sugars after cold storage in S. phureja, S. chacoense and S. goniocalyx is promising for processing.

Strikingly high yields were found in the 1970s among tetraploid derivatives of *S. vernei*, *S. goniocalyx*, *S. phureja*, *S. chacoense* and even *S. demissum*. Apparently many *Solanum* species can rapidly be nobilized suggesting close relationships with cultivated potato.

Hermsen (1977) recommended a systematic research in the wild species *S. chacoense*, *S. vernei*, *S. microdontum* and others with the direct aim of broadening the genetic basis of potato breeding, analogous to the successful adaptation selection programmes with *S. tuberosum* ssp. *andigena* and *S. phureja*/*S. stenotomum*. Peloquin *et al.* (1985) reported that such direct selection in the wild species studied has not been very effective but crosses between particular *S. tuberosum* haploids and a number of diploid wild species produced many high yielding F hybrids. Highly productive F_1's were also obtained from interploidy crosses between potato cultivars and diploid wild clones with first division restitution (FDR).

Conclusion

Several good arguments can be found in this section for using wild species and primitive forms in potato breeding. If these are accepted, questions arise about detecting the best sources for desired traits and about applying the most efficient procedures for their utilization. These questions are the main subject of the following sections.

CHOOSING SOURCES OF DESIRABLE CHARACTERS

Before resorting to related species, the potential of *S. tuberosum* should be carefully considered. When related species are apparently needed, the following criteria for choosing the most suitable should be taken into account.

(i) The degree of relationship with *S. tuberosum*.

Close relationship warrants good crossability, gene transfer via normal recombination and rapid nobilization. Remotely related species can only be recommended if they carry highly valuable genes which are known not to occur in more closely related species. The reasons for such recommendation are obvious. Wide crosses are often hampered by partial or even complete crossability barriers, necessitating either large scale pollination or ploidy manipulation or the use of bridging species or - ultimately - somatic hybridization. Restricted gene recombination limits gene transfer and may require many backcrosses for nobilization. Elimination of chromosomes and loss of genes are not uncommon.

(ii) The ploidy level.

As pointed out by Peloquin (1982) and demonstrated by Peloquin et al. (1985) diploidy is the most efficient ploidy level for interspecific gene transfer and nobilization. Therefore it is highly desirable to breed excellent diploid S. tuberosum clones to use as partners not only in crosses with diploid related species, but also with allopolyploids (= functional diploids). Allotetraploids can readily be crossed with diploid potato but hardly with autotetraploid potato, unless the allotetraploid is first doubled to the octoploid level. The most efficient way to include allotetraploids in a breeding programme at the diploid level is according to the scheme

allotetraploid x advanced diploid → triploid F_1

triploid F_1 x advanced diploid → BC_1 etc.,

unless the triploid F_1 plants are both male and female sterile. It should be realized that diploid and allotetraploid species constitute nearly 90% of all known tuber-bearing Solanum species.

(iii) The type, level, stability and inheritance of resistances or the level and inheritance of other desirable characters in Solanum.

The most valuable resistance is monogenic, dominant immunity. Resistances based on unspecific (e.g. mechanical) barriers to pests and diseases also occur, e.g. insect resistance in S. berthaultii due to glandular trichomes. Durable resistance based on polygenes is common but polygenic resistance as a rule is partial and not easy to handle in breeding, unless it has a high heritability.

It should be pointed out here that many species combine major genes for hypersensitivity to a disease or pest with polygenes for durable resistance. So the polygenes should receive ample attention and not be wasted, as has happened so often in the past.

Based on the criteria mentioned and on experimental evidence, the order of preference in the choice of related species for breeding purposes should be as follows:

- Di(ha)ploid S. tuberosum.
- Diploid and tetraploid primitive forms.
- Diploid wild species from Series Tuberosa.

- Diploid wild species from Series Commersoniana or Megistacroloba.
- Allotetraploid species from Series Tuberosa.
- Allotetraploid Longipedicellata species.
- Hexaploid Tuberosa species.
- Hexaploid Demissa species.
- Diploid species from remotely related Series.
- Polyploid species from remotely related Series.

However, it should be realized that this is a general scheme, into which some species do not fit.

When for a desired trait two or more equivalent sources are available, the choice could be based on the simultaneous presence in the source species of other desirable traits. In this context some complications should be mentioned:

1. One species usually has several desirable characters. For example Hawkes and Hjerting (1969) mentioned 25 in S. chacoense. However, these traits are predominantly scattered over different accessions and individuals.
2. Basically the same holds true for one character: it usually occurs in certain accessions and, within an accession, only in a restricted number of individuals of that accession.
3. Polygenes controlling one character may also be, and usually are, distributed between different individuals.

These facts have implications both for the evaluation of species and for the breeding procedures, as will be explained in the following chapters.

EVALUATION AND "PREBREEDING"

Evaluation should be carried out primarily in the most suitable diploid species, and for those characters which justify the efforts. A sufficient number of accessions per species and individuals per accession should be tested because of the heterogeneity both within species and accessions. A good evaluation will provide information about (i) the occurrence of desired characters in the species, (ii) how they are distributed among and within accessions and (iii) whether different levels or different types of resistance occur. A good evaluation enables the selection of the most suitable individuals to start a "prebreeding" procedure as propounded by Hermsen (1977).

Prebreeding involves a number of breeding activities within a species. The aim is (i) to upgrade the level of polygenic traits, (ii) to combine genes for different components and types of resistance, or genes for different desired traits and (iii) to study the genetics of the desired characters within the species in order to avoid erratic genetic ratios or distributions caused by meiotic irregularities and pre- or post-zygotic interspecific barriers. A reliable genetic analysis provides the material and knowledge needed to choose the best parents for further breeding (nobilization) and to establish the most efficient system for nobilization and gene transfer.

"NOBILIZATION"

The following factors need to be considered for efficient nobilization:

1) feasibility of breeding at the diploid level which, for nobilization, is to be preferred to the tetraploid level.
2) the genetic basis of the characters to be transferred.
3) the availability of closely related source material with the right level and stability of the desired characters.
4) the number of characters which a breeder wants to carry on simultaneously through the generations of nobilization.
5) the frequent need of using more than one species to obtain sufficiently high levels of polygenic characters.

Related to the last factor - more than one species contributing to one polygenic trait - different approaches to nobilization are feasible, of which the "columnar approach" (Hermsen 1974) and the "gene pool approach" are extremes.

In the columnar approach the wild species are nobilized separately and then the products of nobilization intercrossed. The chances of loosing valuable (poly)genes are minimized, but two or more nobilization programmes have to be handled simultaneously for each character.

In the gene pool approach the sources before nobilization are brought together into one population or gene pool which is then enhanced by population breeding or recurrent selection. In this case only one gene pool has to be handled per character, but a large population is needed to prevent loss of genes. A good example of the latter approach is the breeding procedure applied by Hermsen (1985) for late blight resistance from three Mexican species: the diploids S. verrucosum (code V) and S. bulbocastanum (code B) and the allotetraploid S. stoloniferum (code S) according to the scheme:

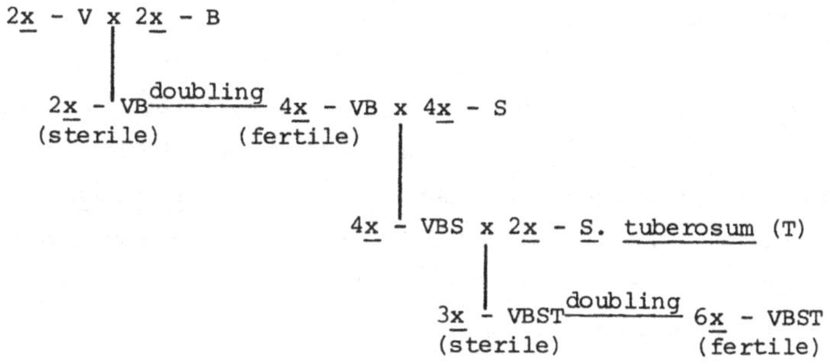

```
2x - V x 2x - B
       |
   2x - VB ---doubling--- 4x - VB x 4x - S
   (sterile)          (fertile)  |
                                 |
                         4x - VBS x 2x - S. tuberosum (T)
                                  |
                          3x - VBST ---doubling--- 6x - VBST
                          (sterile)                (fertile)
```

followed by nobilization of the quadruple species hybrid.

Procedures intermediate between columnar and gene pool approach are often used. The gene pool is usually created after one or more generations of separate enhancement or prebreeding of the source species, especially when these species are remote relatives of *S. tuberosum*. This will briefly be illustrated by four examples.

1. Breeding for bacterial wilt resistance from at least four different sources is being carried out at CIP using an intermediate procedure as described by Schmiediche (1985).
2. Different diploid wild species may, after crosses with selected *S. tuberosum* dihaploids, give rise to surprisingly high-yielding F_1 clones, as reported by Peloquin *et al.* (1985) for nine species of the Series *Tuberosa*, *Commersoniana* and *Megistacroloba*. An efficient breeding scheme would be:
   (i) Selected haploids x 1st species

   F_1

   Selected haploids x 2nd species

   F_1

   Selection for tuber yield and the desired trait(s) from the wild species

   (ii) Intercrossing selected plants from both F_1's

   (iii) Further enhancement of the progeny.
3. A different approach has to be applied when trying to combine late blight resistance from distantly related Mexican species (e.g. *S. bulbocastanum*, *S. pinnatisectum*) with late blight resistance from the closely related South American *Tuberosa* species (e.g. *S. microdontum*, *S. vernei*). In such programmes the Mexican species first have to be made accessible to

hybridization with preselected genotypes from the South American species. Then further enhancement of the gene pool is feasible.

4. A final example is increasing PLRV resistance by combining infection resistance from *S. tuberosum* cultivars with extreme multiplication resistance from the nontuberous Chilean species *S. brevidens* and *S. etuberosum*. The latter species first have to be made accessible via bridging species and the resulting quadruple species hybrids crossed with infection-resistant cultivars (Hermsen 1985).

Basically the same methods can be applied for combining various characters from different species, or from different accessions of one species, and for simultaneous transfer of these characters through the generations of nobilization.

EVALUATION OF THE EBN HYPOTHESIS

The endosperm balance number (EBN) is a hypothetical number assigned to each plant on the basis of its crossing behaviour to a standard species (Johnston *et al.* 1980). Viable endosperm is essential for normal development of the embryo.

In *intraspecific* crosses only one species is involved and the EBN is identical to the number of genomes. It is now established that the endosperm develops normally only when the maternal:paternal genome ratio in the endosperm is 2:1. This is always the case when the parent plants have the same ploidy level. However, this ratio deviates from 2:1 in intraspecific tetraploid-diploid crosses: the endosperm then degenerates and hence the triploid embryos abort.

In *interspecific* crosses this 2:1 genome ratio leads to viable endosperm only if the species genomes have the same "weight" or EBN. If species genomes have different weights, the "effective ploidy", which differs from the real ploidy, should be used in determining the maternal:paternal ratio in the endosperm. The effective ploidy is expressed in EBN units and an interspecific cross produces viable endosperm, and thus is successful only if the maternal:paternal EBN-ratio is 2:1 in the endosperm.

Consequently interspecific crosses are successful only if both parental species have the same EBN.

The EBN-hypothesis may be helpful in predicting the success of interspecific crosses (Ehlenfeldt & Hanneman 1984), but only if the parental EBNs are known. The following objections cast some doubt upon its usefulness.

1. EBNs have to be assigned empirically and *a posteriori*.

2. The hypothesis does not take into account prezygotic barriers which usually evolve between sympatric species (Stebbins 1971) to prevent fertilization by alien pollen.
3. In interspecific crosses, a nearly continuous variation in degree of compatibility, as well as in endosperm quality and quantity, is fairly common. So how can compatibility and incompatibility be distinguished without ambiguity?
4. The hypothesis does not fit for crops such as sugarbeet, watermelon and carnation, where high frequencies of triploids result from intraspecific interploidy crosses.
5. Continuous variation between an optimum result and no result at all is a biological reality not covered by the hypothesis, which therefore is not flexible enough. Why should EBNs always be whole figures and not fractions?

USEFULNESS OF NEW TECHNIQUES IN TISSUE CULTURE AND GENETIC MANIPULATION FOR INTERSPECIFIC HYBRIDIZATION

Embryo culture

Embryos with defective endosperm can be rescued by *in vitro* culture of either dissected embryos or ovules (see Collins *et al.* 1984 for detailed information). Embryo culture is a powerful tool in interspecific and intergeneric hybridization programmes.

Somatic hybridization

In evaluating somatic hybridization in potato, technical problems of fusion, selection and regeneration will not be considered, nor will somaclonal variation. For details of these aspects see Jones (this volume). The following possibilities can be explored using somatic hybridization but not sexual hybridization:

(i) the integral fusion of completely intact (nonreduced) nuclei ($2n$ - FDR gametes from desynaptic clones closely approach this result).

(ii) the production of cytoplasmic hybrids.

(iii) the hybridization of plants that are fully sterile, do not flower, are sublethal, are obligate apomicts or possess a long juvenile phase (trees).

Somatic hybridization undoubtedly broadens the range of species and genera that can be hybridized. However, there are good reasons for having doubts about the usefulness of somatic hybrids between species which cannot be hybridized sexually:

(i) chromosome elimination may occur, leading to sterility.

(ii) if chromosome elimination does not occur, preferential pairing may prevent gene exchange between the parental chromosomes.

(iii) the first backcross to the crop may be at least equally as difficult as producing the F_1 hybrid. Carrying out somatic hybridization twice in succession may raise the number of chromosomes to a level where normal meiosis can no longer be expected.

It may be argued that somatic hybridization provides a shortcut in those cases where, along classical lines, genes can be transferred to cultivars only via bridging species. The diploid species *S. brevidens* is a unique example for comparing sexual and somatic hybridization. Somatic hybrids have been produced between *S. brevidens* and a *S. tuberosum* cultivar (Barsby *et al.* 1984) and between *S. brevidens* and a diploid primitive form (Austin *et al.* 1985). The first somatic hybrid was a completely sterile hexaploid, and so a dead end. The second hybrid was a partially fertile allotetraploid genotype (many plants, displaying some somaclonal variation), which may be useful for further nobilization through sexual crosses.

Classical breeding was carried out with *S. brevidens* (also with its close relative *S. etuberosum*) from 1980-1984 (Hermsen 1985). Two bridging species had to be used before hybridization with potato cultivars was successful. At present thousands of tetraploid true hybrid seeds are available. The classical approach may have taken more time, but has the advantages (i) that late blight resistance was introduced through the bridging species (*S. brevidens* is extremely susceptible); (ii) that there is ample opportunity for selection not only for PLRV resistance (from *S. brevidens*), but also for agronomic characters in the large segregating populations available.

Single gene transfer through bacterial vectors

Excellent vectors are presently available for single gene transfer into plants; such gene transfer has already been realized. However, when considering the agronomically important characters, including durable resistances, very few are controlled by a single gene. Genes for storage proteins, hypersensitivity, immunity to a few viruses, resistance to herbicides and antibiotics are examples of genes which could be used in transforming potato cultivars once they are identified and isolated. An advantage of single gene transfer *in vitro* is that the nobilization phase is superfluous (no backcrosses needed). On the other hand it may be questioned whether polygenes may, in the foreseeable future, be handled in vitro. Nothing is known about the action and products of single minor genes and little about their interaction, their numbers and their distribution on and among chromosomes.

REFERENCES

Austin, S., Baer, M.A. & Helgeson, J.P. (1985). Transfer of resistance to potato leaf roll virus from Solanum brevidens into Solanum tuberosum by somatic fusion. Plant Science 39, 75-82.

Barsby, T.L., Shepard, J.F., Kemble, R.J. & Wong, R. (1984). Somatic hybridization in the genus Solanum: Solanum tuberosum and S. brevidens. Plant Cell Reports 3, 165-7.

Collins, G.B., Taylor, N.L. & Deverna, J.W. (1984). In vitro approaches to interspecific hybridization and chromosome manipulation in crop plants. In Gene Manipulation in Plant Improvement. ed. J.P. Gustafson, Plenum Press, New York. pp. 323-83.

Ehlenfeldt, U.K. & Hanneman, R.E. (1984). The use of endosperm balance number and 2n gametes to transfer exotic germplasm in potato. Theor. Appl. Genet. 68, 155-61.

Hawkes, J.G. (1963). A revision of the tuber-bearing Solanums (second edition). Scott. Pl. Breed. Station Record pp. 76-181.

Hawkes, J.G. & Hjerting, J.P. (1969). The potatoes of Argentina, Brazil, Paraguay and Uruguay. A biosystematic study. Ann. of Bot. Memoir 3, 82-99.

Hermsen, J.G.Th. (1974). The utilization of non-cultivated wild Solanum species. Rept. Plann. Conf. on Utiliz. of Genet. Res. in the Potato, Lima. Peru. pp. 40-6.

Hermsen, J.G.Th. (1977). Incorporation of new germ plasm: wild species. Rept. Plann. Conf. on Utiliz. Genet. Res. in the Potato. Lima, Peru. pp. 81-100.

Hermsen, J.G.Th. (1985). Utilization of wide crosses in potato breeding. Rept. 26th Plann. Conf., 1983, CIP, Lima, Peru. pp. 115-32.

Johnston, S.A., Den Nijs, A.P.M., Peloquin, S.J. & Hanneman, R.E. (1980). The significance of genic balance to endosperm development in interspecific crosses. Theor. Appl. Genet. 57, 5-9.

Peloquin, S.J. (1982). Meiotic mutants in potato-breeding. Stadler Genetics Symposia 14, 1-11.

Peloquin, S.J., Okwuagwu, C.O., Leue, E.F., Hermunstad, S.A., Stelly, D.M., Schroeder, S.H. & Chujoy, J.E. (1985). Use of meiotic mutants in breeding. Rept. 26th Plann. Conf., 1983, CIP, Lima, Peru. pp. 133-41.

Schmiediche, P. (1985). Breeding bacterial wilt (Pseudomonas solanacearum) resistant germplasm. Rept. 26th Plann. Conf., 1983, CIP, Lima, Peru. pp. 45-55.

Simmonds, N.W. (1966). Studies of the tetraploid potatoes. III Progress in the experimental re-creation of the Tuberosum Group. J. Linn. Soc. (Bot.) 59, 279-88.

Stebbins, G.L. (1971). Processes of Organic Evolution, 2nd edition. Englewood Cliffs, New Jersey: Prentice Hall.

ADVANCES AND LIMITATIONS IN THE UTILIZATION OF NEOTUBEROSUM IN POTATO BREEDING

R.L. Plaisted

INTRODUCTION

Dr. N.W. Simmonds began his programme of selection for adaptation of Solanum tuberosum ssp. andigena (Andigena or adg) to the environment of the United Kingdom in 1959 (Simmonds 1966). His motivation was based strongly on the theory that Andigena is a rich source of genetic variability for temperate breeding programmes. He proposed and initiated a project of selection for adaptation within a germplasm collection that has few precedents in other crops. Most efforts to use unadapted germplasm have initiated the process by crossing to adapted genotypes followed by further backcrossing or recurrent selection. Simmonds' scheme was also unique in that it was directed toward the utilization of the total Andigena genotype, not just one or a few genes.

The arguments presented by Dr. Simmonds were one of two reasons for initiating a similar programme in New York in 1963. The second reason is the consequence to US potato breeding of the efforts of Chauncey E. Goodrich with Andigena between 1846 and 1864. In the belief that potato varieties were "running out" and needed an infusion of new parents, he obtained potato varieties from the American consulate in Panama. It is likely that these were Andigena. One of these he called "Rough Purple Chili". From it he produced a seedling "Garnet Chili". Subsequently, this clone gave rise to the variety "Early Rose" in 1861 (Clark & Lombard 1946). It in turn produced the berry from which the "Burbank Seedling" was selected by Luther Burbank in 1876 (Clark & Lombard 1946). This mutated to produce the russet form which is now the most widely grown variety in the United States. Furthermore, the impact of the "Rough Purple Chili" introduction is such that almost all the pedigrees of present day varieties in the United States can be traced to it.

Goodrich's experiences shed further insight into the challenges of introducing new sources of germplasm. In the summary of his potato breeding experience written a few months before his death in May 1865 (Goodrich

(1863), he wrote the following:

"In 1848, I received a variety from Bogota, South America... . This variety was so entirely too late in its maturity as speedily to decline... .

In 1850, I received another variety from the same place, of a little earlier maturity, but it could never be adapted to our climate... .

In 1851, I received eight varieties from Panama, supposed to have been brought from the coast of Chili,... . One of these was exactly like the last sort above noticed. One other was afterwards extensively cultivated and sold by me under the name of Rough Purple Chili. It was the parent of my seedling, the Garnet Chili. The six remaining varieties were all too late for this climate,... .

In December, 1852, a neighbor returning home from the west coast of South America, brought me three varieties. They were entirely too late in maturity, and gradually declined in health.

This record of importation is sufficiently discouraging, one only out of twelve sorts, having had any permanent value.

In 1852 I originated four thousand two hundred and ten sorts, which is considerably more than one fourth of all I have ever originated. Nearly all these families of South American origin were ruinously late, as their parents before them had been, though often the seedlings were an improvement on the parent in time of maturity.

Theoretically, I doubt the propriety of importing sorts from climates very different from our own."

Obviously his own experience did not allow him to foresee the long lasting consequences of his work.

Present experiences with Andigena by Simmonds and Glendinning in the UK and by us in New York seem to reflect the same mixture of discouragement and promise as accompanied Goodrich's effort. The advantage we have is that we can approach the task with more resources and in an orderly fashion, encouraged by the knowledge of the rewards already obtained.

SOURCES OF GERMPLASM AND METHODS APPLIED

The programme to produce an adapted source of Andigena (called Neotuberosum) germplasm in New York was initiated in 1963 using seeds from N.W. Simmonds with one generation of selection in England, and seeds obtained directly from Peru, Ecuador and Colombia. The Commonwealth Potato Collection provided Simmonds with his initial germplasm. A second introduction with an additional cycle of selection was obtained from Simmonds in 1966. Records were maintained of the maternal parents. In a few generations, none of the clones were descended from the direct importations, probably due to the advantage of the initial cycles of selection in the Simmonds source. For this reason, Figure 1 lists only the one origin of germplasm in 1963 and 1966. For three generations the populations were dug by hand and the open pollinated fruits were saved from the plants with the best yield of tubers

Figure 1. Development of the Neotuberosum population at Cornell University.

1963,1966	Introductions from N.W. Simmonds (1 cycle of selection)		
	OP* and CP		
1967	Cycle 2		
	OP		
1968	Cycle 3		
	OP		
1969-1970	Cycle 4	New introductions from South America, IR-1 collection and Commonwealth Potato Collection	
	BP*	OP	
1971-1973	Cycle 5	Cycle 1	
	BP	BP	
1974-1977	Cycle 6	Cycle 2	
	BP	BP	
1978-1980	Cycle 7	Cycle 4	
		BP	
1981-1983		Cycle 5	Tuberosum
		BP	BP
1984-1985		Cycle 6 Neotuberosum	Cycle 6 Hybrids

*OP = Open pollinated
BP = Pollinated with a bulk collection of pollen
CP = Controlled pollination with pollen from one parent

Although this system was efficient, it produced evidence of inbreeding depression that was counterproductive for effective selection. In 1970, we began to reproduce the cycles by collecting pollen in bulk from all the selected clones and pollinating the same clones with this bulk pollen. We also began to transplant seedlings to pots to produce tubers for planting the field trials, rather than transplanting directly to the field. This produced better plants for selection and permitted us to screen for late blight (Phytophthora infestans) and virus susceptibility at the seedling stage in the glasshouse. The consequence was that each generation now required 3 years. This system was practiced for three more cycles.

A survey of the pedigrees after three cycles revealed that our system of selection without regard to lineage had reduced the number of original females to 19. Obviously this was contrary to the goal of the programme, so in 1969 and 1970, 1073 new introductions of Andigena were obtained from South America, the Commonwealth Potato Collection, and the IR-1 collection in the United States. In this material, every effort was made to select at least one clone in each of the original accessions. The first generation was reproduced by use of open pollinated fruits, but thereafter a bulk collection of pollen was used in controlled crosses. The second cycle of this branch of the programme was outcrossed to the sixth cycle of the original population in order to take advantage of the resistance to potato virus Y (PVY) developed in that population and needed in the new population to avoid loss of clones due to virus infection. The resulting generation was labelled as cycle 4. Selections from this generation were crossed again to the more advanced population, this time using clones in cycle 7 selected for late blight resistance. Some of these selections are presented in Figure 2. In 1984, the "fifth" cycle selections were intercrossed using bulk pollen and the same clones were pollinated with a bulk of eleven clones of S. tuberosum ssp. tuberosum (Tuberosum or tbr) with the H_1 gene for resistance to Globodera rostochiensis. These were grown as spaced plants in 1985 and selections made in September after a growing season of 104 days.

RESULTS

Disease resistance

Even though disease resistance was not the primary motivation for involvement with Andigena, it has been one of the earliest contributions. One of the major contributions of Andigena to temperate breeding programmes

Figure 2. Seven Neotuberosum clones obtained after approximately five generations of recurrent phenotypic selection for adaptation to long days.

has been the dominant H_1 gene for resistance to *Globodera rostochiensis*, the potato cyst nematode, discovered by Ellenby in CPC 1673 (Ellenby 1952). Simmonds (Simmonds & Malcomson 1967) selected for resistance to late blight (*Phytophthora infestans*) in Cornwall relying on field epiphytotics. We continued to select for late blight resistance using artificial epiphytotics in the glasshouse and field. Even though Andigena has a reputation for being very susceptible, through recurrent selection with moderate selection pressure, clones with a high degree of general resistance have been selected. We no longer consider it necessary to protect this population with foliar sprays of fungicide at Ithaca, New York.

The third generation of the original population was severely reduced by PVY. The population of Andigena seemed to be far more susceptible than Tuberosum clones. The entire project would have been abandoned had it not been for the discovery of a single dominant gene which gave immunity to the common strain of PVY endemic in these fields (Munoz *et al*. 1975). Subsequent research in Europe (Ross 1982) and South America (Fernandez-Northcote 1982) has shown that this gene is effective against all the strains of PVY used in these tests.

The Andigena population has been screened for resistance to other diseases. Clones resistant to early blight, (*Alternaria solani*), root knot nematodes (*Meloidogyne hapla* and *M. chitwoodi*) and potato virus X (PVX) have been identified.

Combining ability

The primary motivation for selection within Andigena has been to develop an adapted, unrelated source of germplasm to use in hybrid combination with existing Tuberosum varieties. This should provide heterosis for yield. The extent of heterosis should be dependent on the level of adaptation and diversity of the Andigena parents. Munoz (Munoz & Plaisted 1981) investigated the adaptation and combining ability of the six cycles of the population derived from the Simmonds seeds. He used remnant seeds of the CPC introductions in 1969 as a measure of the initial population and remnant seeds of the second to sixth cycles of selection. Trials were conducted in 1977 using transplanted seedlings and in 1978 and 1979 using seed tubers. The averages for the three trials for each cycle are presented in Table 1. The yield level of the clones is still improving, especially in the last cycle when the selection pressure was intense (0.8%). Yield improvement was due to increase in tuber size. The effect of the selection process upon combining ability was evaluated by growing 50 plants representive of each of the five generations, collecting the pollen for each cycle as a bulk, and pollinating four Tuberosum clones. These hybrid seeds were used to produce seedling tubers which were used to plant a yield trial in 1979. The trials also included the cultivar Katahdin and progenies from the same four Tuberosum testers and a bulk of other Tuberosum clones. The results are presented in Table 2. The progress in combining ability tended to parallel the progress in clonal performance. The hybrids of the fourth cycle produced yields approximately the same as the Tuberosum x Tuberosum hybrids, but less than Katahdin. The hybrids of the sixth cycle yielded 126% of the Tuberosum x Tuberosum progenies and 108% of Katahdin. The yield advantage, however, was due to a larger set of tubers. Tuber size of the Tuberosum x Andigena hybrids was still less than the check comparisons. These results are consistent with the data (Table 3) obtained by Cubillos

Table 1. Response to selection for yield within Andigena.

Cycle	Yield (kg/hill)	Tuber number	Mean tuber weight (g)
0	0.32	16	35
2	0.68	15	43
3	0.61	17	36
4	0.74	19	44
5	0.82	15	54
6	1.37	17	74

Table 2. Mean yield, tuber number per hill, and tuber weight of the combining ability trial progenies.

Parental stocks	Yield per hill (kg)	Tuber number per hill	Tuber weight (g)
Progeny means of the Tuberosum clones as females			
Hudson	0.61	7.3	87
Elba	0.53	6.8	80
NY62	0.61	7.9	79
NY60	0.70	7.7	92
Progeny means of the Andigena populations as males			
Cycle 0	0.47	7.7	62
Cycle 2	0.53	7.3	77
Cycle 3	0.60	7.4	82
Cycle 4	0.64	7.5	87
Cycle 6	0.83	7.3	115
Tbr x Tbr	0.66	5.3	129
Katahdin	0.77	3.9	201

& Plaisted (1976) in 1976 using clones from the fourth and fifth cycles of selection. They observed that the Tuberosum x Andigena hybrids yielded 113% of the Tuberosum x Tuberosum hybrid progenies. Again the tuber size of the hybrids was less than the size of the Tuberosum x Tuberosum progenies. Hybrid advantages in similar programmes have been reported by Paxman (1966) to be 50%, by Glendinning (1969) to be 13%, and by Tarn & Tai (1977) to be an average of 21%.

Utilization in variety development

The proof of the value of Neotuberosum in potato breeding is in its ability to produce superior varieties worthy of release. We have used the Neotuberosum (Andigena) clones in our variety development crosses. The extent of this use of the Neotuberosum is presented in Table 4 which summarizes the constitution of the clones grown in 1984. Approximately 65% were Tuberosum x Tuberosum progenies, 20% were 1/4 Andigena (Neotuberosum) and 15% were 1/2 Andigena (Neotuberosum).

The first cycle of selection is based on widely spaced single hills planted with small seedling tubers grown in pots. A single large tuber is saved from the selected hills which is used to plant a widely spaced single hill in the third season. The entire hill of the selected clones is saved from this cycle and these tubers are used to plant an unreplicated

Table 3. Performance of selected Andigena, Tuberosum, and hybrid populations as spaced plants. Ithaca, New York, 1976

	Yield (g) > 4.8 cm	Tubers per plant	Wt. (g) per tuber
Adg x Adg	1254 ± 78	14.1 ± 1.1	110 ± 8
Tbr x Adg	2444 ± 101	14.5 ± 0.6	197 ± 8
Tbr x Tbr	2162 ± 86	9.8 ± 0.3	240 ± 9

Table 4. The use of selected Andigena (Neotuberosum) at all stages of selection in the Cornell variety development programme in 1984.

Year	Stage	Size of generation	Tbr	1/4 Adg	1/2 Adg
			%	%	%
1	Transplants	83 000	85	3	12
2	Seedling hills	65 320	74	19	7
3	Single hills	13 690	47	40	13
4	Unreplicated plots	2 411	59	29	12
5	1st year yield trials	35	60	26	14
6	Advanced yield trials	12	63	11	26

plot the fourth season. The selected clones are included in a replicated yield trial the fifth season. Table 5 presents the percentage selected of each type of clone at each of these stages. The progenies with 50% Neotuberosum are at a disadvantage at all stages of selection. This is due to tuber size, tuber appearance, especially depth of eye, and to some extent skin colour and internal defects. At this stage of selection of the Neotuberosum population, it appears that the progenies which are backcrosses to Tuberosum do as well as the Tuberosum x Tuberosum progenies.

Another perspective is to look at one group of clones. The progenies produced in 1981 were transplanted to pots in 1982, grown as seedling hills in 1983, as single hills from a large tuber in 1984, and as unreplicated plots in 1985. The performance of the three types of progenies at each of the three stages of selection is presented in Table 6. Again the Tuberosum x Neotuberosum progenies are not as selectable as the Neotuberosum x Tuberosum progenies at any of the three stages of selection, but the backcrosses to Tuberosum are at least as good as the Tuberosum x Tuberosum progenies. Presumably the selectability of the hybrids will improve as the tuber appearance and size of the Neotuberosum is improved through further selection.

Vine type

Until 1980, selection was directed towards maturity, tuber yield and type, and disease resistance. Vine type was not a factor in selection in

Table 5. The selectability of three types of progenies at four stages of selection in 1984.

Year	Stage	% Selected		
		Tbr x Tbr	1/4 Adg	1/2 Adg
2	Seedling hills	10	12	9
3	Single hills	12	12	9
4	Unreplicated plots	13	12	8
5	1st year yield trial	29	43	28

Table 6. Percentage of clones selected in the 1981 group of seedlings.

	Initial No. of clones	First	Second	Third	Three year product
Tbr x Tbr	26 700	20	12	30	0.72
Tbr x Adg	14 800	20	10	20	0.24
(Tbr x Adg) x Tbr	20 600	21	11	34	0.78

as much as the vines were always killed prior to harvest and selection. However, it was evident that the vine type of the later generations was more like that of Tuberosum than the earlier generations. Huarte (1984) subjected the most advanced cycle of Neotuberosum to three cycles of selection for vine type. He measured an average correlation between vine score and tuber score of 0.25 and between vine score and total yield of 0.23. This population was very responsive to selection for Tuberosum-like vine type. Table 7 shows that three cycles of selection were sufficient to produce progenies with vine type comparable to the progeny of the variety hybrid, Katahdin x Atlantic.

CONCLUSION

The data presented document the advances made in disease resistance, adaptation, combining ability, and utility in our own variety development programme. The data also document the limitations in selectability using Andigena at the stages of selection available to date. More difficult to document is the impression that this population has just now reached the stage which will enable rapid improvements in the near future. There is one other aspect to consider in the utilization of Andigena in a breeding programme. So far the data presented have been in terms of population means or percentages. Yet the value of Andigena will be in terms of specific genotypes released as varieties. Certainly the history of the impact of Goodrich's Rough Purple Chili and Garnet Chili upon North American variety development is striking. The results of our use of Andigena are much more modest. We have released one variety, Rosa, which is a Tuberosum x Andigena hybrid (Plaisted et al. 1981). Although its prospects are still unclear this demonstrates to some extent that hybrids will be accepted by the growers. Among the 12 clones currently in the early stages of the

Table 7. Effect of selection for vine type.

	Vine Score[a]	Width of distal leaflet (cm)
Original	1.4	3.3
Cycle 1	2.3	4.6
Cycle 2	3.1	5.3
Cycle 3	3.7	6.0
Katahdin x Atlantic	3.5	5.8

[a]Score of 1 is adg-like, score of 5 is tbr-like.

foundation seed programme, three are 25% Andigena. These are not the highest yielding of the 12 clones, but it is interesting that they are the three earliest maturing clones.

REFERENCES

Clark, C.M. & Lombard, P.M. (1946). Descriptions of and key to American potato varieties. USDA Circ. 741.

Cubillos, A.G. & Plaisted, R.L. (1976). Heterosis for yield in hybrids between S. tuberosum ssp. tuberosum and tuberosum ssp. andigena. Amer. Potato J., 53, 143-50.

Ellenby, C. (1952). Resistance to the potato eelworm Heterodera rostochiensis. Woll. Nature, 170, 106.

Fernandez-Northcote, E.N. (1982). Prospects for stability of resistance to potato virus Y. In Proceedings International Congress "Research for the Potato in the Year 2000", ed. W.J. Hooker, p. 82. International Potato Center, Lima, Peru.

Glendinning, D.R. (1969). The performance of progenies obtained by crossing Groups Andigena and Tuberosum of Solanum tuberosum. Eur. Potato J., 12, 13-9.

Goodrich, C.E. (1863). The potato. Its diseases - with incidental remarks on its soils and culture. Transaction of the N.Y. State Agricultural Society, 23, 103-34.

Huarte, Marcelo A. & Plaisted, R.L. (1984). Selection for tuberosum likeness in the vines and in the tubers in a population of neotuberosum. Amer. Potato J., 61, 461-73.

Munoz, F.J. & Plaisted, R.L. (1981). Yield and combining abilities in andigena potatoes after six cycles of recurrent phenotypic selection for adaptation to long day conditions. Amer. Potato J., 58, 469-79.

Munoz, F.J., Plaisted, R.L. & Thurston, H.D. (1975). Resistance to potato virus Y in Solanum tuberosum ssp. Andigena. Amer. Potato J., 52, 107-15.

Paxman, G.J. (1966). Heterosis in Andigena hybrids. Rep. John Innes Inst., 1965, pp. 51-3.

Plaisted, R.L., Thurston, H.D., Sieczka, J.B., Brodie, B.B., Jones, E.D. & Cetas, R.C. (1981). ROSA: A new golden nematode resistant variety for chipping and tablestock. Amer. Potato J., 58, 451-5.

Ross, H. (1982). Major and minor genes in breeding virus resistant varieties. In Proceedings International Congress "Research for the potato in the year 2000", ed. W.J. Hooker, pp. 165-6. Lima: International Potato Center.

Simmonds, N.W. (1966). Studies of the tetraploid potatoes. III. J. Linn. Soc. Bot., 59, 279-88.

Simmonds, N.W. & Malcomson, J.F. (1967). Resistance to late blight in Andigena potatoes. Eur. Potato J., 10, 161-6.

Tarn, T.R. & Tai, G.C.C. (1977). Heterosis and variation of yield components in F_1 hybrids between Group Tuberosum and Group Andigena potatoes. Crop Sci., 17, 517-21.

BREEDING AT THE 2x LEVEL AND SEXUAL POLYPLOIDIZATION

S.A. Hermundstad and S.J. Peloquin

INTRODUCTION

The breeding strategy advocated in this paper is based on genetic diversity. By diversity we mean both the many valuable traits available in the wild and cultivated relatives of the 4x potato (*Solanum tuberosum* ssp. *tuberosum* (hereafter, Tuberosum)), and the allelic diversity present in these relatives that provides the opportunity to approach maximum heterozygosity in developing new 4x cultivars. The breeding strategy contains three essential components; 1) the wild and cultivated relatives are the source of genetic diversity, 2) haploids (2n = 2x = 24) of Tuberosum (2n = 4x = 48) are effective tools in capturing the genetic diversity (putting the germplasm in a usable form), and 3) 2n gametes, gametes with the sporophytic chromosome number, are the basis of an efficient method of transmitting the genetic diversity to the cultivated 4x potato (Peloquin 1982).

An essential for the third component is that there are 2n pollen-producing 2x hybrids from crosses between haploids of Tuberosum and the cultivated 2x species *S. phureja* (hereafter, Phureja) and *S. stenotomum* or the 2x wild species. The 2n pollen can be formed by either the genetic equivalent of first division restitution with crossing-over, FDR-CO (parallel spindles mutant, *ps*) or FDR without crossing-over, FDR-NCO (parallel spindles in combination with the synaptic 3 mutant, *sy* 3). The significance of the mode of 2n pollen formation resides in the genetic consequences. Approximately 80% of the heterozygosity of the 2x parent is transmitted to the 4x progeny following 4x x 2x FDR-CO, and with 4x x 2x FDR-NCO matings, 100% of the heterozygosity and epistasis of the 2x parent is transmitted to the 4x offspring (Peloquin 1983).

The 2x haploid-species hybrids offer the advantage of simpler inheritance patterns (disomic vs. tetrasomic) in synthesizing the desired 2x genotypes for use in incorporating desirable traits from exotic germplasm into the cultivated potato. The large variability for many traits in the haploid-species hybrids also makes them valuable materials for genetic studies at the 2*x* level.

HAPLOID-WILD SPECIES HYBRIDS

The 2*x* clones used to obtain 4*x* progeny from 4*x* x 2*x* and 2*x* x 2*x* matings have been almost exclusively hybrids between Phureja and Tuberosum haploids. The earliest work involved F_1 hybrids between selections of Phureja and the male fertile haploids US-W1 from the cultivar Katahdin and US-W42 from Chippewa. It was in these hybrids that 2*n* pollen formed by FDR-CO was first identified (Mok & Peloquin 1975a). Crosses between particular F_1 hybrids that produced 2*n* pollen were made to obtain a variable group of 2*x* hybrids for use in 4*x* x 2*x* crosses. Fortuitously, it was found that some of the hybrids produced 2*n* pollen by FDR-NCO (Okwuagwu 1981). One disadvantage of using Phureja in crosses with Tuberosum haploids was that the F_1 hybrids were almost exclusively male sterile when Phureja was the male parent. This put a very severe restriction on the haploids which could be used, since less than 3% are male fertile.

The male sterility problem of 2*x* hybrids was circumvented by use of the wild, tuber-bearing species *S. chacoense*. It hybridizes easily with male sterile haploids to give male fertile progeny, permitting the inclusion of a wide range of haploids as potential parents (Leue & Peloquin 1980). The research with haploid Tuberosum x *S. chacoense* hybrids also indicated that tuber yields of the hybrids could exceed those of the haploid parent, tuberization in the hybrid families varied significantly with the haploid parent, and there was large variation in tuberization within some individual families.

The very interesting results with *S. chacoense* stimulated increased research with haploid Tuberosum-wild species hybrids. Forty male sterile haploids, representing gametes of eight 4*x* clones, were crossed with plants from two introductions of each of eight 2*x* wild species; *S. berthaultii* (ber), *S. boliviense* (blv), *S. canasense* (can), *S. infundibuliforme* (ifd), *S. microdontum* (mcd), *S. raphanifolium* (rap), *S. sanctae-rosae* (sct), and *S. tarijense* (tar).

Five exciting and significant results were found by Hermundstad & Peloquin (1985b) in regard to tuberization in the hybrid families:
1) Variation for tuberization within families was large. Plants which had not tuberized were found next to full sibs that had good tuberization. The

lack of a continuous range of expression for tuberization within a family suggests it may be a relatively qualitatively inherited character; 2) Some of the hybrids had tuber yields three to four times that of their haploid parents; the species did not tuberize. A yield trial of 210 selected hybrids; (involving nine wild species), 10 haploids, and four cultivars was conducted at two locations. Many hybrids had tuber yields more than three times that of their haploid parent, and several hybrids outyielded three of the cultivars; 3) Haploids varied significantly in their ability to produce hybrid families with good tuberization. The percentage of plants selected for tuberization within a family varied from 0 to 40 depending on the haploid involved; 4) Haploids from the same 4x parent varied significantly in their ability to produce hybrid families with good tuberization. The percentage of hybrid plants selected for tuberization varied from 9 to 33 in families in which the haploid parent was US-W4139, and from 0 to 6 when US-W721 was the parent; both haploids were extracted from the Wisconsin advanced selection W231; 5) Tuberization in the hybrid families varied with the species parent; ber, can, mcd and tar were better parents than blv, ifd, rap and sct. Limited evidence suggested that plants within a species varied in their ability to produce families with good tuberization. It should now be possible to determine the value of different plant introductions and individual genotypes of a 2x wild species by the tuberization and yield of hybrid offspring obtained from crosses with selected haploids. For the first time we are in a position to evaluate the contributions to yield of different wild species genotypes.

Two of the preceding observations provide the basis of intriguing speculation concerning the genetic relationships between wild species and the cultivated potato. The discrete variation for tuberization within families plus the high yield and good tuber type of many haploid-wild species F_1 hybrid clones could be interpreted as evidence that domestication may not have involved many genes. It will be a fascinating area for future research to determine the number and location of genes involved in the origin of the cultivated potato. The size of F_2 populations necessary to obtain haploid and wild species phenotypes following matings between full sibs from haploid-wild species F_1 hybrids could provide some clues.

Male fertility of haploid Tuberosum-wild species F_1 hybrids depends primarily on the species parent (Hermundstad & Peloquin 1985a). Families with the same haploid parent vary in male fertility according to the species parent; families with the same species parent exhibit similar ratios of fertile to sterile plants. Species parents can be characterised according to the male fertility of the haploid-species hybrids they produce (Table 1).

It is very encouraging that the majority of the species evaluated produce hybrid families with good male fertility. The male sterility observed in some families is best explained on the basis of genetic-cytoplasmic interactions, assuming Tuberosum cytoplasm is sensitive to dominant sterility genes in certain species genotypes. This, however, has not been established; reciprocal crosses with male fertile haploids are necessary to test this hypothesis. Because of the limited number of Plant Introductions and other genotypes of each species used, generalization regarding male sterility of hybrid families should be made cautiously. Extensive studies, involving many plants representing numerous Plant Introductions, must be made to characterize a species.

HAPLOID-PHUREJA VS HAPLOID-WILD SPECIES HYBRIDS

It is interesting to compare haploid Tuberosum-Phureja hybrids (PH) with haploid-wild species hybrids (HS), since preliminary evidence indicates that the HS clones appear to be superior to the PH clones for several traits (Schroeder 1983; Hermundstad & Peloquin 1986). The majority of PH clones have rough tubers due to deep eyes and raised internodes. In contrast the tubers of most HS clones, particularly those involving ber, can and tar are smooth and attractive, often resembling Tuberosum cultivars. The number of tubers per hill of HS clones is significantly lower than that of PH clones. Approximately two-thirds of 210 HS clones had fewer than 15 tubers per hill; most PH clones have two to four times as many tubers per hill. An impressive characteristic of the HS clones is that they closely

Table 1. Male fertility of haploid-species hybrid families.

Species parent	Family fertility*	Species parent	Family fertility
S. berthaultii	F	*S. boliviense*	F, S
S. chacoense	F	*S. sanctae-rosae*	F, F+S, S
S. kurtzianum	F	*S. bukasovii*	F+S
S. spegazzinii	F	*S. canasense*	F+S
S. tarijense	F	*S. raphanifolium*	F+S, S
S. microdontum	F, F+S	*S. infundibuliforme*	S
S. sparsipilum	F, F+S		

*F - male fertile families, S - male sterile families,
F+S - both male fertile and male sterile plants in a family

resemble cultivars for this trait. A striking feature of most HS clones evaluated is the uniform size of the tubers within a hill. This is particularly evident in comparison to almost all PH clones where there is a wide range in tuber size within clones.

Considerable variation among and within HS families occurs for several other tuber characteristics. The range in tuber dormancy for HS hybrids varies from 22 to 207 days from harvest to initiation of sprouting. In comparison, PH hybrids, particularly F_1 hybrids, are almost exclusively short dormancy clones due to the dominance of short dormancy contributed by Phureja. There is also large variation in specific gravity among HS hybrids. The range among 210 hybrids extends from 1.050 to 1.110; it is above 1.085 for 30 hybrids. No PH clones with specific gravity higher than 1.085 have been identified.

An important advantage of HS families involving most 2*x* wild species tested is that they contain male fertile plants for use in further breeding and genetic studies (Hermundstad & Peloquin 1985a). In contrast, when male sterile Tuberosum haploids are crossed with Phureja, most hybrid families are male sterile. Many genotypes, representing a large number of Plant Introductions of Phureja, were crossed with numerous male sterile haploids. The F_1 hybrids were almost exclusively male sterile. This sterility results from the interaction of sensitive cytoplasm of Tuberosum haploids with dominant genes from Phureja (Ross *et al.* 1964). However, male fertile plants have been found in a few haploid Tuberosum x Phureja F_1 families (Carroll 1975). The basis of this fertility can be explained by the presence of dominant fertility restorer genes in the haploid parent. Hybrids between male sterile haploids and the wild species ber, blv, buk, can, chc, ktz, mcd, spl, spg, and tar are male fertile indicating that the wild species genotypes tested are not homozygous for dominant genes that interact with the sensitive cytoplasm of Tuberosum to result in male sterility.

2N POLLEN IN HAPLOID-SPECIES HYBRIDS

The continued exploitation of the 4*x* x 2*x* and 2*x* x 2*x* breeding schemes is dependent on the ability to easily obtain 2*x* haploid-species hybrids that produce 2*n* pollen, and to obtain good seed set following 4x x 2x crosses. The production of 2*n* pollen by these hybrids is dependent on the frequency of the *ps* allele in both species and haploid parents. Plants that produce 2*n* pollen have been found in many 2*x* species evaluated by Quinn *et al.* (1974). More extensive investigations of a few species have determined that the frequency of ps in these species is very high. The frequency of 2n pollen-producing plants among more than 500 plants representing 56 Plant Introductions of three, wild, tuber-bearing species from Argentina, S. gourlayi, S. infundibuliforme and S. speggazzinii, was determined. The frequency of the ps allele was 0.46, 0.37 and 0.29, respectively (Camadro & Peloquin 1980). Recently, similar results have been obtained with other wild, 24-chromosome species (Hermundstad 1984).

The frequency of the ps allele in haploids is related directly to the frequency of this allele in the cultivars from which they are derived. An analysis of 56 USA cultivars revealed a ps gene frequency of 0.69 (Iwanaga & Peloquin 1982). Further, more than 65% of the 4x clones were simplex, Ps ps ps ps. Assuming chromosome segregation for ps in tetraploids, 50% of the haploids would be ps ps and 50% Ps ps. Even if the 4x clone were duplex, Ps Ps ps ps, about 67% of the haploids would be Ps ps and 17% ps ps.

With the relatively high gene frequencies for ps in both wild species and haploids, it should not be difficult to obtain haploid species hybrids that produce 2n pollen. In practice we have found it relatively easy to generate many such hybrids (Hermundstad & Peloquin 1985a). It is important to only use as male parents species plants that produce 2n pollen. Since the majority of haploids are heterozygous Ps ps, haploid-species families in which one-half of the individuals produce 2n pollen can be readily obtained.

The synaptic 3 mutant has not been detected in either wild species or haploid-wild species F_1 hybrids. The value of this mutant is such that considerable effort should be undertaken to transfer it from the Phureja-haploid hybrids to the haploid-wild species hybrids. Hermsen et al. (1985) have outlined in detail the most efficient breeding procedures for accomplishing this task.

Good fruit and seed set from 4x x 2x crosses is an important requirement for breeding new clones, and an absolute necessity if this breeding procedure is to be used for producing a potato crop from true seed. A strong triploid block is present in many tuber-bearing Solanums due to genic imbalance in the endosperms associated with triploid embryos following 4x x 2x crosses (Johnston et al. 1980). As a result almost all the progeny from 4x x 2x crosses are 4x (Hanneman & Peloquin 1968). Because expression of the ps allele is highly variable among clones, seed set following 4x x 2x crosses depends mainly on the frequency of 2n pollen in the 2x parent. An increase in the number of seeds per fruit with increased 2n pollen frequencies was reported by Jacobsen (1980) and Schroeder & Peloquin (1983). Both results revealed a nonlinear relationship between seed set and 2n pollen frequency; frequencies of 2n pollen over 60% did not increase seed set. This was attributed to the increased competitiveness of FDR 2n pollen over n pollen due to gametophyte heterosis, a phenomenon demonstrated by Simon & Peloquin (1976). Further, seed set in 4x x 2x crosses approached that in control 4x x 4x crosses with 2n pollen frequencies as low as 30%. Since functioning of n pollen results in triploids which are rarely recovered, the normal seed set in 4x x 2x crosses with only 30% 2n pollen indicates that the 2n pollen grows faster than n pollen, further evidence for gametophyte heterosis. We have arbitrarily set minimal levels of seeds per fruit from 4x x 2x crosses that permit reasonably efficient use of this breeding scheme. For breeding new clones through conventional selection within families of 100 to 200 individuals we place this minimum at 10-15 seeds per fruit; this level of seed set can be obtained with 2n pollen frequencies as low as 3%. For the production of the crop from true seed, this minimum is 50 seeds per fruit which would require 2n pollen frequencies of about 20%.

RELATIONSHIPS BETWEEN 2X PARENTS AND 4X PROGENY

For the 4x x 2x breeding scheme to be effective, efficient selection among the wide array of 2x parental clones is important. Thus, one must determine correlations between 2x parents and 4x progeny for particular traits. The relationships between 4x progeny and their 4x and 2x parents for maturity, tuber eye depth, and yield has been evaluated by Schroeder (1983). He found that the family means for maturity were highly correlated with the 2x parents, but not with the 4x parents. Similar results were obtained for depth of eye. Most importantly, no significant correlations were found between the total yield of either parent and the total yield of the families. This is particularly unfortunate with regard to the 2x parents, since it was hoped that one could simply evaluate the yields of the 2x clones and use only the best yielders as parents in 4x x 2x crosses. Thus, progeny testing is necessary to determine the parental value for yield of a 2x clone. Preliminary results indicate that a 2x clone must be crossed with at least two cultivars which differ in maturity in order to perform adequate progeny tests (Schroeder 1983).

It was of interest to compare the 4x progeny from 4x x 2x FDR-CO with those from 4x x 2x FDR-NCO. Thus, 20 families from crossing two 4x cultivars with five FDR-CO clones and with five FDR-NCO clones were evaluated (Masson 1985). The within-family variances for total tuber yield, marketable tuber yield, weight of undersized tubers, and number of marketable tubers were significantly higher with FDR-CO. The variance ratio FDR-NCO/FDR-CO was from 0.65 to 0.69 for the four traits. This result is not unexpected, since the gametes from an FDR-NCO clone have identical genotypes. Thus, variation in the 4x progeny is entirely due to the 4x parent. In contrast, there would be considerable variation among the gametes of an FDR-CO clone, since one chiasma is formed in each of the 12 bivalents. Thus, there would be variation from both intra- and inter-chromosomal recombination. However, variation among gametes would be much less than that in a 2x clone without parallel spindles, since with this meiotic mutant all heterozygosity from the centromere to the first crossover in the 2x clone is maintained intact in the gametes, and one-half the parental heterozygosity from the first to second crossover is present in the gametes.

The homogeneity of the male gametes from FDR-NCO 2x clones suggests these clones might be valuable as testers of 4x parents. Following 4x x 2x FDR-NCO crosses, all the variation in the 4x progeny should be attributable to the 4x parent. Thus, the female general combining ability variances were calculated for 4x parents crossed to either 2x FDR-CO or 2x FDR-NCO clones (Masson 1985). For total yield, marketable yield, and weight of undersized tubers these variances were highly significant when the 2x parent was FDR-CO. However, with FDR-NCO 2x clones the same variances were not significant. Thus, the FDR-NCO clones are obviously not good testers of combining ability of 4x parents. The highly heterozygous and homogenous male gametes from FDR-NCO clones appear to prevent detection of the genetic variation in the eggs from 4x parents. The synthesis of relatively homozygous 2x clones that produce a homogenous male gamete population could provide an efficient method of testing the combining ability of 4x parents.

4X PROGENY FROM 4X X 2X CROSSES

Several researchers have reported that tuber yields of particular 4x progenies, resulting from crosses between USA or Canadian 4x cultivars with 2x FDR-CO hybrids, exceeded the yield of the cultivar parents (Mendiburu et al. 1974; Mok & Peloquin 1975b; De Jong & Tai 1977; De Jong et al. 1981; McHale & Lauer 1981). Furthermore, selected 4x clones from 4x x 2x crosses have been released as cultivars (Johnston & Rowberry 1981; International Potato Center 1984).

More recently heterosis for total tuber yield and marketable tuber yield was obtained by Massón (1985, this volume) in 4x families obtained from crosses between European cultivars and 2x clones (haploid-Phureja hybrids). The mean total yield of 49 4x families exceeded the mean total yield of the 4x parents by 19%; the five best families exceeded the mean of the parents by 42%. For marketable yield the mean of all families exceeded the mean of the parents by 10%; the five best families exceeded the mean of the parents by 35%. Similar results were obtained in Wisconsin with 32 families of similar parentage.

The total tuber yield and marketable tuber yield of 32 4x hybrid clones from 4x x 2x crosses, 10 European cultivars, and nine USA cultivars were compared at two locations in Italy, one irrigated and one not irrigated (Concilio 1985, this volume). The mean yield of the 4x hybrid clones under irrigation was 678 q/ha with a range of 212-1207 compared to the European cultivars with a mean of 667 q/ha and a range of 480-870, and the USA cultivars with a mean of 542 q/ha and a range of 340-825. The marketable yield was 86% of the total yield for 4x hybrid clones, 79% for European cultivars, and 76% for USA cultivars. The 4x hybrid clones were superior to European and USA cultivars in both total and marketable yield under non-irrigated conditions. The total mean yield and range in yield was 370 q/ha (157-606) for 4x hybrid clones, 267 (169-353) for European cultivars, and 269 (118-441) for USA cultivars. The marketable yield of the 4x hybrid clones was 87% of total yield as compared to 60% for the European cultivars. These results are particularly encouraging, since the 4x hybrid clones were selected in Wisconsin for tuber type, but not yield before they were tested in Italy. They are also interesting in that the 2x parents of six of the 31 4x hybrid clones are three different haploid-*S. chacoense* hybrids. Further, the six 4x hybrid clones with this parentage had good yields under both irrigated and nonirrigated conditions.

A surprising and important new result is the heterosis for total tuber solids (specific gravity) of 4x families obtained from 4x x 2x crosses. In two separate experiments, one involving USA cultivars and the other European cultivars, the mean total solids of many 4x families exceeded that of the 4x parents. Buso (unpublished) found, in using USA cultivars in a 5 (4x) x 4 (2x) partial diallel, that the amount of total solids in 20 4x families was 112% of that of the 4x parents. More importantly, total solids for the best five families were 124% of those of the 4x parents. In several families the mean of the family was more than 20% solids as compared to

16-18% in the 4x parents and 15-16% in the 2x parents. Very similar results were obtained by Masson (1985) with European cultivars. The mean total solids of 49 4x families exceeded that of the mean of the 4x parents by 11%, and that of the five best families exceeded the 4x parental mean by 22%. Although the basis of heterosis for total solids in 4x progeny from 4x x 2x crosses is not known, selection of hybrids with high specific gravity should provide valuable clones for the processing industry. Genotypes from 4x x 2x crosses could also prove valuable in more basic studies of the genetics, physiology and biochemistry of solids in the potato tuber.

The 4x x 2x breeding scheme is also an important approach for the production of potatoes from true seed (TPS). This approach offers an alternative production method for countries that do not have the conditions to produce virus-free tubers, and cannot afford to buy clean tubers from other countries. The 4x hybrid seed from 4x x 2x crosses is superior to open pollinated seed, which is mainly produced from self-pollination by bumblebees, and hybrid seed obtained from intercrossing cultivars (Macaso & Peloquin 1983; Kidane-Mariam et al. 1985). Families from open pollinated seed were compared to 4x hybrid families obtained from 4x x 2x (haploid-Phureja hybrid) crosses for seed, seedling, plant and tuber characteristics. The 4x hybrids had significantly larger 100-seed weights, better seedling vigour, and they produced about 50% more seedling tubers per unit area in small seedling beds. The vigour and uniformity of the plants from hybrid seed was superior to that of plants from open pollinated seed. The uniformity of the haulms and tubers in a few families approached clonal uniformity, indicating it may be possible in the future to produce certified tubers from true potato seed. Most importantly, the tuber yield of field transplants from hybrid seed exceeds that of those from open pollinated seed by 30-70%. Concilio (1985, this volume) obtained similar results from TPS yield trials at two locations in Wisconsin and one in Italy. He used both European and USA cultivars as 4x parents, and both haploid-Phureja and haploid-S. chacoense hybrids as 2x parents. Interestingly, the four highest yielding families of 28 tested in Italy were hybrids from crosses between European cultivars and haploid-S. chacoense hybrids; these families were also above average for yield in Wisconsin. Further, the tuber type of the 4x hybrid families with haploid-S. chacoense 2x parents was superior to that of 4x hybrids with haploid-Phureja 2x parents.

4X FROM 2X X 2X CROSSES

Several years ago both 4*x* and 2*x* progeny were obtained from crosses between 2*x* haploid-Phureja hybrids that produce 2*n* pollen and 2*n* eggs. The 4*x* progeny were more vigorous and had significantly higher (30-50%) tuber yields than the 2*x* "full sibs" if the 2*n* pollen was formed by FDR (Mendiburu & Peloquin 1977). The mode of 2*n* egg formation was not known at that time, but was assumed to be FDR, because of the heterosis for tuber yield in the 4*x* progeny. Recently, through use of a stain-clearing technique for whole ovules (Stelly *et al.* 1984) it was found that the clones used as females in the 2*x* x 2*x* crosses produced 2*n* eggs by omission of the second meiotic division. This is genetically equivalent to second division restitution (SDR), and it differs from FDR in that less than 40% of the parental heterozygosity is estimated to be transmitted to the gamete. However, SDR 2*n* eggs combined with FDR (FDR-CO) 2*n* pollen resulted in highly heterotic 4*x* progeny with regard to vigour and yield. Similar results were reported by Chujoy (1985). He obtained 4*x* and 2*x* progeny from crosses between haploid-*S. chacoense* hybrids and Phureja-haploid hybrids. Again, the 4*x* progeny significantly outyielded the 2*x* progeny and the 2*x* parental clones. Importantly, cytological analyses revealed that the 2*n* pollen was formed by FDR and the 2*n* eggs by SDR.

The previous results demonstrate that SDR x FDR 2*x* crosses can generate vigorous, high yielding 4*x* progeny. This is a mild surprise, but it is certainly worthy of attempted explanations. From the standpoint of transmission of heterozygosity it is possible to envisage how SDR and FDR can compliment each other. With FDR, 100% of the parental heterozygosity from the centromere to the first crossover and 50% from the first to second crossover is incorporated in the 2*n* gametes; in contrast, with SDR, 0% of the heterozygosity from the centromere to the first crossover and 100% from the first to second crossover is present in the 2*n* gametes. Thus, in regard to transmission of heterozygosity, FDR x FDR is superior to SDR x FDR from the centromere to the first crossover, but from the first to second crossover SDR x FDR is better than FDR x FDR.

It is also interesting to consider these two mating systems in view of the concept that maximum heterozygosity (more than two alleles per locus) is advantageous. For example, crosses between unrelated 2*x* hybrid clones (a1a3 x a2a4) would generate the following expected genotypic frequencies in the 4*x* progeny. For loci near the centromere (no single exchange tetrads) 100% would be tetrallelic following FDR x FDR, and 100%

triallelic after SDR x FDR. In contrast when the locus is very distal to the centromere (100% single exchange tetrads) FDR x FDR gives 25% tetrallelic, 50% triallelic, and 25% diallelic, and SDR x FDR 50% tetrallelic and 50% triallelic. If one assumes that triallelic and tetrallelic loci are more advantageous than monoallelic and diallelic loci, then SDR x FDR could result in slightly better 4x progeny than FDR x FDR. Therefore, it is important to compare 4x progeny from FDR x FDR with those from SDR x FDR to evaluate these *a priori* assumptions.

If possible then, "important loci" located between the centromere and the first crossover should be heterozygous in the FDR parent, and "important loci" between the first and second crossover should similarly be heterozygous in the SDR parent for maximum benefit from SDR x FDR crosses. Although our knowledge of the genetic architecture of the potato is in its infancy, it is possible that one could obtain results in regard to the location of "important loci" that would be relevant. For example, the results of reciprocal crosses between a series of 2x hybrid clones that all produce 2n pollen by FDR and 2n eggs by SDR could be instructive. Differences in yield of 4x progeny from reciprocal crosses may tell us which 2x clones were heterozygous at the most "yield loci" either between the centromere and the first crossover or distal to the first crossover. Identification of such 2x hybrid clones would permit one to synthesize 4x progeny heterozygous for the maximum number of "yield loci", and exploit the full potential of bilaterial sexual polyploidization.

Paper No. 2854 from the Laboratory of Genetics. Research supported by College Agricultural and Life Sciences; International Potato Center; SEA, USDA CRGO 84-CRCR-1-1389; and Frito-Lay, Inc.

REFERENCES

Camadro, E.L. & Peloquin, S.J. (1980). The occurrence and frequency of 2n pollen in three diploid Solanums from northwest Argentina. Theor. Appl. Genet., 56, 11-5.

Carroll, C.P. (1975). The inheritance and expression of sterility in hybrids of dihaploid and cultivated diploid potatoes. Genetica, 45, 149-62.

Chujoy, J.E. (1985). Tuber yields of 2x and 4x progeny from 2x x 2x crosses in potato; barriers to interspecific hybridization between *Solanum chacoense* and *S. commersonii*. Ph.D. Thesis, University of Wisconsin-Madison.

Concilio, L. (1985). Evaluation of yield and other agronomic characteristics of TPS families and advanced clones from different breeding schemes. M.S. Thesis, University of Wisconsin-Madison.

De Jong, H. & Tai, G.C.C. (1977). Analysis of tetraploid-diploid hybrids in cultivated potatoes. Potato Res., 20, 111-21.

De Jong, H., Tai, G.C.C., Russell, W.A. & Proudfoot, K.G. (1981). Yield potential and genotype-environment interactions of tetraploid-diploid (4x - 2x) potato hybrids. Amer. Potato J., 58, 191-9.

Hanneman, R.E. & Peloquin, S.J. (1968). Ploidy levels of progeny from diploid-tetraploid crosses in potato. Amer. Potato J., 45, 255-61.

Hermsen, J.E.Th., Ramana, M.S. & Jongedyk, E. (1985). Apomictic approach to introduce uniformity and vigor into progenies from True Potato Seed (TPS). In Report of Planning Conference on Present and Future Strategies for Potato Breeding and Improvement, pp. 99-114. Lima: International Potato Center.

Hermundstad, S.A. (1984). Production and evaluation of haploid Tuberosum-wild species hybrids. M.S. Thesis, University of Wisconsin-Madison.

Hermundstad, S.A. & Peloquin, S.J. (1985a). Male fertility and 2n pollen production in haploid-wild species hybrids. Amer. Potato J., 62, 479-87.

Hermundstad, S.A. & Peloquin, S.J. (1985b). Germplasm enhancement with potato haploids. J. Hered. (in press).

Hermundstad, S.A. & Peloquin, S.J. (1986). Tuber yield and tuber traits of haploid-wild species F_1 hybrids. Potato Res. (in press).

International Potato Center (1984). Potatoes for the Developing World. Lima, Peru.

Iwanaga, M. & Peloquin, S.J. (1982). Origin and evolution of cultivated tetraploid potatoes via 2n gametes. Theor. Appl. Genet., 61, 161-9.

Jacobsen, E. (1980). Diplandroid formation and its importance for the seed set in 4x x 2x crosses in potato. Z. Pflanzensuhtg., 84, 240-9.

Johnston, S.A., den Nijs, T.P.N., Peloquin, S.J. & Hanneman, R.E. (1980). The significance of genic balance to endosperm development in interspecific crosses. Theor. Appl. Genet., 57, 5-9.

Johnston, G.R. & Rowberry, R.G. (1981). Yukon gold: a new yellow-fleshed, medium early, high quality table and french fry cultivar. Amer. Potato J., 58, 241-4.

Kidane-Mariam, H.M., Arndt, G.C., Macaso, A.C. & Peloquin, S.J. (1985). Comparisons between 4x x 2x hybrid and open-pollinated true-potato-seed families. Potato Res., 28, 35-42.

Leue, E.F. & Peloquin, S.J. (1980). Selection for 2n gametes and tuberization in Solanum chacoense. Amer. Potato J., 57, 189-95.

Macaso, A.C. & Peloquin, S.J. (1983). Tuber yields of families from open pollinated and hybrid true potato seed. Amer. Potato J., 60, 645-51.

Masson, M.F. (1985). Mapping, combining abilities, heritabilities and heterosis with 4x x 2x crosses in potatoes. Ph.D. Thesis, University of Wisconsin-Madison.

McHale, N.A. & Lauer, F.I. (1981). Breeding value of 2n pollen diploid from hybrids and Phureja in 4x - 2x crosses in potatoes. Amer. Potato J., 58, 365-74.

Mendiburu, A.O., Peloquin, S.J. & Mok, D.W.S. Potato breeding with haploids and 2n gametes. In Haploids in Higher Plants, ed. K.J. Kasha, pp. 249-58. Guelph: Guelph University Press.

Mendiburu, A.O. & Peloquin, S.J. (1977). Bilateral sexual polyploidization in potatoes. Euphytica, 26, 573-83.

Mok, D.W.S. & Peloquin, S.J. (1975a). Three mechanisms of 2n pollen formation in diploid potatoes. Can. J. Genet. Cytol., 17, 217-25.

Mok, D.W.S. & Peloquin, S.J. (1975b). Breeding value of 2n pollen (diplandroids) in 4x - 2x crosses in potatoes. Theor. Appl. Genet., 46, 307-14.

Okwuagwu, C.O. (1981). Phenotypic evaluation and cytological analysis of 24-chromosome hybrids for analytical breeding in potatoes. Ph.D. Thesis, University of Wisconsin-Madison.

Peloquin, S.J. (1982). Meiotic mutants in potato breeding. In Stadler Genetic Symposium, ed. G. Redei, 14, 99-109.

Peloquin, S.J. (1983). Genetic engineering with meiotic mutants. In Pollen: Biology and Implications for Plant Breeding, eds. D.L. Mulcahy and E. Ottaviano, pp. 311-6. New York: Elsevier.

Quinn, A.A., Mok, D.W.S. & Peloquin, S.J. (1974). Distribution and significance of diplandroids among the diploid Solanums. Amer. Potato J., 51, 16-21.

Ross, R.W., Peloquin, S.J. & Hougas, R.W. (1964). Fertility of hybrids from *Solanum phureja* and haploid *S. tuberosum* matings. Eur. Potato J., 7, 81-9.

Schroeder, S.A. (1983). Parental value of 2x, 2n pollen clones and 4x cultivars in 4x x 2x crosses in potato. Ph.D. Thesis, University of Wisconsin-Madison.

Schroeder, S.A. & Peloquin, S.J. (1983). Seed set in 4x x 2x crosses as related to 2n pollen frequency. Amer. Potato J., 60, 527-36.

HAPLOIDS EXTRACTED FROM FOUR EUROPEAN POTATO VARIETIES

L. Frusciante and S.J. Peloquin

INTRODUCTION

It is well known that haploids extracted from Solanum tuberosum ssp. tuberosum provide unique opportunities for germplasm transfer and genetic manipulation.

Haploid extraction in potato has become a routine method since Hougas & Peloquin (1957) showed that haploids are relatively easy to obtain from 4*x* x 2*x* crosses.

Haploids can be easily hybridized with most 24-chromosome tuber-bearing species. The hybrids obtained are vigorous, fertile and have an improved tuberization under long-day conditions. Hermundstad (1984) found that many of the haploid tuberosum-species hybrids outyielded their haploid parents as well as some 4*x* cultivars.

Results on the extraction of haploids from varieties adapted to Italian conditions are reported in this paper.

MATERIALS AND METHODS

Four varieties widely grown in Italy (Desiree, Jaerla, Primura and Sirtema) were crossed with pollinator S. phureja, clone PI 1.22. All crosses were done using the technique described by Peloquin & Hougas (1958). Seedlings from the crosses were grown in the glasshouse and root-tip chromosomes were counted.

RESULTS AND DISCUSSION

The results of the pollination, in terms of fruit set, seeds, seedlings, haploid frequency and haploids per 100 fruits are presented in Table 1.

A total of 863 pollinations were made which resulted in 286 fruits, 329 seeds and 250 seedlings. More than 14% of the seedlings were found to be haploids with an average of 12.1 haploids per 100 fruit.

The ability of the clone PI 1.22 to induce haploids with the four parents used was confirmed.

Differences in the frequency of haploid production were found among the four varieties. According to Hougas et al. (1964) these can be attributed to

1. different frequencies of recessive lethal genes,
2. different abilities of the haploids to develop,
3. different influences of the endosperm on developing seeds.

Table 1. Results of the crosses between four 4x cultivars and 2x Phureja clone PI 1.22.

Cultivar	Pollinated flowers	Fruits/ pollinated flower	Seeds	Seedlings	Haploids/ seedling	Haploids/ 100 fruit
	no.	%	no.	no.	%	no.
Desiree	302	20.8	46	31	32.2	15.9
Jaerla	185	36.7	112	92	6.5	8.8
Primura	320	39.0	135	97	17.5	13.6
Sirtema	56	53.5	36	30	10.0	10.0
Total	863	37.5	329	250	14.4	12.1

REFERENCES

Hermundstad, S.A. (1984). Production and evolution of haploid tuberosum-wild species hybrids. M.S. thesis, University of Wisconsin.

Hougas, R.W. & Peloquin, S.J. (1957). A haploid plant of the potato variety Katahdin. Nature, 180, 1209-10.

Hougas, R.W., Peloquin, S.J. & Gabert, A.C. (1964). Effect of seed-parent and pollinator on frequency of haploids in Solanum tuberosum. Crop Sci., 4, 593-5.

Peloquin, S.J. & Hougas, R.W. (1958). The use of decapitation in inter-specific hybridization in Solanum. Amer. Potato J., 35, 726 (abst.).

HETEROSIS FOR TUBER YIELDS AND TOTAL SOLIDS CONTENT IN 4*x* x 2*x* FDR-CO CROSSES IN POTATO

Marc F. Masson and Stanley J. Peloquin

INTRODUCTION

Peloquin (1982) proposed five breeding schemes for the utilization of primitive cultivars and wild relatives of the potato, *Solanum tuberosum* ssp. *tuberosum* (Tuberosum) (2*n* = 4*x* = 48). Of these breeding schemes, 4*x* x 2*x* crosses (unilateral sexual tetraploidization), using diploids which produce 2*n* gametes, allow the breeder to obtain almost entirely 4*x* progenies, due to the strong triploid block existing in potato and its wild relatives (Johnston *et al*. 1980).

Tetraploid offspring with high tuber yields, good vine vigour and significant heterosis have been reported from 4*x* x 2*x* crosses (Hanneman & Peloquin 1969; Mok & Peloquin 1975; Kidane-Mariam & Peloquin 1974, 1975; Mendiburu & Peloquin 1977; De Jong & Tai 1977; De Jong *et al*. 1981; McHale & Lauer 1981; Schroeder 1983). In their crosses the 2*x* clones, used as males, produced 2*n* pollen by first division restitution (FDR).

This paper presents results obtained from 4*x* progenies of 4*x* x 2*x* FDR crosses. The experiment was carried out in France during 1983 and three major economic traits were studied: total tuber yield, marketable tuber yield and total solids.

MATERIALS AND METHODS

During the spring of 1981, 4*x* x 2*x* crosses were made in the glasshouses of the GIE Germicopa Recherche et Création, at Chateauneuf du Faou, France. Six European cultivars (2*n* = 4*x* = 48) - Claustar, Spunta, BF 15, Sirtema, Charlotte and Desiree - were used as seed parents. US-W 5295.7 and five progeny clones of US-W 5295.7 x US-W 5337.3 from the University of Wisconsin potato programme were used as male parents. These 2*x* haploid *Tuberosum* x *S. phureja* (Phureja) hybrid clones were characterized as producing 2*n* pollen by a FDR mechanism, in which crossing-over does occur (FDR-CO). Seeds from all the 36 crosses of a partial diallel mating design were obtained.

The seeds were sown in glasshouses during September, 1981 at Chateauneuf du Faou and the tubers harvested during February, 1982. The tubers were stored at 3°C until 26 April, when they were planted in the field for multiplication. Two tubers (>35-45 mm) of 22 genotypes per cross were reserved for field trials in 1983.

On 15 April 1983, tubers were planted in the field at Chateauneuf du Faou, in a split plot design. The trial had two replications, in which females were laid out as whole plots, with males forming the subplots. The six 4x cultivars used as female parents were included in this trial. Double row plots were used, each plot planted with 22 different genotypes for the 4x progenies and with 22 tubers for the 4x parents. Spacing was 65 cm between rows and 33 cm within the rows. The trial was harvested in the first week of September, 140 days after planting.

Data were collected on a plot basis for (1) total tuber yield (in kg), (2) marketable tuber yield (weight of tubers >35-45 mm, in kg), (3) total solids (determined by the weight-in-air to weight-in-water ratio). In addition, the number of stems per hill was counted by taking at random 10 hills per plot in one replication.

Analysis of variance was computed with SAS at the Computer Center of the University of Wisconsin, Madison, Wisconsin, USA. Calculations were made on raw data for total tuber yield and marketable yield and after arcsin transformation for total solids.

RESULTS AND DISCUSSION

The analyses of variance are given in Table 1. For total and marketable yields there are significant female and male effects, whereas the female x male interaction is not significant. In contrast, for total

Table 1. Analyses of Variance for the Chateauneuf trial (see text for details).

		Probability values for variance ratios		
	df	Total yield	Marketable yield	Total solids
Female	5	0.004 **	0.002 **	0.18 NS
Female x Rep.	6	0.31 NS	0.33 NS	0.0002 **
Male	5	0.004 ***	0.0001 ***	0.0001 ***
Female x Male	25	0.08 NS	0.13 NS	0.0009 ***
CV		8.45	10.65	1.64

*, **, *** = significant at p = 0.05, 0.01 and 0.001 respectively.

solids content, the female effect is not significant but male and female x male interaction effects are highly significant. The average number of stems per hill was 8.5 for 4x progenies and only 5.3 for their 4x parents.

Values for heterosis over the better parent are presented in Table 2. They range from 1.10 to 1.19 when calculated on the basis of the overall progeny means and from 1.22 to 1.42 on the basis of the average of the five best 4x progenies from 4x x 2x FDR-CO crosses.

These results for tuber yields confirmed those of previous investigations with 4x x 2x FDR crosses (Hanneman & Peloquin 1969; Mok & Peloquin 1975; Kidane-Mariam & Peloquin 1974, 1975; Mendiburu & Peloquin 1977; De Jong & Tai 1977; McHale & Lauer 1981; Schroeder 1983) and results obtained in Hancock, Wisconsin, USA with the same material (Masson 1985). Although the observed heterotic effect could be explained by the higher number of stems per hill in the 4x progenies (8.5 *vs* 5.3 for the 4x parents), the results obtained for marketable tuber yield (tubers >35/45 mm) at a rather close spacing indicate that the number of tubers per plant does not account for all the yield increase. This significant heterosis can be a reflection of either the importance of the high level of heterozygosity of the 4x progenies or the effect of complementary genes present in the different parents (Phureja and Tuberosum). Chujoy (1985), working with 2x haploid Tuberosum x *S. chacoense* hybrids (HC) and 2x Phureja x haploid Tuberosum hybrids (PH), obtained large differences for total yield between 2x and 4x progenies from HC x PH, PH x PH and HC x HC crosses. Tetraploid progenies outyielded diploid ones.

Table 2. High-parent heterosis in the Chateauneuf trial. Mean values are presented for 36 4x progenies (prog.), and for the five best 4x progenies from 4x x 2x crosses with six 4x European female cultivars and six 2x FDR-CO males.

	4x parents*	4x prog.*	Five best 4x prog.	4x prog. / 4x parents	Five best 4x prog. / 4x parents
Total tuber yield	0.67	0.80	0.95	1.19	1.42
Marketable tuber yield	0.60	0.66	0.83	1.10	1.38
Total solids[x]	16.1	17.9	19.6	1.11	1.22

* Expressed in Kg/plant^{-1} for total and marketable tuber yield.
x In the same field the 4x cultivar Bintje had a mean of 18.6 for total solids.

In Chujoy's 2x x 2x crosses, male parents at least formed 2n pollen by FDR, so the resultant 4x progeny genotypes have more heterozygosity than their 2x full sibs. Consequently, these results indicate the importance of heterozygosity for tuber yield, compared with complementary gene action.

No comparable data are available for total solids content. However, for tuber yield, the female x male interaction is highly significant and confirms Tai's data (1976). Thus the two hypotheses, heterozygosity or complementary gene action, are equally possible.

Implementation of the 4x x 2x breeding scheme would be further expedited if correlations for major traits occurred between the performances of 2x FDR parents and those of their 4x progenies. For tuber yield, a lack of correlation has been found by several authors (Mok & Peloquin 1975; De Jong & Tai 1977; McHale & Lauer 1981; Veilleux & Lauer 1981; Schroeder 1983; Masson 1985). As a consequence the breeder has to perform 4x progeny testing in order to choose his 2x parents. On the other hand, good agreement between 2x FDR clones and their 4x progenies has been obtained for general tuber appearance and eye depth in 4x x 2x crosses (Quinn 1973; De Jong & Tai 1977; Schroeder 1983). In addition to these findings, several authors have found evidence, in 2x x 2x crosses, for qualitative inheritance of early tuber bulking and vine maturity under long-day conditions (Leue 1983; Schroeder 1983; Hermundstad 1984; Masson & Peloquin, unpublished). In their studies, a two-gene model accounts for almost all the variation observed for tuberization and vine maturity.

Consequently, the breeder can easily put germplasm of the wild species into a form that will tuberize at the 2x level upon crossing with Tuberosum haploids. This will allow him to capture all the valuable genes present in wild species but absent in the tetraploid commercial potato (e.g., late blight, (*Phytophthora infestans*), virus and nematode (*Globodera* spp.) resistance). Then by using 2x FDR clones in 2x x 2x or 4x x 4x crosses he will be able to transfer these genes as a group to the 4x level.

REFERENCES

Chujoy, J.E. (1985). Tuber yields of 2x and 4x progeny from 2x x 2x crosses in Potato; Barriers to interspecific hybridization between *Solanum chacoense* Bitt. and *S. commersonii* DUN. Ph.D. Thesis, University of Wisconsin-Madison.

De Jong, H. & Tai, G.C.C. (1977). Analysis of 4x-2x hybrids in cultivated potatoes. Potato Res., 20, 111-21.

De Jong, H., Tai, G.C.C., Russel, W.A., Johnston, G.R. & Proudfoot, K.G. (1981). Yield potential and genotype-environment interactions of tetraploid-diploid (4x-2x) potato hybrids. Amer. Potato J., 58, 191-9.

Hanneman, R.E. Jr., & Peloquin, S.J. (1969). Use of Phureja and haploids to enhance the yield of cultivated tetraploid potatoes. Amer. Potato J., 46, 436 (Abstract).

Hermundstad, S.A. (1984). Production and evaluation of haploid Tuberosum-wild species hybrids. M.S. Thesis. University of Wisconsin-Madison.

Johnston, S.A., den Nijs, T.P.M., Peloquin, S.J. & Hanneman, R.E. Jr. (1980). The significance of genic balance to endosperm development in interspecific crosses. Theor. Appl. Genet., 57, 5-9.

Kidane-Mariam, H. & Peloquin, S.J. (1974). The effect of direction of hybridization (4x x 2x vs. 2x x 4x) on yield of cultivated potatoes. Amer. Potato J., 51, 330-6.

Kidane-Mariam, H. & Peloquin, S.J. (1975). Method of diplandroid formation and yield of progeny from reciprocal (4x - 2x) crosses. J. Amer. Soc. Hort. Sci., 100, 602-3.

Leue, E.F. (1983). The use of haploids, 2n gametes and the topiary mutant in the adaptation of wild Solanum germplasm and its incorporation into Tuberosum. Ph.D. Thesis, University of Wisconsin-Madison.

Masson, M.F. (1985). Mapping, combining abilities, heritability and heterosis with 4x x 2x crosses in Potato. Ph.D. Thesis, University of Wisconsin-Madison.

McHale, N.A. & Lauer, F.I. (1981). Breeding value of 2n pollen diploid from hybrids and Phureja in 4x-2x crosses in potatoes. Amer. Potato J., 58, 365-74.

Mendiburu, A.O. & Peloquin, S.J. (1977). The significance of 2n gametes in potato breeding. Theor. Appl. Genet., 49, 53-61.

Mok, D.W.S. & Peloquin, S.J. (1975). Breeding value of 2n pollen (diplandroids) in 4x-2x crosses in potatoes. Theor. Appl. Genet., 46, 307-14.

Peloquin, S.J. (1982). Meiotic mutants in potato breeding. In Stadler Genetic Symposium. Vol. 14, pp. 99-109. University of Missouri.

Quinn, A.A. (1973). Use of experimental tetraploids in potato breeding. Ph.D. Thesis. University of Wisconsin-Madison.

Schroeder, S.H. (1983). Parental value of 2x, 2n pollen clones and 4x cultivars in 4x x 2x crosses in potato. Ph.D. Thesis, University of Wisconsin-Madison.

Tai, G.C.C. (1976). Estimation of general and specific combining abilities in potato. Can. J. Genet. Cyto., 18, 463-70.

Veilleux, R.E. & Lauer, F.I. (1981). Variation for 2n pollen production in clones of Solanum phureja Juz. and Buk. Theor. Appl. Genet., 59, 95-100.

VARIABILITY OF F_1 PROGENY DERIVED FROM INTERPLOIDY ($4\underline{x}$ x $2\underline{x}$) CROSSING

J. Jakubiec

INTRODUCTION

Diploid potato clones which produce unreduced gametes merit particular attention because of their value in genetic research and in the breeding of new cultivars. There are various mechanisms that lead to the production of unreduced ($2\underline{n}$) gametes; and these have diverse genetic consequences. The value of diploid clones and their utilization depends on the production mechanism (Mendiburu *et al*. 1970; Mendiburu & Peloquin 1979). Clones forming unreduced gametes via First Division Restitution (FDR) can be used in the following ways:-

1) To transfer genes from diploid species to tetraploid cultivars since they cross easily with both diploids and tetraploids (Hanneman & Peloquin 1967, 1968; Jakubiec & Suska 1981).

2) To study the inheritance of characters in the tetraploid progeny and the combining ability of the maternal parent (Jakubiec & Narkiewicz 1981), because they form homogenous gametes.

3) To transfer heterozygous blocks of alleles to the F_1 progeny by crossing selected diploids with tetraploids, and thus ensuring a high level of heterosis (de Jong *et al*. 1981).

4) To produce F_1 seeds which are genotypically uniform by utilizing the ability of clones to produce female and male unreduced gametes. This provides the opportunity of producing cultivars from true seed (TPS) (Anon 1979; Jackson, this volume).

In this paper the following are discussed:-

1) The characteristics of 'NG' clones which produce unreduced gametes.

2) The fluctuation of big pollen production during the growing season.

3) Variation in tuber yield and tuber size in F_1 progeny from $4\underline{x}$ x $2\underline{x}$ crosses.

MATERIALS AND METHODS

The mechanism leading to the formation of 2n gametes was identified by studying microsporogenesis. The viability of pollen grains was determined prior to pollination; numbers of flowers and seeds per berry were also recorded. F_1 seeds were sown in the glasshouse and the resulting seedlings were planted in field plots in a hexagonal design with two replicates. A number of characters were measured including the shape, colour, eye depth, yield and starch content of tubers.

RESULTS

Various abnormalities in the frequency of 2n gametes were observed. Clones NG_1 and NG_3 formed 2n gametes as a result of parallel spindles (ps) and fused spindles in metaphase II. Clone NG_2 produced 2n pollen and big pollen as a result of ps and premature cytokinesis. When diploid NG clones were used as a source of pollen, the frequency of big pollen formation and the viability of the grains fluctuated depending on the plant growth stage. The highest frequency of big pollen occurred at the commencement of the flowering phase. Successful hybridization was affected by the quantity of big pollen produced. The diploid parent significantly affected the number of seeds produced per pollinated flower and the number of seeds per berry, clone NG_1 being better in these respects than NG_2. However, the mean frequency of big pollen production did not differ significantly. Successful hybridization was independent of the maternal parent. When considering the F_1 progenies of 4x x 2x crosses we assumed that the 2n gametes were homogenous (produced by FDR) and therefore that variation in characters measured within and between families resulted from the heterozygosity of the 4x maternal parent. Tuber yields and starch content for 10 families are shown in Table 1.

Variation in tuber yield was similar for six of the families (1, 3, 4, 6, 8 & 10 - see Figure 1). There were significant differences in starch content between families and some of them showed considerable variation for this character (Table 1). Differences in the morphology of tubers, their mean size and shape, colour of skin and depth of eyes, were also significant. NG clones with small tubers transferred this character to their progeny. Four crosses produced clones which had very uniform tubers and high marketable yield. The maternal parents of these crosses were Dekema, Pilica, OL-21851 and OL-22127.

Table 1. Tuber yield and starch content for families derived from 4x x 2x crosses

Family No.	Cross	Tuber yield $\bar{x}$	range	SD	CV(%)	Starch content (%) $\bar{x}$	range
1	Atol x NGl	78.56	18-181	37.65	47.9	14.17	10.0-17.6
2	Certa x NGl	89.79	12-244	47.98	53.4	15.54	10.6-24.5
3	Dekama x NGl	76.26	15-188	38.28	50.2	14.88	11.2-22.8
4	Duet x NGl	97.57	15-185	40.84	41.9	16.27	7.0-20.0
5	Falka x NGl	54.05	12-126	25.98	48.1	16.62	13.2-20.4
6	Narew x NGl	70.54	12-156	35.96	51.0	17.75	13.0-20.4
7	Pilica x NGl	99.55	18-175	33.85	34.0	16.60	11.3-21.0
8	OL21851 x NGl	99.46	30-245	39.30	39.5	17.37	13.9-26.1
9	OL22127 x NGl	111.11	27-285	56.72	51.0	14.74	8.6-19.0
10	OL22128 x NGl	90.17	20-200	36.39	40.4	16.28	8.1-21.2
Uran		181.00		27.10	29.9	14.63	
LSD (p=0.05)		12.29				0.64	

SD = standard deviation
CV = coefficient of variation

Figure 1. Range of variation in tuber yield per plant for different families derived from 4x x 2x crosses.

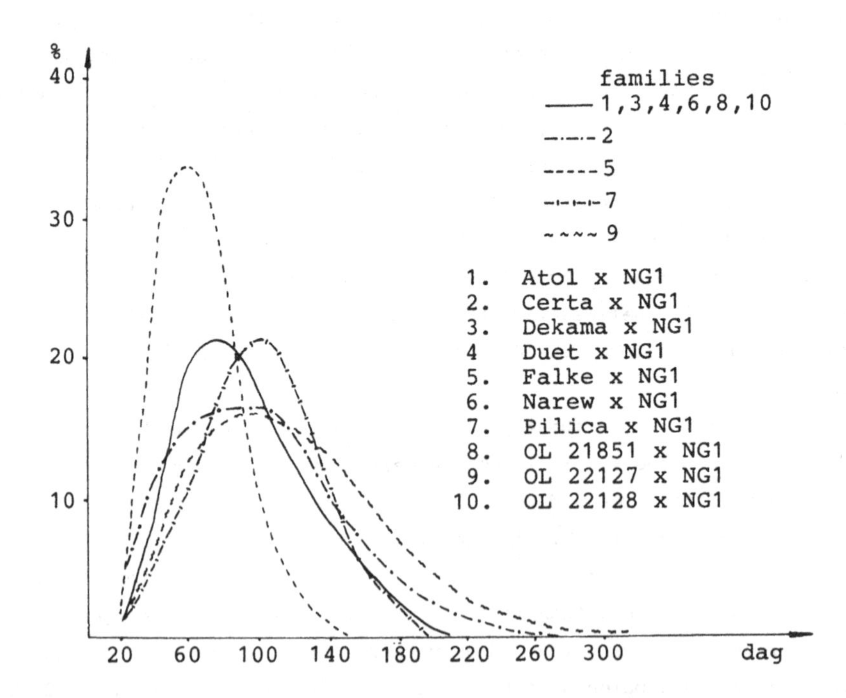

CONCLUSION

Breeding at the diploid level will play an important role in variety improvement for a whole range of characters in the very near future, because disomic inheritance simplifies breeding methodology. Diploids will also play an important role in the production of TPS. Sowing true seed instead of planting seed tubers is especially important in regions where tuber seed production is difficult.

REFERENCES

Anon. (1979). International Potato Center Ann. Rep. 1978, pp. 43-4.

Hanneman, Jr, R.E. & Peloquin, S.J. (1967). Crossability of 24-chromosome potato hybrids with 48-chromosome cultivars. Eur. Potato J., 10, 62-73.

Hanneman Jr., R.E. & Peloquin, S.J. (1968). Ploidy levels of progeny from diploid-tetraploid crosses in the potato. Am. Potato J., 45, 255-61.

Jakubiec, J. & Suska, M. (1981). Mieszance miedzygatunkowe donorem odpornosci na choroby przechowalnicze ziemniaka. Hod. Rosl. Aklim. Nas., 25, 43-53.

Jakubiec, J. & Narkiewicz, M. (1981). Poszukiwanie form diplandronicznych i diplogynicznych w potomstwie krzyzowek klonow diploidalnych z tetraploidalnymi odmianami Solanum tuberosum. Hod. Rosl. Aklim. Nas., 25, 87-96.

de Jong, H., Tai, G.C.C., Russell, W.A., Johnston, G.R. & Proudfoot, K.G. (1981). Yield potential and genotype-environment interactions of tetraploid-diploid (4x - 2x) potato hybrids. Am. Potato J., 58, 191-9.

Mendiburu, A.O., Peloquin, S.J. & Hanneman Jr., R.E. (1970). The significance of 2n gametes in potato breeding. Am. Potato J., 47, 356-7.

Mendiburu, A.O. & Peloquin, S.J. (1979). Gene-centromere mapping by 4x-2x matings in potatoes. Theor. Appl. Genet., 54, 177-80.

UNREDUCED GAMETES IN THE BREEDING OF POTATOES AT THE DIPLOID LEVEL

E. Zimnoch-Guzowska and I. Wasilewicz

At the Institute for Potato Research, the final goal in breeding at the diploid level is to obtain parental forms possessing a range of valuable characters (high yielding ability, good tuber shape, resistance to various pathogens) associated with the ability to produce first division restitution (FDR), 2*n* gametes. For several years our diploids have been selected for their ability to produce 2*n* gametes.

The following criteria have been used:

1. More than 5% big pollen grains (BP).
   Up until 1985, from over 900 clones and 2100 first year seedlings, 293 clones and 1100 seedlings had been selected. The character was observed most frequently in hybrids derived from *Solanum phureja* and *S. chacoense*. It was confirmed that environmental conditions influenced the frequency of BP. If both parents produced 2*n* gametes, a higher percentage of BP was found in their progeny (Table 1).
2. Seed set in 4*x*-2*x* or 2*x*-4*x* crosses.
   Of 73 hybrids tested, 13 produced over 10 seeds per berry in 4*x*-2*x* crosses.
3. Presence of cytological abnormalities at meiosis indicating the formation of 2*n* gametes.

Table 1. The incidence of big pollen grains (BP) in parents producing 2*n* gametes and in their progeny.

	Parents		Progeny	Progeny
	Male	Female	mean	range
DG 79-455 x DG 79-986	8	5	11.8	0-33.7
DG 79-986 x H_2 439	5	10	23.8	0-80.0
DG 79-465 x DG 79-986	1	5	8.5	0-33.0

Among cytologically evaluated clones, deviations indicating either FDR or SDR (second division restitution) were observed. Recently a FDR clone, DG 81-68 has been selected.

Evaluation of three tetraploid F_1 progenies of 4x-2x crosses

In 1985 heterosis and the extent of variation for some characters were evaluated in three F_1 progenies obtained from crossing Polish cultivars with the 2*x* FDR clone DG 81-68. These progenies were: Bryza x DG 81-68 (B), Wilga x DG 81-68 (W), and Certa x DG 81-68 (C). The F_1 were planted in two replicates with two plants per plot; 6 plots of each parent were planted in each replicate. The following characters were evaluated: number of tubers per plant, mean tuber weight (g), tuber yield per plant (dag) and starch content (%).

The mean number of tubers per plant was significantly (*p*=0.01) higher than the midparent value for each of the F_1 progenies but this was coupled with a reduction in mean tuber weight, which was significant (*p*=0.01) for progenies B and C (Table 2). There was a significantly higher

Table 2. Comparison of parents and their progenies.

		F_1 progenies	
	B+	W	C
Number of tubers/plant			
mean of parents: (4*x*)	12.3	10.7	15.4
(2*x*)	9.1	9.1	10.6
midparent value	10.7	9.9	13.0
mean of F_1 hybrids	21.4**	16.5**	28.5**
Mean tuber weight (g)			
mean of parents: (4*x*)	121.4	113.2	112.4
(2*x*)	31.7	31.7	29.7
midparent value	76.6**	72.5	71.1**
mean of F_1 hybrids	51.7	63.4	39.9
Tuber yield/plant (dag)			
mean of parents: (4*x*)	135.1	112.1	181.6
(2*x*)	28.4	28.5	31.3
midparent value	81.7	70.3	106.5
mean of F_1 hybrids	105.9**	95.9*	105.0
Starch content (%)			
mean of parents: (4*x*)	12.0	12.3	15.2
(2*x*)	15.4	15.4	15.6
midparent value	13.7	13.8	15.4**
mean of F_1 hybrids	13.3	13.3	13.5
Number of F_1 hybrids	80	77	39

*,** = significant at the 5% and 1% respectively between midparent value and mean of F_1 hybrids.
+ B;W;C - see text for details.

mean tuber yield per plant for F_1 progenies B and W ($\underline{p}$=0.01; 0.05 respectively) than their midparent values. Hence, tuber size was more than compensated for by tuber number. The mean percentage starch content for all three F_1 progenies was slightly lower than the respective midparent value; this was significant ($\underline{p}$=0.05) for progeny C.

DESYNAPSIS AND FDR 2N-EGG FORMATION IN POTATO: ITS SIGNIFICANCE TO THE EXPERIMENTAL INDUCTION OF DIPLOSPORIC APOMIXIS IN POTATO

E. Jongedijk

INTRODUCTION

In the cultivated potato, meiotic polyploidization by means of 2*n*-gametes has received considerable attention, both in relation to the development of more efficient alternatives to current breeding programmes (Mendiburu *et al*. 1974; Peloquin 1982; Hermsen 1984b) and the production of well-performing and relatively uniform hybrid progenies for the new technology of growing potatoes from true seeds (Peloquin 1983). As to the latter, the possibility of using 2*n*-egg formation in attempts to induce gametophytic apomixis has recently been contemplated (Hermsen 1980; Iwanaga 1982; Hermsen *et al*. 1985).

In gametophytic apomixis an unreduced embryo sac is formed that can be of either diplosporic (sexual) or aposporic (somatic) origin (Rutishauser 1967). As there are strong suggestions from earlier literature (Powers 1945; Petrov 1970; Asker 1980; Hermsen 1980; Matzk 1982) that both aposporic and diplosporic apomixis comprise distinct and genetically regulated elements, the experimental induction of gametophytic apomixis might well be achieved by a combination of them.

The elements of diplosporic apomixis that can be distinguished are a strongly reduced crossing over in megasporogenesis, the formation of unreduced megaspores and embryo sacs, and parthenogenetic development of the unreduced egg cell. In aposporic apomixis, parthenogenetic development should be combined with the development of a somatic cell of the ovule into an unreduced embryo sac. In potato, the genetically controlled elements of displosporic apomixis are available (Hermsen 1980; Hermsen *et al*. 1985). With apospory, however, the aforementioned development of a somatic cell, as claimed to occur in diploid potato hybrids (Iwanaga 1980, 1982), has not yet been reported (Jongedijk 1985). Therefore, at this stage, the experimental induction of diplosporic apomixis in potato appears to offer the best prospects.

REDUCED CROSSING OVER AND 2N-MEGASPORE FORMATION

Various abnormal meiotic events may lead to 2n-gamete formation. Depending on the genetic consequences, however, only two distinct types of 2n-gametes can be distinguished: first-division-restitution (FDR) and second-division-restitution (SDR) gametes. In highly heterozygous crops FDR gametes, when compared to SDR, preserve a much larger amount of parental heterozygosity and epistasis, and thus more strongly resemble the parental genotype from which they derive (Mendiburu et al. 1974; Hermsen 1984a). Therefore, the first step in the induction of diplosporic apomixis in potato is to combine FDR megaspore formation with a reduction in crossing over. In potato, the latter is a prerequisite for FDR megaspore formation (Jongedijk 1985).

Crossing over may be restricted by the action of mutant genes controlling synapsis (asynapsis and desynapsis) or as a result of limited chromosome homology. In these cases only FDR gametes are expected to be functional (Ramanna 1983; Jongedijk 1985). With a complete lack of crossing over, such FDR gametes will preserve the parental genotype intact, as is the case with diplospory. Though the combination of asynapsis (no chromosome pairing and thus no gene recombination) with FDR megaspore formation might thus be preferred (Hermsen 1980), desynapsis or genome divergency may also suffice if gene recombination is strongly reduced and predominantly restricted to the chromosome ends.

PRESENT RESULTS AND FUTURE STRATEGY

For the experimental induction of diplosporic apomixis in potato, genetically controlled asynapsis and desynapsis are of special interest. As genes controlling asynapsis are still unknown in potato, our experiments are focussed on desynapsis instead. By test crossing (Jongedijk 1983), several diploid, and recently also tetraploid, genotypes combining desynapsis and FDR megaspore formation have been selected.

The single recessive gene (ds) for desynapsis (Ramanna 1983) was found to be identical to the gene sy_3 reported by Okwuagwu & Peloquin (1981), and is expressed at both microsporogenesis (Okwuagwu & Peloquin 1981; Ramanna 1983) and megasporogenesis (Jongedijk 1983). The single recessive gene sy_1 (Iwanaga & Peloquin 1979), expressed in megasporogenesis only, was found to be nonexistent. Though the Ds/ds locus is characterized by normal pachytene pairing and separation of bivalent chromosomes at diakinesis, genetic analysis with marker genes indicated a severe reduction in crossing over. In attempts to induce diplosporic apomixis the locus therefore appears to be a valuable alternative to true asynapsis.

In contrast to earlier findings (Iwanaga & Peloquin 1979), FDR megaspores were formed predominantly through pseudohomotypic division in all selected genotypes. From some of them, parthenogenetic progeny could be obtained via pseudogamy using marked diploid and tetraploid *Solanum phureja* pollinators.

Though the viability of experimental induction of diplosporic apomixis in potato has now been illustrated, many things remain to be studied. Maximizing frequencies of FDR megaspore formation and pseudogamous parthenogenesis still require extensive fundamental research. In addition, apomictic seed production needs to be simplified. The latter might be achieved by introducing genes for pseudogamy into self-compatible desynaptic genotypes with FDR megaspore formation or by using cytoplasmic male sterility and insect pollination.

REFERENCES

Asker, S. (1980). Gametophytic apomixis: elements and genetic regulation. Hereditas, 93, 277-93.

Hermsen, J.G.Th. (1980). Breeding for apomixis in potato: pursuing a utopian scheme? Euphytica, 29, 595-607.

Hermsen, J.G.Th. (1984a). Mechanisms and genetic implications of 2n-gamete formation. Iowa State J. Res., 58, 421-34.

Hermsen, J.G.Th. (1984b). The potential of meiotic polyploidization in breeding allogamous crops. Iowa State J. Res. 58, 435-48.

Hermsen, J.G.Th., Ramanna, M.S. & Jongedijk, E. (1985). Apomictic approach to introduce uniformity and vigour into progenies from true potato seed (TPS). *In* Report XXVI Planning Conference 'Present and future strategies for potato breeding and improvement', pp. 99-114. Lima: CIP.

Iwanaga, M. & Peloquin, S.J. (1979). Synaptic mutant affecting only megasporogenesis in potatoes. J. Heredity, 70, 385-9.

Iwanaga, M. (1980). Diplogynoid formation in diploid potatoes. Ph.D. Thesis, University of Wisconsin, Madison.

Iwanaga, M. (1982). Chemical induction of aposporous apomictic seed production. *In* Proceedings International Congress 'Research for the potato in the year 2000', pp. 104-5. Lima: CIP.

Jongedijk, E. (1983). Selection for first division restitution 2n-egg formation in diploid potatoes. Potato Res., 26, 399 (Abstract).

Jongedijk, E. (1985). The pattern of megasporogenesis and megagametogenesis in diploid *Solanum* species hybrids; its relevance to the origin of 2n-eggs and the induction of apomixis. Euphytica, 34, 599-611.

Matzk, F. (1982). Vorstellungen über potentielle Wege zur Apomixis bei Getreide. Arch. Züchtungsforsch., 12, 183-95.

Mendiburu, A.O., Peloquin, S.J. & Mok, D.W.S. (1974). Potato breeding with haploids and 2n gametes. *In* Haploids in Higher Plants, ed. K. Kasha, pp. 249-58. Guelphi: University of Guelph.

Okwuagwu, C.O. & Peloquin, S.J. (1981). A method of transferring the intact parental genotype to the offspring via meiotic mutants. Am. Potato J., 58, 512 (Abstract).

Peloquin, S.J. (1982). Meiotic mutants in potato breeding. Stadler Genetics Symposia, 14, 1-11.

Peloquin, S.J. (1983). New approaches to breeding for the potato of the year 2000. In Proceedings International Congress 'Research for the potato in the year 2000', pp. 32-4. Lima: CIP.

Petrov, D.F. (1970). Genetically regulated apomixis as a method of fixing heterosis and its significance in breeding. In Apomixis and Breeding, ed. S.S. Khoklov, translated from Russian by B.R. Sharma, pp. 18-28. New Delhi: Amerind Publishing Co.

Powers, L. (1945). Fertilization without reduction in guayule (Parthenium argentatum GRAY) and hypothesis as to the evaluation of apomixis and polyploidy. Genetics, 30, 323-46.

Ramanna, M.S. (1983). First division restitution gametes through fertile desynaptic mutants of potato. Euphytica, 32, 337-50.

Rutishauser, A. (1967). Fortpflanzungsmodus und Meiose apomiktischer Blütenpflanzen. Protoplasmatologia, Band VI F3. Vienna: Springer Verlag.

UTILIZING WILD POTATO SPECIES VIA *SOLANUM PHUREJA* CROSSES

R.N. Estrada

Wild potato species are an immense source of genetic variability which has been relatively little used. Many diploid and tetraploid wild species are compatible with diploid clones of *Solanum phureja* cultivated on the Andes of South America. The resulting hybrids can have various levels of ploidy:

a) diploid, 2*n* = 24 (wild diploid x cultivated diploid)

b) triploid, 2*n* = 36 (wild tetraploid x cultivated diploid)

c) tetraploid, 2*n* = 48 (wild tetraploid x cultivated diploid producing 2*n* gametes.

Gene transfer from wild to cultivated clones in the case of a) diploid and b) triploid hybrids is assured by utilizing them as male parents, which may produce 2*n* pollen grains by First division restitution or Second division restitution in the meiotic process, in order to fertilize tetraploid cultivars or clones of *S. tuberosum* ssp. *tuberosum* or spp. *andigena*. In the case c), in which the hybrids produced are tetraploid, gene transfer is more simple since the hybrids can be used either as males or females in backcrosses to *tuberosum* or *andigena* tetraploid clones.

By this method, hybrids were obtained from crosses between *S. phureja* clones and 12 species (*S. acaule*, *S. andreanum*, *S. boliviense*, *S. bulbocastanum*, *S. chacoense*, *S. colombianum*, *S. microdontum*, *S. sanctae-rosae*, *S. sogarandinum*, *S. stoloniferum*, *S. toralapanum* and *S. vernei*) belonging to seven taxonomic series (Figure 1).

Several of the F_1 hybrids have already been used successfully in backcrosses to tetraploid cultivars or to advanced clones (Table 1).

The utilization of *S. phureja* and wild species by this method makes the transfer of valuable genes from wild species into cultivars more feasible (Estrada 1984; Hermunstad & Peloquin 1985; Quinn *et al.* 1974; Peloquin & Mendiburu 1972).

Using the conventional method, the chromosomes of the wild species had to be doubled by the use of colchicine. This led to higher ploidy levels (4x or more) and therefore increased genetic complexity, requiring more backcrosses to the tetraploid cultivars and much higher populations for the selection of valuable genes.

Table 1. Species hybrids, their pollen viability and crossability.

Clone	Parents	Stained* pollen %	Compatible with 4x
acaphu	*S. acaule* x *S. phureja*	60	yes
80-655-3	*S. stoloniferum* x *S. phureja*	40	yes
84-611-1	*S. chacoense* x *S. phureja*	70	yes
84-616-4	*S. sanctae-rosae* x *S. brevidens*	60	yes
84-618-1	*S. sanctae-rosae* x *S. phureja*	90	yes
84-604-1	*S. boliviense* x *S. phureja*	90	yes
84-621-3	*S. sogarandinum* x *S. phureja*	60	yes
84-627-4	*S. vernei* x *S. phureja*	70	yes

* The ability of pollen grains to take up stain is an indication of viability.

Figure 1. Tubers obtained from the hybrids *S. vernei* x *S. phureja* (left), and *S. sogarandinum* x *S. phureja* (right).

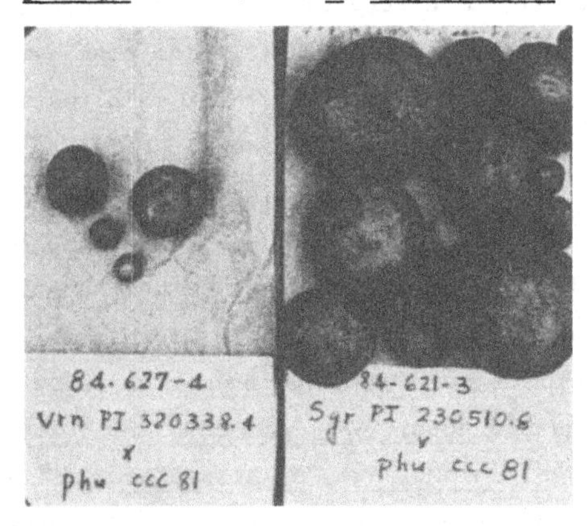

REFERENCES

Estrada, R.N. (1984). Acaphu: a tetraploid, fertile breeding line, selected from *S. acaule* x *S. phureja* cross. Am. Potato J., 61, 1-7.

Hermunstad, S.A. & Peloquin, S.J. (1985). Tuber yield and tuber characteristics of haploid-wild species hybrids. Abst. Potato Assoc. Am. 69th Ann. Meet.

Peloquin, S.J. & Mendiburu, A.O. (1972). Sexual polyploidization in relation to breeding and evolution. Am. Potato J., 49, 363.

Quinn, A.A., Mok, D.W.S. & Peloquin, S.J. (1974). Distribution and significance of diploandroids among the diploid solanums. Am. Potato J., 51, 16-21.

THE USE OF DIPLOID <u>SOLANUM PHUREJA</u> GERMPLASM

C.P. Carroll

INTRODUCTION

Until very recently the contribution of the diploid edible potatoes of South America to breeding in the Northern Hemisphere has been limited to a few, unimproved genotypes. To examine the potential of this material, a substantial and widely based sample should be acclimatized to higher latitudes, extensively screened for fungal and viral disease resistances, and suitable breeding strategies devised (bearing in mind its diploid nature) for combining it with a range of modern *Solanum tuberosum* ssp. *tuberosum* (Tuberosum) parents. Any general advantages which could reasonably be expected from the new material, such as more rapid attainment of particular breeding goals at the diploid level, the presence of unexploited additive variation for yield components, and of heterotic effects in hybrids should be utilized as fully as possible. This paper describes briefly a programme for the use of *S. phureja* (Phureja) germplasm at Pentlandfield, based upon these requirements.

METHODS

Work on the improvement and subsequent utilization of diploid potatoes began in 1967 with a mass-selection scheme, which was continued until 1979 (Carroll 1982). In 1968 crossing was started to produce first-generation hybrids with dihaploids of Tuberosum. Further crossing and selection took place amongst the hybrid material, and a pedigree scheme for Phureja was initiated, using individual selections from the mass population. Direct crosses between Phureja and Tuberosum cultivars began on a small scale in 1973 using the "diplandroid" system to produce tetraploid offspring. Several hundred pollinations of 2*x* males with 4*x* females produced hardly any progeny based on "diplogynoids" (Sudheer (Carroll) 1977). Production of 4*x* F_1 hybrids has been steadily expanded using a wide range of diploid and 4*x* cultivar parents. Crosses between these hybrids enable useful disease resistances to be combined, and highly heterozygous material to be generated by crossing clones of diverse parentage. Backcrossing to Tuberosum parents is in progress to study the effects of dilution of the Phureja input, which is likely to occur when hybrid material is introduced to conventional cultivar breeding programmes.

RESULTS

Yield improvement of Phureja

Progress in yield under mass selection will depend upon reserves of general combining ability (GCA); De,Maine (1978) was able to demonstrate the differences in GCA for yield in the first clonal year among Phureja clones drawn from the mass population. Yield improvement has continued; in recent trials elite diploid clones grown as spaced plants have exceeded the yield of cv. Pentland Crown controls by as much as 100%. However, results from randomized field experiments at commercial plant spacing give a fairer picture of the present performance. In 1982, 12 elite Phureja clones were included in a replicated trial at 0.3m spacing with five modern cultivars for comparison. The mean yield per plant for all Phureja entries was 0.73 kg compared with 0.96 kg for the cultivars. Phureja yields ranged from 0.19 kg for a 1977 selection to 0.99 kg for one selected in 1981. In general, 2*x* hybrids of Phureja with dihaploid Tuberosum show slightly higher yields, and have larger tubers, in our trials than pure Phureja. It is this type of 2*x* hybrid which has been used by Peloquin and his collaborators in developing high yielding material at the 4*x* level (see Mok & Peloquin 1975; Kidane-Mariam *et al*. 1985). If Phureja itself is used as the diplandroid-producing parent, the tetraploid F_1 will contain 50% of new, non-Tuberosum germplasm rather than 25%; in subsequent crossing at the 4*x* level any advantage consequent upon hybridity might be expected to decline less rapidly.

Yield and characteristics of 4x hybrids

In a 1985 trial, unselected progenies (25 clones each) from crosses between the cultivars Pentland Crown and Cara and two elite Phureja selections were examined in the second clonal year. Mean progeny yields greater than those of the higher parent were demonstrated for the crosses involving Pentland Crown, but not for those with Cara. Although the demonstration of this kind of heterosis appears to depend on which 4*x* parents are chosen, it is relatively easy to select individual F_1 hybrids with yields exceeding those of current cultivars. In a recent replicated trial at commercial spacing (0.3m), 21 4x hybrids were compared with five cultivars (Table 1).

Tuber sizes were in the Tuberosum range, giving marketable yields around 90%. In the past, Phureja has been regarded as a possible genetic source of earliness. In fact, early tuber bulking is frequently found in the 4x hybrids, but it is not accompanied by early haulm senescence: such clones continue growth, giving very large yields at a maincrop lift. Under adverse conditions (drought, late frosts), they can perform better than conventional first earlies.

Disease resistance in Phureja

The stocks of Phureja, derived from mass selection, have been screened for a variety of disease resistances in both laboratory tests and field trials. As a result, certain diploid clones which combine high resistance to potato leafroll virus and potato virus Y are being evaluated in the SCRI virus resistance breeding programme. Phureja material has shown good resistance to *Phytophthora infestans* in field exposure trials in Cornwall and the West of Scotland, despite an inconsistent response to standard laboratory tests. Phureja also possesses useful resistances to tuber diseases; 48 clones out of 102 tested for resistance to common scab (*Streptomyces* spp.) over three seasons were rated very resistant. Extremely few were very susceptible. High levels of gangrene (*Phoma exigua* var. *foveata*) resistance have also been identified. Although the number of clones tested against soft rot (*Erwinia carotovora*) has been small, several are very difficult to infect and are currently being examined for their blackleg (*Erwinia carotovora*) reaction. Screening of 4x hybrids suggests that resistances in Phureja can be transferred effectively to the tetraploid level.

Table 1. Comparison between 4x hybrids and current cultivars

Group	Yield per nine plant plot (kg)	Comparison
4x x F_1 hybrids	11.28)	F value = 19.9;
Tuberosum cultivars	8.61)	$p < 0.001$

REFERENCES

Carroll, C.P. (1982). A mass selection method for the acclimatization and improvement of edible diploid potatoes in the United Kingdom. J. Agric. Sci., Camb., 99, 631-40.

De,Maine, M.J. (1978). The inheritance, in diploid potatoes, of yield and of rate of sprout growth during storage. Potato Res., 21, 163-70.

Kidane-Mariam, H.M., Arndt, G.C., Macaso-Khwaja, A.C. & Peloquin, S.J. (1985). Comparisons between 4x and 2x hybrid and open-pollinated true-potato-seed families. Potato Res., 28, 35-42.

Mok, D.W.S. & Peloquin, S.J. (1975). Breeding value of 2n pollen (diplandroids) in tetraploid x diploid crosses in potatoes. Theor. Appl. Genet., 4, 307-14.

Sudheer (Carroll, C.P.) (1977). The use of hybridization between diploid American and tetraploid European potatoes in a breeding programme. In Interspecific Hybridization in Plant Breeding. Proc. 8th Cong. Eucarpia, Madrid.

ADVANCES IN POPULATION BREEDING AND ITS POTENTIAL IMPACT ON THE EFFICIENCY OF BREEDING POTATOES FOR DEVELOPING COUNTRIES

H.A. Mendoza

INTRODUCTION

The mandate of the International Potato Center (CIP) is to increase potato productivity by helping to develop varieties better adapted to the growing conditions of developing countries and improving agronomic, seed production, and storage technologies.

Since the foundation of CIP, its scientists have constantly visited and consulted with the research staff of national programmes to analyse biotic and abiotic yield constraints related to the performance of their commercial cultivars. Using this information, research needs for breeding programmes have been established for the developing countries associated with each of the seven regions of CIP. The regional network has been developed to coordinate CIP's research involvement through the world. Furthermore, Planning Conferences have been organized on a regular basis with the participation of breeding experts from leading institutions throughout the world, CIP breeding staff, and scientists from developing countries. At these conferences, research problems are discussed and breeding priorities defined. CIP's main contribution to potato production has been in germplasm improvement and the distribution of genetic materials. These materials are distributed as selected clones with specific adaptation and resistance to or tolerance of climatic stresses, pests and diseases. CIP also distributes advanced populations with a wide genetic diversity segregating for the same characters as the clonal materials. National programmes have to evaluate this advanced germplasm and select varieties adapted to their local conditions and suitable for their needs. They will eventually release their own varieties from CIP's genetic material. As an additional activity, CIP is selecting true potato seed (TPS) progenies to be utilized in commercial potato production. This represents a radical departure from traditional potato production by seed tubers and is beginning to revolutionize potato production in some developing countries.

Besides assembling its own breeding team, CIP has established research contracts with leading research institutions in the developed and developing world in order to create a breeding programme which has a wide base with regard to expertise, philosophy and genetic resources.

The major element of CIP's breeding strategy is to maximize the utilization of the genetic variability available in the World Potato Collection, which consists of a large number of both primitive cultivated and wild Solanum species. In addition, the genetic resources developed by the research contracts are very important.

CIP has four experimental stations in Peru located between latitudes 6° and 12°S and ranging in altitude from 175 m to 3300 m above sea level. These sites provide a wide range of environmental variation in their biotic and abiotic components. This enables researchers to adequately assess and select breeding populations. Moreover, CIP has established a regional research network with scientists stationed in each of the seven strategic regions in the developing world. The regional scientists actively cooperate in the final testing of selected germplasm which is to be transferred to national programmes of the countries linked to each region.

Because of its worldwide responsibilities, CIP decided on an innovative breeding approach which has advantages over traditional potato breeding methods when producing advanced germplasm to fit the needs of countries of the tropical and subtropical regions of the world. This is the population breeding strategy, based on phenotypic recurrent selection in the early stages of the breeding programme with the addition of progeny testing in the later stages. This approach, which is in essence a multidisciplinary effort, fulfills the following basic strategic objectives: a) to maintain a wide genetic diversity, b) to increase the frequency of genes controlling desirable attributes and c) to stimulate the recombination of these desirable genes.

Recurrent selection with progeny testing permits the selection of parental material with high general combining ability (GCA) for the traits being considered. Therefore, this method allows additive genetic variability to be exploited in addition to nonadditive genetic variability. The latter has been more generally used in traditional breeding schemes. When considering breeding for the utilization of true potato seed (TPS), the selection of progenitors with high GCA for yield, tuber uniformity and other traits is a very efficient strategy to obtain progenies with high performance for these characters.

From the advanced germplasm that CIP has developed and distributed as selected clones, as well as from segregating populations, a collection of new high performance commercial varieties has been selected and released, and these varieties are currently cultivated in various countries of the developing world. The varieties are in general characterized by their tolerance of climatic stresses and resistance to one or more of the important pests or diseases present in these specific countries.

BIOTIC AND ABIOTIC YIELD CONSTRAINTS AFFECTING THE POTATO IN THE DEVELOPING WORLD

The potato crop is vulnerable to attack by a large number of pests and diseases. To date, 23 viruses, 38 fungi, six bacteria, two mycoplasms, one viroid, 68 nematode species and 128 insects and related pests have been reported, totalling 266 pathogens and pests. Some of these are more destructive than others and are distributed worldwide, i.e. late blight (*Phytophthora infestans*), potato leafroll virus (PLRV) and potato virus Y (PVY). In tropical zones, there are additional plant parasites of major economic importance such as bacterial wilt (*Pseudomonas solanacearum*), blackleg (*Erwinia* spp.), root knot nematodes (*Meloidogyne* spp.), cyst nematodes (*Globodera* spp.), tuber moth (*Phthorimaea operculella*) and other related species of the family Gelechiidae. There are also diseases such as early blight (*Alternaria solani*), wilts (*Fusarium* spp.), leaf blight (*Choanephora cucurbitae*) and charcoal rot (*Macrophomina phaseoli*), which may have a devastating effect on the crop but are confined to certain geographical zones. In addition to these biotic constraints, the potato crop in the developing world is frequently threatened by climatic stresses such as hail, frost, drought and heat.

Most of the commercial varieties presently grown in the developing world have been bred in temperate countries where they yield well. Late blight and/or virus resistance have not had a high priority in these temperate countries since these diseases can be effectively controlled by pesticides and certified-seed programmes. When these varieties are cultivated in tropical and subtropical countries, where the use of pesticides is limited and seed programmes are either not sufficiently effective or do not exist, their yields are seriously reduced. The presence of diseases different from those in temperate climates might further reduce yield and quality of the crop.

BREEDING PRIORITIES AND PROGRAMME ORGANIZATION

CIP's breeding programme is largely aimed at combining pest and disease resistances, together with tolerances of climatic stresses in order to provide developing countries with high performance genetic materials which have stability across environments. There are three major components to the programme:

1. Breeding for hot environments. The priority research projects in this area are: a) breeding for earliness and heat tolerance and b) breeding for combined resistance to bacterial wilt and root knot nematode.
2. Breeding for cool environments. The priority research projects in this area are: a) breeding for tolerance to frost, and b) breeding for resistance to cyst nematodes.
3. General breeding projects: a) Adaptation and utilization of potato populations. This project produces early maturing parental clones with a wide range of adaptation to be used in other projects, including TPS breeding. Breeding for resistance to *Alternaria* and *Erwinia* is also included. b) Breeding for resistance to late blight. c) Breeding for resistance to PLRV and PVY. d) Ploidy level manipulations using haploids and 2*n* gametes. This project concentrates on combining disease and pest resistance at the diploid level and transfer of this genetic variability to the tetraploid level via 4*x*-2*x* matings. e) Breeding for insect resistance, principally tuber moth.

In spite of the fact that breeding for specific resistances is carried out in the form of individual projects, which may create the impression of a "column breeding system", all projects are closely linked to the main objectives, i.e. breeding for hot environments and breeding for cool environments. The breeding programme is conducted by an interdisciplinary group of scientists who cooperate in the various phases of the work.

POPULATION BREEDING PHILOSOPHY AND STRATEGY

The gene pool of the genus *Solanum* contains both primitive cultivated and wild species. It is well known that this pool contains valuable genes controlling adaptation to the environment, yield *per se* and resistance to or tolerance of climatic stresses, pests and diseases. Throughout evolution, introgression among species has occurred and there is evidence that such introgression is still taking place. However, in certain cases geographical isolation has prevented gene flow, and valuable genes are still confined within species which inhabit certain geographical areas. Incompatibility and other crossing barriers, especially among species belonging to different taxonomic series, has certainly limited introgression. In spite of the presence of unreduced gametes in diploid species, ploidy differences may also have limited the gene flow between diploid and tetraploid species to a certain extent. For these reasons it is highly unlikely that naturally occurring desirable combinations will be found. This is particularly true for genes conferring disease and stress resistance which would be present at an extremely low frequency. Therefore, valuable genetic attributes will still be confined within species or closely related biosystematic groups. For this reason, it might be stated that the frequency of desirable genes is extremely low and that these genes are held in small segments of the *Solanum* gene pool.

In order to bring together the variability contained in the genus *Solanum* in both its wild and primitive cultivated species, to increase gene frequencies and to stimulate recombination of desirable alleles while maintaining a wide genetic diversity in the population, two preconditions were necessary: 1) An adequate breeding scheme, 2) seedling screening methods to test for resistance to pests, diseases and stresses.

1). Adequate breeding scheme: CIP's breeding objective is not to release varieties but to conduct a programme of germplasm improvement and to distribute to developing countries advanced genetic materials with resistance to pests and diseases. This goal had to be achieved as fast as possible. Several breeding strategies were considered in order to be able to manage the very large potato gene pool. Traditional potato breeding strategies such as the back cross method or "nobilization", the "pedigree" method, and the "column breeding" method (see also Hermsen, this volume) were discarded for three reasons: a) because of their slowness, b) because of their orientation towards variety release and hence their utilization of the nonadditive type of gene action, c) because these methods are not easily adapted to the breeding of large populations. A more dynamic strategy utilizing both the additive and nonadditive types of gene action (for yield and other quantitatively inherited characteristics) became necessary. At the beginning it was decided to carry out a population breeding scheme based on phenotypic recurrent selection. However, as basic genetic information was gained about the nature of the genetic variability present in the breeding populations, the breeding scheme was modified, and progeny testing was included. With this readjustment a significant response to selection for several traits was achieved. Progeny testing permits the selection of parental clones with

high GCA for various characters. The selection of such parents is very important for CIP, since it allows several objectives to be achieved: a) a high percentage of the progenies sent to the regions are eligible for variety selection, b) as a consequence of a) the sample size sent to the regions can be decreased without diminishing the probability of selecting varieties, c) the performance of crosses between parents with high GCA can be predicted and d) in the TPS breeding programme, it is fundamental to have parental clones with high GCA for yield and tuber uniformity in order to obtain high yielding progenies with tubers uniform in size, colour and shape.
2) Seedling screening techniques: The availability of these techniques is fundamental for an efficient population breeding scheme aimed at the accumulation of resistance genes. The rationale behind this scheme is to eliminate susceptible genotypes from the population at an early stage, and then to select for adaptation, yield and other agronomic characteristics from this resistant material. In the very early stages of the population development, screening was carried out for resistance to a single disease or stress factor. As the selection cycles progressed double or triple screenings were carried out. Seedling screens are carried out for resistance to the following diseases or stress factors: late blight, early blight, bacterial wilt, root knot nematode, PVY, PLRV, potato virus x (PVX), frost, and aluminium toxicity.

BREEDING METHODOLOGY

The amount of genetic variability in the breeding populations for traits such as yield, earliness, disease resistance and stress tolerance, is assessed in each selection cycle using the mating designs North Carolina I and II. Besides providing basic genetic information, these designs ensure that good clones are selected at harvest and then re-evaluated. Those maintaining good characteristics are included in the group of potential progenitors or selected clones. These genotypes are then progeny tested to gain information about their parental value. The methods most commonly utilized to evaluate parental value are: a) pollinations with bulked pollen from clones of different origins, (this method provides information on the GCA of the female clones), b) line x tester analysis and c) complete and partial diallel designs. These methods provide information on both GCA and specific combining ability for the selected parents.

BREEDING RESULTS

In the 10 years during which the population breeding system has been operating, consistent progress has been made in increasing the frequency of genes controlling several useful attributes.

In the populations bred for hot environments, the first need was to select for heat tolerance and earliness. The frequency of selected clones in the base population was only 0.06%. After seven cycles of selection the frequency of early-maturing heat-tolerant clones was increased to 20-30% depending on individual progenies. As a result of this work, the early maturing and heat tolerant clones DTO-2, DTO-28, DTO-33, N561.5, LT-1, LT-2, LT-4 (also immune to PVY), LT-5, LT-6 and LT-7, have been distributed for regional testing. The clone LT-8, also immune to PVY and PVX, is at present being meristem cultured and will soon be distributed. A group of parental clones

In the breeding project for bacterial wilt resistance involving a wide-based diploid population generated by intercrossing the wild *Solanum* species *S. sparsipilum*, *S. chacoense* and *S. microdontum* and the cultivated species *S. phureja*, the frequency of resistant clones was 30%. After two cycles of recurrent selection with progeny testing the frequency was raised to 60-65%.

At the tetraploid level, significant progress has also been made. The tetraploid population which had *S. phureja* as the only source of resistance had 15-20% resistant clones. The tetraploid population derived from the widely based diploid populations mentioned above had 60-75% resistant clones. A significant number of these clones were also heat toler-
from the widely based diploid populations mentioned above had 60-75% resistant clones. A significant number of these clones are also heat tolerant since the tetraploid materials used in the 4*x*-2*x* transfers were progenitors selected for high GCA for earliness and heat tolerance.

In the breeding project for resistance to *Meloidogyne* spp., the base population derived from cultivated and wild diploid species had 3.5% resistant clones. After four cycles of recurrent selection the frequency of resistance has been increased to 35.5%. This resistance has already been transferred to the heat-tolerant early-maturing population using 4*x*-2*x* matings.

In the programme for PVY immunity, a heat-tolerant early-maturing population, comprising 126 progenies (each with 150 seedlings) was evaluated in 1985 and gave the following segregation ratios of PVY immunity to susceptibility: 48 progenies, 1:1; 27 progenies, 3:1; 26 progenies, 5:1;

and 25 progenies 11:1. This demonstrates the high frequency of the gene controlling PVY immunity.

Progress has also been made in breeding for resistance to early blight. Although this work has only been started recently, several diallel and line x tester experiments have permitted the determination of a group of progenitors with a high GCA for resistance to this disease. The best progenitors found are LT-7, 377892.7, 378676.6, 7XY.1, BL2.9, and Murca. Most of them are also early maturing and heat tolerant. In spite of the fact that the resistance is positively correlated with lateness, a few early maturing progenies with resistant individuals have also been found. These progenies are: Maine 47 x 378015.16, CFR-69.1 x LT-7, Beauvais x LT-7, Maine 53 x 377892.7, BR63.65 x 378676.6, 575049 x LT-7 and some others. A narrow sense heritability estimate of h2 = 0.7 has been obtained.

In the breeding project for field resistance to late blight, initial sources of field resistance present in S. tuberosum ssp. andigena, S. phureja and Mexican germplasm derived from S. tuberosum were improved through four cycles of recurrent selection. The frequency of clones with resistance, good agronomic type, earliness and adaptation to longer daylengths was considerably improved. A low frequency of resistance in the initial stages was increased to an average of 20% based on seedling screening and field selection after four cycles of recurrent selection. The maturity period was reduced from 180 days to 120 on average, and the frequency of adaptibility to long days (40°S) has been increased to an average of 30% of the population. Field testing the screened population for resistance to late blight at the seedling stage indicated 70-80% resistant clones after 4 years of testing in Colombia, and 80% of the Colombia selections were resistant at Toluca, Mexico. Selection for agronomic characters has also been carried out in each cycle, so that by the fourth cycle a higher frequency of desirable genotypes was available for selection.

The breeding population for tolerance of frost has undergone five cycles of recurrent selection since the beginning of the project in 1973. Sources of frost tolerance from wild as well as from cultivated species were selected for improvement in frost tolerance, agronomic characters and yield, as well as adaptation to longer daylengths. Since 1980, a sub-population has been separated and specifically selected for earliness and adaptation to longer daylengths. Results have indicated an average frequency of 30% of clones with tolerance of frost in the base population (seedling screening at -4°C in a growth chamber) and of approximately 12% in the sub-population which is being improved for earliness and adaptation to long days. A sample of the latter tested under natural long days (40°S) indicated a frequency of approximately 18% of genotypes that were able to tuberize under long days. Selected clones outyielded the locally adapted cultivars in a 120 days growing period. Seven outstanding frost tolerant clones have been selected from the overall population as potential varieties by National Programmes.

Since 1978, considerable progress has been made in research on the utilization of TPS in commercial potato production. A great deal of information has been obtained on types of genetic variability, heritabilities and phenotypic and genotypic correlations between characters relevant for TPS utilization. Comparisons of the performance of various types of TPS populations demonstrated that hybrid progenies were superior to either open-pollinated or self-pollinated progenies, indicating that even the rather small amount of inbreeding present in the two latter progenies was depressing yield. However, in these experiments 160 hybrid progenies were included in contrast to only 20 open-pollinated and 20 selfed progenies. The latter two samples may not have been large enough and therefore a larger experiment has recently been done. A random sample of 50 clones was taken from an Andigena population, and from each clone four types of progenies were obtained: a) selfed (S), b) open-pollinated (OP), c) hybrid by crossing to a 10-clone pollen bulk (B_{10}), and d) hybrid by crossing to a 20 clone pollen bulk (B_{20}). In total, 200 progenies were obtained, 50 of each type. These populations were evaluated in replicated field trials for two seasons at Huancayo. Preliminary analysis of data indicated that in about 50% of the clones there were no differences in yield between their OP, B_{10}, and B_{20} progenies. In the rest of the clones the inbreeding depression ranged from slight to very severe. These results suggest that carefully selected OP progenies may give satisfactory yields at a much lower cost than would be the case with hybrid TPS.

In the use of TPS for commercial potato production, the most important achievement has been the identification of a number of clones with high GCA for yield and tuber uniformity which are also mostly early maturing. Among the most important are: LT-7, 378015.13, 378015.16, R-128.6, Atlantic, Atzimba, Maine 28, Santo Amor, 7xy.1 and C83.1199. Progenies such as Maine 28 x 378015.13, C83.119 x Santo Amor, C83.511 x LT-7 have recently been found to give outstanding yield and uniformity. These are in addition to other progenies previously identified by the TPS breeding project, such as Atzimba x LT-7, Atzimba x DTO-28, Atzimba x 7xy.1 and Atzimba x R-128.6. At present, late blight resistance and PVY immunity are being introduced into the TPS progenies. The progeny Atzimba x 7XY.1 already combines resistance to late blight, PVY and early blight and the progeny India 1035 x LT-7 has similar attributes.

As a result of the population breeding work carried out at CIP, a number of genetic materials sent to developing countries for testing and utilization have reached various stages of development ranging from advanced trials to the releasing of varieties. Table 1 summarizes the present status of these materials.

Table 1. Main attributes of CIP clones at various stages of selection in developing countries

Region	Selection Stage			
	Advanced Selection Trials	Variety Trials	Released	Named
I	4*(110)**Frost, Cyst, LB, BW, PVY, PVX	2(32)Frost,Cyst, LB, BW	1(1)PLRV	3(6)BW,LB, PLRV
II	1(50) Early, LB,BW, EB, PVY, PVX	1(5) Early, HT	-	1(1) LB
III	5(87) Early,LB,BW	7(49) LB, BW	4(13)BW, PVY,HT, Early,LB	3(6)BW,LB
IV	1(50) Early,PLRV, PVY,PVX,EB	1(2) Early,LB	-	-
V	-	-	-	2(2)BW,LB, HT
VI	1(11) LB,BW,HT, Early	4(43)BW,LB,PLRV, Wart,HT	1(4)LB, Wart, PLRV,PVY	2(1)BW,LB
VII	2(4) LB,BW	2(4) LB	-	2(4)BW,LB

* No. of countries
** No. of clones

LB = late blight resistance
BW = bacterial wilt resistant
Cyst = cyst nematode resistant
Wart = wart resistant

Early = maturing early
EB = early blight resist.
HT = heat tolerant
Frost = frost tolerant
PVX) PVY) PLRV) = resistant to these viruses

CONCLUSION

There is no doubt that CIP's innovative approach to potato breeding has led to the development of superior potato germplasm in less than 15 years. In traditional programmes, this is the time needed to develop one or two varieties. Two remarkable developments within this brief period merit special mention; the development of potato germplasm for the hot humid tropics and the utilization of TPS. None of this could have been achieved with traditional methods. In the next 5 years we shall see the release of scores of new varieties based on CIP's breeding populations.

TRUE POTATO SEED

BREEDING STRATEGIES FOR TRUE POTATO SEED

Michael T. Jackson

INTRODUCTION

The utilization of true potato seed (TPS) for potato production has developed rapidly in many parts of the world since 1978, when TPS was adopted as a principal research programme at the International Potato Center (CIP) in Peru. By 1984, the range of TPS activities had become extensive (International Potato Center 1984). In 34 countries, research on TPS was being conducted at the experiment station level, while 10 countries were involved in on-farm research. TPS was being used at that time by farmers in Sri Lanka, the People's Republic of China, Rwanda, Samoa and in the Philippines.

Much of the early work on TPS agronomy was carried out using open pollinated seed collected from cultivars which produced large quantities of true seed. One of the problems with this true seed was the heterogeneity of some of the progenies, because of the high heterozygosity of the mother plants, which could not be compared with clones in terms of uniformity. Initially it was felt that complete uniformity might not be necessary for potatoes from TPS. Researchers at CIP had been encouraged in this belief by the fact that they had observed farmers in Costa Rica (Central America) mixing white-skinned and red-skinned varieties at harvest. It was thought that this might also be typical of the situation in other countries. Subsequent studies have indicated however that uniformity is important, not only with regard to phenotype in the field, but also with regard to maturity, pest and disease resistance and cooking quality. Uniformity must be given due consideration in any breeding programme if potatoes from TPS are to achieve their potential in many countries.

This review of breeding for TPS covers some of the genetical studies which have formed the basis of breeding research, the breeding schemes which are currently having most impact, namely the use of hybrid progenies from $4x$ x $2x$ crosses and open pollinated progenies, as well as longer term research on inbreeding and apomixis. Aspects of seed production are discussed in relation to breeding priorities as well as some points concerning the types of potatoes to be raised from true seed.

THE NATURE OF GENETIC VARIABILITY FOR TPS TRAITS

When research commenced in the late 1970s, little was known about the genetic basis and inheritance of traits such as berry number, per cent seed germination and transplant survival, which were considered to be important for potatoes grown from true seed. Furthermore characters such as seedling vigour and uniformity had never been previously considered in the context of a field crop of potatoes, as all breeding efforts prior to this time had concentrated on the production of clonally propagated varieties. The determination of the heritabilities of these traits in different progenies has provided a basis for much of the breeding research.

The estimation of genetical parameters for TPS traits has been undertaken in CIP and at Cornell University in the United States. The results have differed depending upon the nature of the genotypes used in each experiment. For example, Mendoza (1980) and Thompson *et al.* (1983), using a North Carolina Design I (Comstock & Robinson 1952) to analyse a Neotuberosum population, found no additive genetic variance for yield *per se*, although estimates for additive variance for the components of yield, such as tuber number and tuber size, were high. The high negative correlation between these characters indicated that selection for an increase in one should decrease the other. Tuber size was positively correlated with days to maturity, size uniformity and eye depth, whereas tuber number was negatively correlated with these characters. Tuber size gave a higher estimate of narrow sense heritability (h^2) than tuber number, but a lower estimate of the additive genetic by environment interaction variance. Hence response to selection for tuber size should be more rapid than for tuber number. Heritability estimates for seedling vigour and uniformity in the nursery and seedling vigour at 60 days were high enough to indicate that these characters could be improved by selection. Transplant survival, on the other hand, gave a low estimate of heritability. Nonadditive variance was the most important component in the control of uniformity of colour.

In terms of seed production, the positive genetic correlations between number of berries and transplant survival, tuber size, uniformity of tuber size, and depth of eyes in this material are advantageous because selection for improvement in these characters should result in increased seed production. The relatively high estimates of heritabilities and low estimates of genotype by environment interactions for tuber size and other tuber yield and quality related traits indicate that individual plant selection should be considered instead of family selection. Thompson et al. (1983) also suggested that the significant estimate of nonadditive variance for yield favours F_1 hybrids as the optimum type of progeny for TPS production.

In another set of experiments using a North Carolina Design II (Comstock & Robinson 1952) with a CIP breeding population comprising Solanum tuberosum ssp. tuberosum and ssp. andigena, tuberosum x andigena hybrids, as well as tuberosum x phureja hybrids. Thompson & Mendoza (1984) determined the genetic parameters for 11 traits. Narrow sense heritability estimates were above 0.52 for berry number, tuber number, tuber size, yield, tuber smoothness and uniformity of tuber colour. No additive genetic variance was found for uniformity of tuber size. Significant estimates of non-additive variance were also obtained for berry number, yield, tuber smoothness, uniformity of tuber colour and uniformity of tuber size. What is interesting about these results is the high estimate of h^2 for yield, in contrast to the estimates determined for the Neotuberosum material in earlier experiments (Thompson et al. 1983). With respect to berry number, tuber number and tuber size, there was coincidence in heritability estimates between the different experiments. The high genetic correlations between yield-related traits form the basis for the adoption of a selection index method of improvement. Based on the material studied and the use of a selection index, they suggested that improvement in yield, tuber smoothness, and uniformity of colour and shape should be rapid at the locations used in their study.

Mendoza (1985) has described the performance of different types of progeny developed from the material used by Thompson & Mendoza (1984). The parental clones used in the N.C. Design II were selfed and bulk pollinated; open pollinated seed was also collected. Multilines were created by mixing equal numbers of seeds of four of the crosses in each set. With respect to per cent germination, transplant survival, number of plants harvested and yields, the hybrid progenies were better than the selfed or open pollinated progenies, which were considered to be partially inbred. With regard to tuber weight, there were no differences amongst the different progenies. Furthermore, the selfed and open pollinated progenies were equal to or better than the hybrids in tuber uniformity and depth of eyes. There were no significant differences in tuber weight, but the hybrid progenies yielded more because of the greater number of tubers per plant.

BREEDING FOR POTATOES FROM TRUE POTATO SEED

The production of acceptable TPS progenies is a compromise between the ease of their production, either as open pollinated progenies or as hybrids, and the cost of their production. Several breeding strategies have been outlined to achieve uniformity in TPS progenies. They include (i) the use of open pollinated seed from selected tetraploid clones; (ii) 4*x* x 2*x* hybridization with unreduced gametes from First Division Restitution (FDR) in the diploids; (iii) inbreeding; and (iv) exploitation of various aspects of apomixis in potatoes.

Hybrids vs. open pollinated TPS

The most detailed information from breeding studies for TPS to date have come from Peloquin and his associates in Wisconsin. Two of the breeding schemes outlined by Peloquin (1983) are based on unilateral and bilateral sexual polyploidization, which have been shown to be efficient methods of exploiting diploid species for widening the genetic base of tetraploid varieties, but which have also been applied successfully for producing superior TPS progenies. Peloquin (1983) has demonstrated that the genetic diversity for both valuable qualitative and quantitative traits and the allelic variation necessary for maximum heterozygosity, can be transferred almost intact to such progenies. Unreduced gametes produced by FDR do transmit a considerable proportion (80%) of the heterozygosity and epistasis of the parents to the offspring. The superiority of the hybrid progenies is interpreted however in terms of heterozygote advantage. Most comparisons have been made between 4*x* x 2*x* hybrid and open pollinated progenies. In almost all cases, hybrid progenies have outyielded cultivars and open pollinated TPS from tetraploid clones (see chapters by Hermundstad & Peloquin and others, this volume). Macaso-Khwaja & Peloquin (1983) stated that the low yields of 4*x* open pollinated progenies were probably an effect of inbreeding depression. The increase in homozygosity and decrease in intra and interallelic interactions, due to the reduction in number of alleles per locus as a result of selfing, were the basis of low yields. In fact, Hermsen (1983) has discounted the use of homozygous lines in breeding true seed potatoes, suggesting that they cannot have the vigour of hybrids which is commonly attributed to the high frequency of tetra-allelic loci and favourable epistatic effects.

Kidane-Mariam et al. (1985a) compared 30 TPS families from 4x x 2x crosses and open pollination, for tuber yield and plant uniformity. The hybrid families were from crosses between 4x clones and 2x Phureja-haploid Tuberosum hybrids producing pollen by FDR. The open pollinated (OP) families were from three categories of 4x parents: (a) DTs - 4x derived from 4x x 2x crosses; (b) advanced clones which were highly male fertile; and (c) advanced clones known to possess variable male fertility in which pollen stainability ranged from <3% to 10-30% during the flowering period.

The average tuber yield of the hybrids was 28% higher than that of the highest yielding OP group and about 48% higher than the combined mean yield of the three groups of OP families. The heterotic response of the hybrid 4x x 2x progenies was interpreted on the basis of the mode of 2n pollen formation (FDR), and the level of heterozygosity transmitted to the progeny. On the basis of these results, Kidane-Mariam et al. (1985a) state that the 4x x 2x approach is the best breeding scheme for the development of TPS progenies. Nevertheless the yield and other characteristics of some of the OP families was sufficiently high to warrant further study, and because of the substantially lower cost of seed production (about 10-20% of the cost of hybrid seed) their use for potato production would be economically advantageous in many situations, even though yields are lower on average than those of hybrids.

While acknowledging the breeding value of the 4x x 2x progenies, Hermsen (1983) has expressed some concern about the narrow genetic base of the diploids used by Peloquin and others, because these clones trace back to a few ssp. *tuberosum* dihaploids and a few clones from *S. phureja*. He advised that these points should be borne in mind when considering parental materials because these should be complementary for horticultural traits and resistances to pests and diseases, as well as abiotic stress conditions. A further point which should also be considered is the glycoalkaloid content of some of the hybrid progenies from 4x x 2x crosses, since some wild species are included in these breeding schemes.

The use of locally-adapted tetraploid cultivars for the generation of TPS progenies has been examined by Kidane-Mariam et al. (1985b). The performances of 262 single and bulk cross hybrids and four families from open pollination were evaluated at three locations in Peru. Significant differences were observed among families in tuber yield, uniformity and transplant survival in the field. Hybrid TPS families gave higher yields and more uniform tubers than families from open pollination. This difference was attributed to inbreeding depression due to selfing in the open pollinated materials.

Hybrid TPS progenies from intermating selected tetraploid stocks can produce good yields and have satisfactory uniformity. However, the significant family-environment interaction for tuber yield in this study indicated that if TPS families are generated from 4x x 4x crosses, then it is important to determine whether the parental stocks have a good level of adaptation to the specific locality where the TPS materials are to be grown.

Open pollinated seed and inbreeding

Farmers could produce TPS themselves by collecting seed resulting from natural pollinations in their own fields, and in fact some may have no reasonable alternative but to save open pollinated seed. However, the reluctance of some potato breeders to accept open pollinated progenies probably rests on the assumption that such materials will become substantially inbred through selfing, and that this should be avoided at all costs.

The quantity and quality of seed from open pollination is related to parental genotype, but also to pollinator activity. Atlin (1985) has discussed the effects of collecting open pollinated seed over several generations and argued that it would lead to inbreeding. There is some evidence to suggest however, that potato populations may in general be more outbred than expected from some pollination studies, where high selfing rates were observed (Glendinning 1976; White 1983). Kidane-Mariam et al. (1985 a) compared first, second and third generation open pollinated progenies from several potato clones, and found no significant difference for yield. The explanation they gave for this was that the individual plants serving as seed parents for the next generation were either the most heterozygous of the selfed progeny or were the product of outcrossing. The indications from this study and one carried out in San Ramon, Peru (Atlin 1985) were that inbred individuals were much less viable and fertile than individuals resulting from cross pollination. It also appeared that inbreeding depression was more severe under conditions of stress, causing even low levels of inbreeding to result in large yield declines. Furthermore, S_1 and S_2 progenies produced less pollen per flower and pollen of a poorer quality than did parental hybrids, suggesting that inbred parents contribute little to the pool of male gametes in a mixed population, and that the level of selfing was very much lower than the average estimate of 80% obtained by the use of genetic markers.

In these experiments using open pollinated seed, the mixtures of clones have been considered more or less as synthetic varieties composed of a number of parents, because of the observed absence of a decline in yield between first and second generations of natural pollination (Atlin 1985). Some inbreeding is still to be expected in such populations, but could be reduced by using methods of male sterility to develop synthetic varieties with low selfing rates. Cytoplasmic sterility as proposed by Brown (1984) or that recently identified in the progeny of Atzimba x IVP-35, in which abundant stainable pollen is produced but which does not function in fertilization would be one way of achieving this.

Another option which has received little attention is to develop clones which are tolerant of inbreeding and from which open pollinated seed could be collected for the production of homozygous, uniform potatoes, without any decline in yield over generations. In a recent paper, Jackson et al. (1985) have proposed an inbreeding strategy for the production of TPS, through single seed descent in diploid potatoes. Inbreeding may be exploited either through the production of F_1 hybrids between inbred lines (Yashina & Pershutina 1971), a strategy in which heterozygosity is favoured at the expense of ease of seed production, or by the use of autogamous inbred lines in which gametic uniformity is due to homozygosity. These methods are compromises between maximizing heterozygosity, maintaining the necessary level of gametic uniformity and the relative ease of seed production. The choice between these approaches is dependent on whether heterosis is related to heterozygosity.

Jinks & Lawrence (1983) have questioned the widespread belief that the best phenotypes, particularly for yield, are produced by heterozygotes rather than by homozygotes. In outbreeding crops, breeders attempt to avoid inbreeding depression, which is expressed as a reduction in vigour, fertility and yield. The usual explanation of inbreeding depression is that it is due to the fixation of unfavourable or deleterious recessives. Inbreeding increases the frequency of loci which are homozygous and some will become homozygous for these deleterious recessives. Jinks & Lawrence (1983) have indicated that inbreeding depression is due to the presence of genes in the base population that display dominance, or dominance and epistatic effects, and that control characters of primary interest to the breeder such as yield.

Heterosis is the converse of inbreeding depression. Its genetical base is still the subject of some debate. The two main genetic models of heterosis are the overdominance model (Hull 1945), favoured by Mendoza & Haynes (1974) to explain the genetic control of yield in autotetraploid potatoes, and the dominance model (Williams 1959; Sinha & Khanna 1975). The overdominance model, which proposes that heterozygosity is intrinsically advantageous, has been expanded to a multilocus model by Li (1967), and to include multiallelic effects by Mendoza & Haynes (1974). Equally, in support of the dominance model, it has been shown that heterosis may result from additive x additive and additive x dominance effects at a few loci (Seyffert & Forkman 1976), or from linkage and linkage disequilibrium (Sved 1972; Arunchalam 1977). In recent theoretical work on the genetical basis of heterosis, Jinks (1981; 1983) has argued that heterosis is not dependent on heterozygosity _per se_, but on the genic content of the individual, and therefore that heterosis may be 'fixed' in homozygous recombinants produced through inbreeding. It is suggested that heterosis results from linkage disequilibrium of genes with dominance and epistatic effects. Although Jinks & Lawrence (1983) do not dispute that overdominance may occur, they argue that there are few substantiated cases of major genes showing overdominance.

They further point out that the effect of selection in cultivated species will be to raise the frequency of genes for favourable expression of the selected character(s), that is, those genes which display dominance in the desired direction, and therefore, when such material is inbred, the resultant inbreeding depression is proportional to the response that has previously been obtained by selection. Jinks & Lawrence (1983) further argue that although in such populations heterozygotes, on average, may be superior in fitness terms to homozygotes because of dominance, the objective of plant breeding is to identify individuals whose performance is well above average, and that these individuals, in the absence of overdominance, are as likely to be homozygotes as heterozygotes. That is, the apparent correlation between yield and heterozygosity is spurious.

With regard to potatoes, different lines seem to behave differently to inbreeding (Krantz & Hutchins 1929; Krantz 1946; Pushkarnath 1960). Trinkler _et al_. (1976) have stated that they were able to select inbred lines which showed little or no inbreeding depression, and Trinkler _et al_. (1980) found no inbreeding depression when comparing the performance of secondary inbred lines. As Atlin (1985) has stated, most inbreeding studies have not continued past the S_1 generation, so full evaluations of its application to potatoes have not been made. The research project of Jackson _et al_. (1985) should contribute valuable data in this respect.

Exploiting apomixis in potatoes

Hermsen (1983) put forward the idea that autonomous apomixis could be introduced into potatoes, as another way of producing genetically uniform potatoes from true seed. Gametophytic apomixis is characterized by apospory and diplospory. The existence of these mechanisms in potatoes, and their choice for breeding true seed potatoes, has been the subject of some controversy.

Peloquin (1983) has considered the use of apospory in two breeding schemes. In one of these, crosses are made between diploid hybrids. In one 2*x* hybrid, there must be a high frequency of 2*n* eggs which are highly heterozygous. Such gametes can be formed either through apospory, or meiotically with no crossing-over followed by FDR. In the other, highly heterozygous male gametes must be formed at meiosis in which there is no crossing-over followed by FDR. Consequently all the male and female gametes would have the same gentoypes as their respective sporophytes. Meiotic mutants for 2*n* pollen which fulfil these criteria have already been identified. The sy_3 mutant (which produces only univalents at the first division of meiosis) and *ps* (parallel spindles) at the second division have been combined, and consequently 100% of the parental genotype is transmitted to the male gametes. The production of 2*n* eggs through apospory or FDR with no crossing-over, and their identification, are the principal constraints to the development of this breeding scheme.

In the other scheme, the formation of seeds asexually is envisaged on desirable 4*x* hybrid clones. In order to achieve this, 4*x* seed may be formed either by apospory or by no crossing-over followed by FDR; fertilization can be circumvented by parthenogenesis or pseudogamy. Irikura (in Peloquin 1983) reported that apomictic seeds were induced in diploid and tetraploid potato cultivars with 2,4-D sprays on emasculated flowers. The chemical induction of aposporous apomictic seed, i.e. clonal true seed, has also been discussed by Iwanaga (1983).

Jongedijk (this volume) has discussed the experimental induction of diplosporic apomixis, through desynapsis and FDR 2*n* egg formation. He states that the induction of diplosporic apomixis in potato appears to offer the best prospects because the genetically controlled elements of the system are already available. These include strongly reduced crossing-over in megasporogenesis, the formation of unreduced megaspores and embryo sacs, and the parthenogenetic development of the unreduced egg cell.

Production of true seed

True seed production on a commercial scale is dependent upon the flowering characteristics of the parental material, the quality of TPS produced and the economics of TPS production (Upadhya *et al.* 1985). Genetical studies have shown a correlation between berry number and the components of yield (Thompson & Mendoza 1984), and Dayal *et al.* (1984) have shown that there is a positive and significant correlation between 1000-seed weight, tuber yield per plant and tuber number per plant.

Almekinders & Wiersema (1985) have shown that the position of the inflorescence on potato plants is related to seed weight. More stems per plant had a favourable effect on TPS production through reduced flowering period, fewer axillary branches and a larger proportion of berries originating from first and second position inflorescences. By decreasing the number of flowers per inflorescence, berry size and 100-seed weight could be increased. It is also clear from the work of Pallais *et al.* (1985) that potatoes grown for the production of true seed have specific needs, such as nutritional requirements, that differ from those grown for commercial production.

The production of hybrid seed is a feature of TPS research which needs particularly careful consideration. Upadhya *et al.* (1985) and Pallais *et al.* (1985) have given figures for the hand emasculation and hand pollination of potato flowers in the field. In India, for example, it has been calculated that the cost of production of 100 g of hybrid true seed (sufficient to plant 1 ha) would be about US$6.00, if emasculation was carried out, but less than US$2.00 if flowers are pollinated without emasculation. The quality of seed produced is also dependent upon the pollen load. Pollination of the stigma three times within the receptivity period of 18-36 hours produced the largest number and size of berries, as well as the highest quantity and quality of true seed (Upadhya *et al.* 1985).

Although the use of hand labour for pollination appears feasible in certain situations, it is clearly not in others. What alternatives are there? Under field conditions, potatoes are pollinated by bees. Since potatoes do not produce nectar or other food 'rewards', they depend mainly on their pollen content to attract pollinators. White (1983) has reported that bee pollination is important for the production of open pollinated seed. In a series of field experiments at Huancayo (3200 m) in central Peru, bees were observed to forage within single clones for short periods, but also to switch occasionally to different clones. Consequently much of the seed from open pollinated clones resulted primarily from self or intraclonal pollinations. For the production of hybrid seed or synthetic varieties using natural pollination, White (1983) suggested that studies should be undertaken to determine the effects of border rows and planting patterns on rates of cross pollination.

Given that a considerable amount of self pollination occurs naturally in the field, the use of bees to produce hybrid seeds will only be effective if one of the parents is male sterile. Bees rarely visit sterile flowers because of the lack of pollen. Consequently incorporation through breeding of the tetrad sterility identified by Brown (1984) offers promise in this respect. Bees visit flowers of the tetrad sterile clone and can pollinate these flowers with fertile pollen from an adjacent male fertile clone. The flowers of the tetrad sterile clone look normal and shed copious amounts of tetrad pollen when vibrated. The production of hybrid seed, or even the efficient production of open pollinated seed is also dependent upon a greater understanding of the ecology of the pollinators, *Bombus* spp., and why they sometimes fail to pollinate.

A TPS ideotype

Agronomic comparisons between potatoes grown from TPS and from seed tubers have little justification. While the use of seedling tubers, as indicated by Wiersema (1983; 1985) gives farmers many of the advantages of TPS while still handling seed tubers, the use of TPS for direct production of a ware crop presents several problems. A TPS seedling is a one-stemmed plant and does not have the food reserves in tubers upon which rapid early growth of clonally propagated potatoes is based. TPS progenies often take longer to reach maturity. Although total yields from some of the 4*x* x 2*x* hybrid progenies reported by Macaso-Khwaja & Peloquin (1983) are high, the important feature to consider is marketable yield per unit area.

Satisfactory yields could also be obtained from TPS seedlings which produce only a small number of tubers per plant, but all of which fall within the marketable size. Such a plant would channel all its productivity into this small number of tubers, rather than into many of which only a small proportion would be of marketable size. Variation in tuber size affects both total and marketable yield, and is due to several factors including (i) space per plant; (ii) stems per plant; (iii) size of stems; (iv) date of emergence; and (v) tuber sizes on one stem. When seedlings are transplanted, the first four of these factors would be reduced or eliminated. The only factor of significance would be due to differences between genotypes in producing a range of tuber sizes, and this could be improved by choice of suitable parental material. Total yield would then be manipulated through planting density. What is clear is that the use of TPS presents potato researchers with a unique opportunity to develop new ways of growing the potato; that is, it would be a mistake to think solely in terms of how the crop is currently raised from seed tubers.

CONCLUSIONS

Considering that virtually nothing was known about growing potatoes from TPS before the late 1970s, the fact that true seed progenies are not only now being evaluated under experimental conditions but also being utilized by farmers is an indication of the rapid development of this technology. The close association of breeders, geneticists, agronomists and physiologists in breeding TPS progenies has been the basis of the rapid progress which has been made in such a short period of time.

REFERENCES

Almekinders, C. & Wiersema, S.G. (1985). TPS production. True Potato Seed (TPS) Letter, 6, no. 1, 1-2.

Arunchalam, V. (1977). Heterosis for characters governed by two genes. J. Genet., 63, 15-24.

Atlin, G. (1985). Farmer maintenance of TPS varieties. In Report of a Planning Conference on "Innovative Methods for Propagating Potatoes", held at Lima, Peru, December 10-14, 1984, pp. 39-62.

Brown, C.R. (1984). Tetrad sterility: a cytoplasmic-genic male-sterility attractive to bumblebees. EAPR - Abstracts of Conference Papers, 9th Triennial Conference, held at Interlaken, Switzerland, July 1-6, 1984, pp. 101-2.

Comstock, R.E. & Robinson, H.F. (1952). Estimation of average dominance of genes. In: Heterosis, ed. J.W. Gower, pp. 494-516. Ames: Iowa State College Press.

Dayal, T.R., Upadhya, M.D. & Chaturvedi, B.N. (1984). Correlation studies on 1000-true seed weight, tuber yield and other morphological traits in potato (Solanum tuberosum L.). Potato Res., 27, 185-8.

Glendinning, D.R. (1976). Neotuberosum: new potato breeding material. 4. The breeding system of Neo-tuberosum and the structure and composition of the Neo-tuberosum gene pool. Potato Res., 19, 27-36.

Hermsen, J.G.Th. (1980). Breeding for apomixis in potato: pursuing a utopian scheme? Euphytica, 29, 595-607.

Hermsen, J.G.Th. (1983). New approaches to breeding for the potato in the year 2000. In Research for the Potato in the Year 2000, ed. W.J. Hooker, pp. 29-32. Lima: International Potato Center.

Hull, E.H. (1945). Recurrent selection and specific combining ability in corn. J. Am. Soc. Agron., 37, 134-45.

International Potato Center. (1984). Potatoes for the Developing World. Lima, Peru.

Iwanaga, M. (1983). Chemical induction of aposporous apomictic seed production. In Research for the Potato in the Year 2000, ed. W.J. Hooker, pp. 104-5, Lima: International Potato Center.

Jackson, M.T., Taylor, L. & Thomson, A.J. (1985). Inbreeding and true potato seed production. In Report of a Planning Conference on "Innovative Methods for Propagating Potatoes", held at Lima, Peru, December 10-14, 1984, pp. 169-79.

Jinks, J.L. (1981). The genetic framework of plant breeding. Phil. Trans. Roy. Soc. Lond. B., 292, 407-19.

Jinks, J.L. (1983). Biometrical genetics of heterosis. In Heterosis. Monographs on Theoretical and Applied Genetics. Vol. 6, ed. R. Frankel. Berlin: Springer-Verlag.

Jinks, J.L. & Lawrence, M.J. (1983). The genetical basis of inbreeding depression and heterosis: its implications for plant and animal breeding. PORIM Research Bulletin, Kuala Lumpur, Malaysia.

Kidane-Mariam, H.M., Arndt, G.C., Macaso, A.C. & Peloquin, S.J. (1985a). Hybrids vs. open pollinated TPS families. In Report of a Planning Conference on "Innovative Methods for Propagating Potatoes", held at Lima, Peru, December 10-14, 1984, pp. 25-37.

Kidane-Mariam, H.M., Mendoza, H.A. & Wissar, R.O. (1985b). Performances of true potato seed families derived from intermating tetraploid parental lines. Am. Potato J., 62, 643-52.

Krantz, F.A. (1946). Potato breeding methods. III. A suggested procedure for potato breeding. Tech. Bull. Minn. Agric. Exp. Stn. No. 173.

Krantz, F.A. & Hutchins, A.E. (1929). Potato breeding methods. II. Selection in inbred lines. Tech. Bull. Minn. Agric. Exp. Stn. No. 58.

Li, C.C. (1967). Genetic equilibrium under selection. Biometrics, 23, 397-484.

Macaso-Khwaja, A.C. & Peloquin, S.J. (1983). Tuber yields of families from open pollinated and hybrid true potato seed. Am. Potato J., 60, 645-51.

Mendoza, H.A. (1980). Preliminary results on yield and uniformity of potatoes grown from true seed. In Report of a Planning Conference on "Production of Potatoes from True Seed", held at Manila, Philippines, September 13-15, 1979, pp. 156-72.

Mendoza, H.A. (1985). Selection of uniform progenies to use TPS in commercial potato production. In Report of a Planning Conference on "Innovative Methods for Propagating Potatoes", held at Lima, Peru, December 10-14, 1984, pp. 5-16.

Mendoza, H.A. & Haynes, F.L. (1974). Genetic basis of heterosis for yield in the autotetraploid potato. Theor. Appl. Genet. 45, 21-5.

Pallais, N., Fong, N. & Berrios, D. (1985). Research on the physiology of potato sexual seed production. In Report of a Planning Conference on "Innovative Methods for Propagating Potatoes", held at Lima, Peru, December 10-14, 1984, pp. 149-68.

Peloquin, S.J. (1983). New approaches to breeding for the potato in the year 2000. In Research for the Potato in the Year 2000, ed. W.J. Hooker, pp. 32-4. Lima: International Potato Center.

Pushkarnath. (1961). Potato breeding and genetics in India. Indian J. Genet. Pl. Breed., 21, 77-86.

Seyffert, W. & Forkman, G. (1976). Simulation of quantitative characters by genes with biochemically definable action. VIII. Observation and discussion of non-linear relationships. In Population Genetics and Ecology, ed. S. Karlin & E. Nevo. New York: Academic Press.

Sinha, S.K. & Khanna, R. (1975). Physiological, biochemical and genetic basis of heterosis. Adv. Agron., 27, 123-74.
Sved, J.A. (1972). Heterosis at the level of the chromosome and at the level of the gene. Theor. Popn. Biol., 3, 491-506.
Thompson, P.G. & Mendoza, H.A. (1984). Genetic variance estimates in a heterogeneous potato population propagated from true seed (TPS). Am. Potato J., 61, 697-702.
Thompson, P.G., Mendoza, H.A. & Plaisted, R.L. (1983). Estimation of genetic parameters for characters related to potato propagation by true seed (TPS) in an andigena population. Am. Potato J., 60, 393-401.
Trinkler, Yu.G., Denisova, I.B. & Mikhalev, E.V. (1980). Enhancement of some characters following crosses in some inbred lines of potato. Trudy Gor'kovskogo Sel'skokhozyaistvennogo Instituta, 147, 73-6.
Trinkler, Yu.G., Kalachev, V.D. & Matenkova, T.V. (1976). The behaviour of inbred lines in breeding potato propagated by seed. Doklady Vsesoyuznoi Ordena Lenina Akademii Sel'skokhozyaistvennykh Nauk Imeni V.I. Lenina, 4, 7-8.
Upadhya, M.D., Thakur, K.C., Asha Juneja & Kadian, M.S. (1985). True potato seed production: flowering, quality and economics. In Report of a Planning Conference on "Innovative Methods for Propagating Potatoes", held at Lima, Peru, December 10-14, 1984, pp. 117-47.
White, J.W. (1983). Pollination of potatoes under natural conditions. CIP Circular, 11. no. 2, 1-2.
Wiersema, S.G. (1983). Potato seed-tuber production from true seed. In Research for the Potato in the Year 2000, ed. W.J. Hooker, pp. 32-4. Lima: International Potato Center.
Wiersema, S.G. (1985). Production and utilisation of seed tubers derived from true potato seed (TPS). In Report of a Planning Conference on "Innovative Methods for Propagating Potatoes", held at Lima, Peru, December 10-14, 1984, pp. 95-116.
Williams, W. (1959). Heterosis and the genetics of complex characters. Nature, 184, 527-30.
Yashina, I.M. & Pershutina, O.A. (1971). The use of inbred forms of the S_1 and S_2 generations to obtain high-yielding potato bybrids. Trudy Nauchno-Issledovateliskogo Instituta Kartofelnogo Khozyaistva, 8, 44-9.

EVALUATION OF YIELD AND OTHER AGRONOMIC CHARACTERISTICS OF TRUE POTATO SEED FAMILIES AND ADVANCED CLONES FROM DIFFERENT BREEDING SCHEMES

Luigi Concilio and S.J. Peloquin

True potato seed (TPS) might be considered for use in particular agronomic and economic situations in developed as well as in developing countries.

To investigate this possibility, 38 hybrid TPS families from two different groups of crosses involving 4*x* and 2*x* *Solanum phureja*-haploid *S. tuberosum* hybrids were evaluated in replicated trials at two locations in the USA and one in Italy. One group of crosses comprised 32 4*x* x 2*x* dihaploid tetraploid (DT) hybrids, the other comprised six 4*x* x DT hybrids; these were compared with open pollinated (OP) families derived from six 4*x* clones. Families from the 4*x* x 2*x* crosses were the highest yielding (Table 1), and also had the best plant vigour and uniformity. The yield performance and the high degree of uniformity could be attributed to the large amount of heterozygosity transmitted from the 2*x* parent to the 4*x* progeny by the 2*n* pollen, formed *via* first division restitution (FDR).

Table 1. Mean yields for three groups of TPS families grown at two locations in the USA and at one in Italy

Family group	Number of families (USA)	Yield (q/ha) Hancock (USA)	Rhinelander (USA)	Number of families (Italy)	Imola (Italy)
DT (4*x* clone x 2*x* hybrid)	32	214 (122-289)	120 (61-198)	18	141 (37-374)
4*x* clone x DT	6	201 (137-250)	120 (63-183)	2	63 (54-69)
4*x* clone OP	6	148 (127-187)	67 (40-94)	8	66 (34-84)

Figures in brackets = range of yield
DT = dihaploid tetraploid
OP = open pollinated.

In another experiment planned to evaluate the agronomic performance of potato clones derived from different breeding schemes, a random sample of 72 advanced clones from the University of Wisconsin breeding programme was grown in Italy. The clones had been selected only for tuber shape and colour, and were evaluated for tuber yield and other characteristics in irrigated and unirrigated conditions. These clones were compared with 19 commercial cultivars; nine from the USA and 10 from Europe. The 72 clones were from two types of crosses; 31 were DTs (4x clone x 2x hybrid) and 41 were DT x 4x clone. The total tuber yield of the 4x x 2x clones was higher than of the other hybrid group and the two groups of commercial cultivars. The most encouraging results were obtained in unirrigated conditions from the 4x x 2x clones; these outyielded the European cultivars both in total yield (+39%) and in marketable yield (+101%) (Table 2).

Table 2. Mean total and marketable tuber yields for the four groups of clones grown under irrigated and unirrigated conditions at Imola (Italy).

	No. of clones	Yield (q/ha) Irrigated		Unirrigated	
		total	marketable	total	marketable
DT (4x clone x 2x hybrid)	31	678.4 212.3-1206.7*	580.6 (85.6) 180.6-1108.1	370.3 157.1-606.2	323.0(87.2) 129.3-560.1
DT x 4x clone	41	521.9 309.4-734.2	444.7 (85.2) 238.5-671.8	278.6 126.5-446.8	244.7(87.8) 121.7-429.4
USA cultivars	9	542.0 340.0-825.0	409.5 (75.6) 273.7-725.0	268.5 118.3-441.2	215.4(80.2) 101.3-398.0
European cultivars	10	667.8 479.7-870.1	526.6 (78.9) 370.7-786.6	266.9 169.1-353.4	160.5(60.1) 93.9-259.5

Figures in brackets = percentage of total yield

* yield range

EVALUATION OF TRUE POTATO SEED FAMILIES OBTAINED FROM DIFFERENT BREEDING SCHEMES IN THE SOUTH OF ITALY

L. Frusciante, S.J. Peloquin and A. Leone

INTRODUCTION

The use of true potato (TPS) seed for potato production has been considered a good alternative to seed tubers, since it minimizes the transmission of viruses and other pathogens. Using TPS also means that the total crop is available for consumption and consequently potatoes are grown at a relatively lower cost. In addition, storage and transfer of TPS is generally easy and inexpensive compared with seed tubers.

To further develop the use of TPS, research is needed to identify breeding methods which will generate high yielding and uniform families and to produce hybrid seed at low cost by using natural pollinators. These objectives can be obtained by using the breeding schemes proposed by Peloquin (1979, 1983).

The purpose of this study is to explore the possibility of using TPS in the south of Italy.

MATERIALS AND METHODS

Sixty-four TPS families obtained from different types of crosses and open pollinations were evaluated in a trial near Naples.

The 64 families consisted of 23 Dts (4*x* progenies from crosses between 4*x* cultivars and 2*x* hybrids); 15 4*x* progenies from crosses between 4*x* European cultivars and 4*x* American cultivars; 11 4*x* progenies from cultivars x Dts and 15 4*x* progenies from open pollinated (OP) cultivars. The diploids used as male parents were *Solanum phureja*-haploid *S. tuberosum* and *S. chacoense*-haploid *S. tuberosum* hybrids, which formed 2*n* pollen by parallel spindles (first division restitution or FDR).

RESULTS AND DISCUSSION

Large differences were found among the families for yield, tuber uniformity and plant vigour (Table 1); on average the hybrids behaved better than OP families and the best result was obtained from 4x x 2x crosses. The highest yields were also obtained from TPS derived from 4x x 2x crosses (Table 2), confirming that the 4x x 2x FDR breeding method is the most promising at present.

Table 1. Yield (q/ha), plant vigour and tuber uniformity of 64 families from four different types of cross.

Type of cross	Number of families	Mean yield	Plant vigour*	Tuber uniformity*
4x x 2x	23	260 (152-452)	2.65	2.34
4x x 4x	15	256 (115-318)	2.25	2.20
Dt x 4x	11	240 (144-331)	1.81	2.04
OP	15	152 (60-239)	1.63	1.80

*Rating ranged from 1 (poor) to 3 (very good)

Table 2. Yield (q/ha) of the best 8 families for each of the four types of cross and Student's t values for each comparison.

Type of Cross	Mean yield	4x x 2x	4x x 4x	Dt x 4x	OP
4x x 2x	347	-	2.89**	3.39**	7.37**
4x x 4x	294		-	0.84NS	5.72**
Dt x 4x	273			-	4.44**
OP	194			-	-

NS Not significant
** Significant at $p = 0.01$

REFERENCES

Peloquin, S.J. (1979). Breeding methods for achieving phenotypic uniformity. In Production of Potatoes from True Seed. Report of a Planning Conference, Manila, Philippines, pp. 151-5. Peru: International Potato Center.

Peloquin, S.J. (1983). New approaches to breeding for the potato of the year 2000. In Research for the Potato in the Year 2000, ed. W.J. Hooker, pp. 32-4. Peru: International Potato Center.

POTATO PRODUCTION FROM TRUE POTATO SEED IN ITALY

L. Martinetti

INTRODUCTION

At present, potato production from true seed (TPS) has advantages in developing countries, where market quality is not too important. However, this technology might even be adopted by developing countries, if progenies with improved quality were available. At the Institute of Agronomy, University of Milan, the possibility of growing potatoes from true seed in Italy is being evaluated. It is well known that production of high quality seed tubers is difficult to achieve in Italy, because of unfavourable weather conditions; TPS technology could solve this problem, besides saving money and making crop establishment, storage and transport less of a problem.

MATERIAL AND METHODS

Four progenies selected by the International Potato Center (CIP) were studied: open pollinated DTO-28 and DTO-33, and hybrids Atzimba X R-128.6 and Atzimba X 7xy-1. A preliminary trial was carried out in the glasshouse. Seeds were sown on 4 April 1985 in two different substrates in peat pots; one substrate was prepared with peat and sand (1:1 v/v), the other one with peat and soil (1:1 v/v). Both substrates were fertilized with 100 mg N, 300 mg P_2O_5 and 100 mg K_2O. A split-plot design with three replications was used. Seedlings at the 4 to 5 leaf stage were transplanted in the field, in ridges 75 cm apart. Groups of four seedlings were placed in the ridge 30 cm apart, so that there were 4.4 hills per m^2. No herbicide was used: weeds were mechanically removed. Tubers were hand lifted on 23 October.

RESULTS

The germination of seedlings was similar in both substrates. DTO-28 had the lowest percentage germination (58.6%), hybrid progenies the highest (75%). Seedlings were about the same height 35 days after sowing, but there were differences in the average number of seedling leaves: DTO-28 had the lowest average number, with 4.1 leaves per plant.

The number of days between transplanting and flowering was similar for all progenies; on average 45 days. The flowering stage was rather long in the hybrid progenies. The percentage of plants flowering was significantly lower for DTO-28 and DTO-30 (7.7% and 11.2% respectively) than for Atzimba XR-128.6 and Atzimba X 7xy-1 (50.0% and 75.7% respectively). Some variability was found in flower colour; colours generally ranged from purple to white and only Atzimba X 7xy-1 showed almost exclusively white flowers. No berries were observed.

Effects of daylength, light intensity, temperature and relative humidity on flowering and fruit set should be further studied; under the conditions in which the trial was carried out, only hybrid progenies had a satisfactory flower set, but none were suitable for seed production. At the flowering stage, both hybrid progenies were significantly taller (by an average 43 cm) than open pollinated ones. No differences occurred among progenies in the number of days from transplanting to agronomic maturity (150 days). The growing period was probably abnormally prolonged because of the cold and rainy spring and the hot and dry summer of 1985 in Italy.

Tuber shape was mostly round-oval, except in Atzimba X 7xy-1 which was mainly oval. Shape was often irregular, partially because the eyes were frequently very deep. Skin colour was yellow in all the progenies and sometimes particoloured red around the eyes. Flesh colour was white in DTO-28, white to light yellow in DTO-33, yellow to white in Atzimba X R-128.6 and white to yellow in the other hybrid. Uniformity was generally satisfactory. Some tubers were slightly affected by common scab (<u>Streptomyces</u> spp.), growth cracks, and second growth. No significant differences in specific gravity occurred; this averaged 1.029 which is rather low and does not satisfy the demands of Italian market.

Total yield was almost the same for all the progenies (average 24.88 t/ha) and satisfactory in comparison to traditional tuber-grown crops. Nevertheless, tubers tended to be small. The percentage of tubers with diameters <30 mm was high, particularly in DTO-33 and Atzimba X 7xy-1 (36.82 and 36.65% respectively). Yield of middle-sized tubers (with diameters of 30-50 mm) was significantly higher in Atzimba X R-128.6 (68.17%) than in DTO-33 and Atzimba X 7xy-1 (55.68 and 57.41% respectively). Yield of large tubers (diameters >50 mm) was almost the same in all the progenies (average 10.57%). So overall the percentage of marketable yield was rather low for our requirements.

We may conclude that the yielding capacity of the progenies tested was satisfactory. It has yet to be determined if this would be true in a shorter growing season. It is also necessary to determine if tuber size can be improved by more appropriate agronomic treatment, for example by changing the plant density. A considerable amount of lateral branch growth was observed from each main stem; therefore, by increasing plant spacing in the ridge larger tubers may be produced.

This recent research suggests that TPS technology is worthy of further study in Italy where, perhaps, it could be profitably adopted in the future.

FIELD SEEDING OF TRUE POTATO SEED IN A BREEDING PROGRAMME

M.W. Martin

INTRODUCTION

A method has been developed for growing commercial-type potatoes from true potato seed (TPS) sown directly into fields. The productivity, quality and uniformity of such crops are usually inferior to those of tuber-grown crops; however, plants superior in many important characteristics can be selected for subsequent clonal evaluation. Direct-seeded TPS populations are exposed to diseases, pests and cultural or environmental stresses and selected for resistance. The full-grown, mature tubers produced can be selected for type, productivity, quality, handling, storability and processing characteristics. Tubers from selected seedlings provide normal-size seed-pieces that are planted as single tubers or in replicated plots for evaluation in the first clonal generation. This first clonal generation is again exposed to a wide range of selection pressures. This contrasts with single hills normally grown from pot-grown tubers where over 99% of the genetic potential of segregating populations is discarded based on cosmetic attributes of tubers from single hills grown from tiny, variable-sized seed tubers under minimal stresses and selection pressures. This new method provides more information about each clone and increases the chances of finding and saving valuable genotypes.

PROCEDURE FOR DIRECT SEEDING TPS

During the past nine years crops have been grown annually from TPS sown directly into fields in eastern Washington and Oregon (Martin 1978a, 1983a,b). They have been grown on soils ranging from light sandy, through various loams to heavy clay, using either sprinkle or furrow irrigation. Machinery, soil preparation, fertilization, planting, irrigation, chemical weed control and cultural methods used for direct-seeded tomatoes are generally successful with TPS. Some changes are needed because of the smaller size and slower germination and growth of potato seedlings and their need to be ridged up before tuber formation.

SELECTING IN DIRECT-SEEDED TPS POPULATIONS

Many characteristics needed in potatoes can be selected in direct-seeded TPS populations and subsequent early clonal generations by applying appropriate selection pressures (Martin 1978b, 1979, 1981, 1983b, 1984). TPS is seeded into soils infested with *Verticillium*, common and deep-pitted scab (*Streptomyces* spp.), Columbia root knot nematode (*Meloidogyne* sp.) and Colorado potato beetles (*Leptinotarsa decemlineata*). Cultural conditions and locations are used which provide severe exposures to early blight (*Alternaria solani*), *Sclerotinia* wilt (*Sclerotinia sclerotiorum*) and powdery mildew (*Erysiphe cichoracearum*). TPS rows are rub or spray inoculated with potato virus Y and interplanted with rows of tubers infected with potato leafroll virus. The bases of plants expressing resistance to these foliar diseases or pests are sprayed with fluorescent red paint. They are thus easily identified during harvest with a flatbed digger, which places plants and tubers together on the soil surface. At harvest tubers of desired shape, skin type, number, size and yield, and free from cracks and growth abnormalities are selected. Planned overwatering during tuber initiation and interruptions in irrigation during bulking expose genetic weaknesses and promote stress-induced interior and exterior tuber maladies so that clones with such weaknesses can be eliminated. We select for high specific gravity, low sugars (light colour on frying), bruise resistance, long dormancy and good storability by screening seedling tubers and first generation clones through a 1.080 specific gravity salt solution, frying a plug from each tuber, bruising each tuber with a motor-driven thumper, and placing the tubers on screen trays in an environment inducive to sprouting and rotting.

REFERENCES

Martin, M.W. (1978a). Field seeding of true potato seed. Amer. Potato J., 55, 385. (Abstr.).

Martin, M.W. (1978b). Potato improvement by breeding. Proc. Washington State Potato Conf., 17, 107-9.

Martin, M.W. (1979). Mass selection for potato disease resistance in field-grown seedlings. Amer. Potato J., 56, 472-3. (Abstr.)

Martin, M.W. (1981). Mass screening for early dying and scab resistance by direct seeding into disease infested fields. Amer. Potato J., 58, 510. (Abstr.)

Martin, M.W. (1983a). Techniques for successful field seeding of true potato seed. Amer. Potato J., 60, 245-59.

Martin, M.W. (1983b). Field production of potatoes from true seed and its use in a breeding programme. Potato Res., 26, 219-27.

Martin, M.W. (1984). First generation selection in field grown potato seedlings. Amer. Potato J., 61, 529. (Abstr.)

UNCONVENTIONAL BREEDING METHODS

RECENT PROGRESS IN MOLECULAR BIOLOGY AND ITS POSSIBLE IMPACT ON POTATO BREEDING: AN OVERVIEW

R.B. Flavell

Potatoes are now beginning to receive special attention from molecular biologists and tissue culture experts who are committed to introducing new genes from the test tube and establishing the principles of gene regulation. This is because the potato is one of the few crop species that is infected by *Agrobacterium tumefaciens*, the organism which can transfer new genes into chromosomes of many dicotyledonous species; plants can also be regenerated from the single cells that have received the new genes, although this latter property is still in need of considerable improvement for routine genetic engineering of the crop. The status of current research is covered by other contributions to this volume, in particular the papers by Ooms and Blau *et al*.

Potato breeders can be excited that their economically important crop has become the experimental organism of a new group of scientists eager to try out new techniques and ask fundamental questions on the frontiers of academic research. It is inescapable that potato breeding will be influenced by this research.

The introduction of new genes into potatoes is in its infancy but progress in learning how to do it will surely be rapid over the next few years because some of the knowledge (and many of the genes) being gained from studies on tobacco is transferable to potato. There are unlikely to be difficulties in transferring DNA into potato protoplasts by direct DNA uptake procedures (Shillito *et al*. 1985) or to cells by infection with *Agrobacterium tumefaciens* but the efficiency of regenerating transformed plants from these cells needs to be improved. Progress in the introduction of new genes into tobacco plants and in the analysis of the transformed plants has accelerated considerably as the efficiency and speed of plant regeneration has been improved (Thomas & Hall 1985). It is now possible to obtain a rooted tobacco plantlet carrying new genes in its chromosomes within a month of infecting tissue explants with the appropriate *Agrobacterium tumefaciens*.

Considerable effort is now being devoted to learning how plant genes are regulated and how to engineer the directed expression of genes in a plant. Blau et al. (this volume) address this aspect. Progress has been dramatic over the past 2 years or so. New information is emerging almost monthly from around the world. Nevertheless, the problems of the control of plant development are so complex that it will be a long time before a full understanding of the control of even simple differentiation systems is available. Current research is motivated, of course, by a desire to understand gene expression but it is also clearly understood that to produce modified genotypes with commercial value it is likely to be necessary to engineer an inserted gene such that it works only in certain tissues or cells or at a specific level. One of the highlights of results so far is the redesigning of bacterial genes so that they are expressed in cells of a plant (Bevan et al. 1983; Herrera-Estrella et al. 1983; Fraley et al. 1983). The redesigning involved the joining of the bacterial coding sequence to the regulatory elements of a gene known to be expressed in plants. In subsequent experiments when the 5' regulatory element was replaced by that from the ribulose bisphosphate carboxylase gene of pea, expression of the bacterial coding sequence was brought under the control of light via the phytochrome response (Herrera-Estrella et al. 1984). These experiments showed, perhaps surprisingly, that the pea regulatory DNA contains sequences recognized by regulatory molecules in tobacco cells.

In other experiments, the DNA encoding the first few amino acids of the ribulose bisphosphate carboxylase precursor protein which are known to be responsible for the transfer of the protein into the chloroplast, was joined to the bacterial coding sequence, so that the short amino acid sequence became part of the bacterial protein sequence. After insertion of this gene into tobacco plants the bacterial protein was found in the chloroplasts (Schreier et al. 1985). This illustrates how new enzyme functions can be engineered into specific cell compartments under specialized regulation.

Storage protein genes from bean (Phaseolus vulgaris) have been transferred into tobacco and found to be expressed strongly in the developing seed but not elsewhere, showing that the regulatory signals on the bean gene defining tissue-specific expression are recognized also by developmental regulatory systems in tobacco (Sengupta-Gopalan et al. 1985). This kind of conservation of regulatory mechanisms is very heartening for the genetic engineer who wishes to use genes isolated from one plant to manipulate another, such as potato.

Even though it is possible to insert genes into tobacco, potato and some other crop plants, evidence from tobacco and petunia suggests that the level of activity of the inserted gene varies from plant to plant. This may be due to effects imposed on the gene from neighbouring chromosomal regions or due to uncontrolled methylation of the gene - it is known that methylation of specific cytosine bases can "silence" genes. This variation in expression may, from some points of view, be a nuisance to molecular biologists wanting to learn what sequences control the extent of gene expression. On the other hand, it enables plants with a diversity of gene expression to be surveyed by the geneticist or breeder and the appropriate genotype selected.

Now that the ability to insert genes into potato is being established, the important question of what genes to insert becomes even more urgent. Here breeders and those who know the problems of the crop and its breeding systems have an opportunity to guide the molecular and cell biologists. It is important to remember, however, that only one or a few genes can be inserted into potato using the currently envisaged techniques. Furthermore, methods for deleting or replacing genes have not yet been developed. To date, molecular biologists have taken note of the need to improve virus resistance in potato and have initiated experiments to try and achieve this long-term aim. Also, as described in the paper by Ooms (this volume), modifications to plant phenotype have been made by the introduction of genes involved in hormone production.

Undoubtedly many useful changes could be introduced into potato by manipulation of the expression of existing potato genes concerned with regulating plant development. The function and molecular characteristics of such genes are unknown. How then will they be isolated? One way to isolate dominant genes is to first inactivate them with a transposable element which has been characterized at the molecular level (Fedoroff et al. 1984). To do this it will be necessary to find, by phenotypic observation, a plant with the desired gene inactivated by the insertion of the transposable element. The element can then be repurified from the mutant plant together with the adjacent gene. The latter DNA sequence can then be used to isolate nonmutant forms of the gene from other plants. If this kind of experimental approach is to be used to isolate useful potato genes, it is necessary to have plants which contain active, defined transposable elements. None have yet been described in potato so molecular biologists are attempting to insert such purified elements from maize to see if they will transpose in their new host plant. Such experiments seem worthwhile in view of the recent finding (B. Baker, J. Schell and N. Fedoroff, in press) that the maize element 'Activator' transposes in tobacco cells.

From research in molecular biology new methods of screening germplasm will emerge. It is possible to find unique pieces of potato DNA that reside in chromosomes closely linked to genes which specify traits which are difficult or expensive to assay by conventional methods. The trait can then be followed in breeding programmes by DNA hybridization assays using the DNA sequence that is closely linked to the desired gene. Where a mutation in or around only one allele of a pair allows or prevents cutting of isolated DNA by a restriction endonuclease, then the DNA fragments carrying each allele will be of different length and easily separated by electrophoresis. When a large number of DNA sequences have been characterized from all over the chromosome complement, then in theory it should be possible to follow many specific chromosome segments through breeding programmes. Although this could be a powerful method for screening germplasm, it is not easily adapted to the large populations that breeders may wish to screen. To overcome the problems associated with isolating DNA from each plant, it should be possible to insert active, or potentially active, genes into the parental potato lines and find the plants with the marker gene in the appropriate position. The germplasm can then be screened for the desirable allele by following the active marker gene. Perhaps in the future the germplasm of breeding programmes will contain many such sequences placed there by the molecular geneticist to help the breeder improve the efficiency of screening germplasm. Certainly, the genes that the molecular geneticists insert will often be linked to a selectable marker so that the new genes will be followed easily by selection or screening systems in the laboratory or the field.

As these sorts of screening aids are developed in response to breeders' needs, breeders will incorporate techniques of molecular biology into their activities. This has already happened in the breeding programme at the Plant Breeding Institute, where breeders are using DNA hybridization assays routinely to screen for the presence/absence of viruses in breeding material (Baulcombe *et al.* 1984; Boulton *et al.*, this volume). The adoption of these techniques has improved the reliability of the screening tests and probably reduced costs.

In the USA a patent application has been upheld recently that establishes the principle that a new trait (and the genes which give it) can be patented. If this become established practice, breeders will need to know which patented genes are in their material. It will often be more convenient to check this using DNA-DNA hybridization techniques than to assay the trait in the field. When many different useful genes, inserted by genetic engineering and patented, are present in elite germplasm, breeding programmes, distinctness testing and variety release will obviously become much more complicated and will certainly involve molecular biologists.

REFERENCES

Baulcombe, D.C., Flavell, R.B., Boulton, R.E. & Jellis, G.J. (1984). The use of cloned hybridisation probes to detect viral infections in a potato breeding programme. In Genetic Manipulation of Plants and its Application to Agriculture, ed. P. Lee, pp. 185-95. Oxford: Oxford University Press.

Bevan, M.W., Flavell, R.B. & Chilton, M.-D. (1983). A chimaeric antibiotic resistance gene as a selectable marker for plant cell transformation. Nature, 304, 184-7.

Fedoroff, N., Furtek, D. & Nelson, O. (1984). Cloning of the Bronze locus in maize by a simple and generalisable procedure using the transposable controlling element. Ac. Proc. Natl. Acad. Sci. USA, 81, 3825-39.

Fraley, R.T., Rogers, S.G., Horsch, R.B., Sanders, P.R., Flick, J.S., Adams, S.P., Bittner, M.L., Brand, L.A., Fink, C.L., Fry, J.S., Galluppi, G.R., Goldberg, S.B., Hoffman, N.L. & Woo, S.C. (1983). Expression of bacterial genes in plant cells. Proc. Natl. Acad. Sci. USA, 80, 4803-7.

Herrera-Estrella, L., De Block, M., Messens, E., Hernalsteens, J.-P., Van Montagu, M.T. & Schell, J. (1983). Chimaeric genes as dominant selectable markers in plant cells. EMBO J., 2, 987-95.

Herrera-Estrella, L., Van den Broeck, G., Maenhaut, R., Van Montagu, M., Schell, J., Timko, M. & Cashmore, A. (1984). Light inducable and chloroplast-associated expression of a chimaeric gene introduced into Nicotiana tabacum using a Ti plasmid vector. Nature, 310, 115-20.

Schreier, P.H., Seftor, E.A., Schell, J. & Bohnert, H.J. (1985). The use of nuclear encoded sequences to direct the light-regulated synthesis and transport of a foreign protein into plant chloroplasts. EMBO J., 4, 25-32.

Sengupta-Gopalan, C., Reichert, N.A., Barker, R.F., Hall, T.C. & Kemp, J.D. (1985). Developmentally regulated expression of the bean p-phaseolin gene in tobacco seed. Proc. Natl. Acad. Sco. USA, 82, 3320-24.

Shillito, R.D., Saul, M.W., Paszowski, J., Muller, M. & Potrykus, I. (1985). High efficiency direct gene transfer to plants. Bio/Technology, 3, 1099-103.

Thomas, T.L. & Hall, T.C. (1985). Gene transfer and expression in plants: implications and potential. BioEssays, 3, 149-53.

COMBINED APPLICATION OF CLASSICAL AND UNCONVENTIONAL TECHNIQUES IN BREEDING FOR DISEASE RESISTANT POTATOES

G. Wenzel, S.C. Debnath, R. Schuchmann and B. Foroughi-Wehr

INTRODUCTION

All breeding strategies, classic as well as unconventional, aim at the production of better varieties. Today this means in particular, higher levels of disease resistance. This aim is most easily achieved when both classical and unconventional techniques complement each other and increase the efficiency of the breeding processes, i.e. creation of variation, hybridization and selection. New variation may be obtained *in vitro* through somaclonal variation or by making use of natural meiotic segregation. For the production of clones with resistance to *Phytophthora* and *Fusarium*, spontaneous *in vitro* mutation can be used; this is, however, only applicable when the induction of variation is coupled with a powerful *in vitro* screening system. From abiotic cultures of the two fungi mentioned, exotoxins can be extracted and used for selection. This process may be also used for hybrid selection, when both fusion partners have a different resistance. In addition, for hybrid selection, hybrid vigour of the fusion products may be used. The regeneration of potatoes via anther or isolated microspore culture (Wenzel *et al.* 1982; Uhrig 1985) provides the basis for the use of meiotic segregation *in vitro*. Haploids can be produced from tetraploids and from dihaploids. The use of such haploids should facilitate the incorporation of virus resistance, and of other characters, regardless of whether they are monogenically or polygenically inherited.

METHODS AND RESULTS

Plant material

For most experiments dihaploid potatoes were used as starting material. The plants were grown in glasshouses under semicontrolled conditions and in the field at three different locations: 1) Köln-Vogelsang with a high natural virus pressure because of high aphid populations (up to 1000 aphids per leaf), 2) Grünbach, an area without commercial potato production and 3) Scharnhorst, an area with few aphids.

Regeneration ability

Regeneration of plantlets from protoplasts and/or microspores is essential for genetic manipulation. Not all clones give good results, measured as the frequency of regeneration *in vitro*, but selection for the character "regeneration ability" can be successful. This was demonstrated in experiments aimed at the regeneration of isolated microspores from tetraploids (Table 1) and for several clones from which protoplast regeneration was intended (Table 2). In regenerants, particularly from protoplasts, a wide range of phenotypic variation and altered ploidy levels (polyploidy,

Table 1. Regeneration ability of microspores on three different media; AH clones have been passed once through anther culture (Uhrig 1985).

	Medium containing (mgl^{-1})		
Anther donor	1.0 BAP 0.1 IAA	0.5 BAP 0.2 IAA	2.0 BAP -
with preselection for regeneration ability			
AH 84.2316/7.1	248	288	440
AH 84.2316/8.1	103	25	25
AH 84.2306/15.1	534	-	1923
without preselection for regeneration ability			
H 81.691/3	0	0	0
H 83.325/1	0	3	2

Table 2. Regeneration ability of several dihaploid potato clones after protoplast isolation; protoplast yield is measured on a 1-9 scale; 1 = very few viable protoplasts.

Clones	Protoplast yield	Survival %	Division %	Calli/ plate	Shoots %
AH 78.5111	8 ab	33 b	11 b	335 b	65 ab
AH 82.4521/6	9 a	33 b	11 b	344 b	67 a
AH 82.4551/6	3 e	30 b	9 b	270 c	52 cd
AH 84.4568/6	8 ab	30 b	10 b	347 b	64 ab
AH 78.6147	5 cde	23 c	8 b	266 c	45 d
AH 82.4552/6	3 e	13 e	3 e	0 d	0 f
AH 84.4566/6	6 bcd	22 cd	7 b	303 bc	56 bc
H 75.1207/7	7 abc	47 a	31 a	578 a	25 e
PH 83.044	4 de	17 de	8 b	317 bc	25 e

Numbers followed by common letters are statistically identical at 5% level of significance.

mixoploidy and aneuploidy) have been reported in protoplast-derived potato plants (Binding et al. 1978; Sree Ramulu et al. 1983). These variations are generally detrimental and should be reduced for the benefit of genetic manipulation. Therefore, experiments have been carried out to select the clones most responsive to protoplast regeneration and to optimize the isolation and regeneration procedure. Clones with good regeneration ability and clones possessing resistance were crossed, the F_1 family was screened for resistance, and then the resistant clones served as donor material. The behaviour of such F_1 hybrid families indicated that the capacity for producing macroscopic structures from microspores is under genetic control. It is, however, difficult to define the mode of inheritance. It could be demonstrated that regeneration ability falls into two distinct classes, indicating a not too complex mode of inheritance (Figure 1).

Eight dihaploid clones, seven developed from anther culture and a protoclone from H 75.1207/7, namely PH 83.044, were used in this study (Table 2). Protoplasts were isolated from 14-day-old axenic shoot cultures using the method of Binding et al. (1978). The plating efficiency was defined as the number of dividing cells expressed as a percentage of the initial number of plated protoplasts. After 14 days, the developing cell

Figure 1. Regeneration ability of several clones of one F_1 family; the parents have either a high (HI) or a very low (H3-704) regeneration ability.

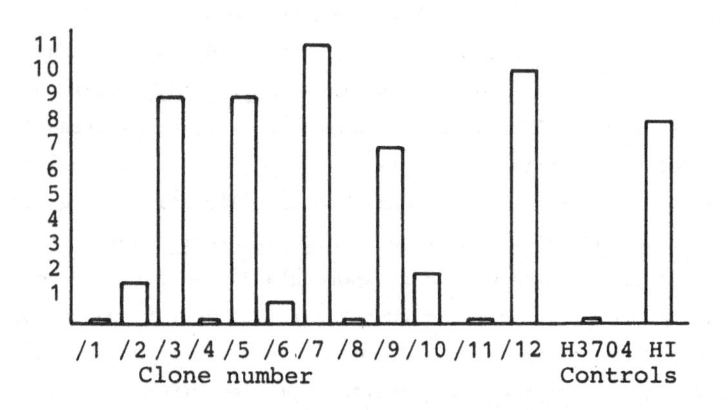

aggregates were cultured for another 2 weeks in the same growth chamber in a semi-solid (0.2% agar) V-KM medium (Binding et al. 1978) by diluting with an equal volume of a 30°C V-KM medium supplemented with 0.4% agarose at 550 mOsm kg^{-1}.

Protoplasts were obtained from all clones studied. Analysis of variance for protoplast yield and survival, division of surviving cells and regeneration of calli showed highly significant mean squares (p >0.01) indicating significant differences between clones for all characters studied. The relative performance for each clone is presented in Table 2. The clones AH 778.5111, AH 82.4521/6 and AH 84.4568/6 were statistically identical for all five characters. These three clones were superior to clone AH 82.4551/6 for protoplast yield, number of calli per plate and percentage of shoots regenerated. Clones AH 78.6147 and AH 84.4566/6 gave statistically similar results for all characters except percentage of shoot regeneration, for which the latter was superior. These two clones, however, were better than AH 82.4552/6 for all traits studied. Clone H 75.1207/7, which was used as control, was statistically superior to its protoclone PH 83.044 for all traits studied except for percentage of protoplast survival and division, and number of calli per plate was highest for clone H 75.1207/7. However, the percentage of shoot regeneration in this clone was low.

SELECTION FOR RESISTANCE TO DISEASES WITHOUT USING SEXUAL HYBRIDIZATION

Screening for *Phytophthora* resistance in callus systems

In her early experiments on the selection of *Phytophthora*-resistant potato clones, Behnke (1979) used unfractionated liquid fungal growth medium, which contained exotoxins responsible for the virulent reaction. Calli were placed on a medium containing *Phytophthora* toxin at a concentration that induced 90% mortality. Survivors were transferred five times to identical toxin-containing media before regeneration of plants was induced. It was found that selected clones had significantly smaller local lesions following mechanical inoculation with *Phytophthora* spores, than unselected control plants. The number of sporangia formed later did not differ from the controls, which means that the infection rate was reduced but not the growth rate of the fungus or the sporulation rate. This resistance was unspecific and is probably quantitatively inherited. Of the regenerated plants, 34 were field tested under natural infection pressure over a 5-year period. These experiments clearly demonstrated that the selected clones were inferior to the starting material in many characters, particularly in resistance to other diseases. Probably so many mutations occurred during the callus phase, most of them deleterious, that regardless of a possible increase in resistance to Phytophthora, these selected clones were of no practical value.

In the present work purified toxins were used, extracted from the fungal growth medium by ultrafiltration. The molecular weight of the active compound is about 20 000 daltons (Stolle & Schöber 1984). Toxin effectiveness was determined by measuring the increase in fresh weight of shoot tip cultures at different toxin concentrations after 3 weeks of abiotic growth. This allows a constant toxin concentration to be used in all further experiments. Calli with genetically different degrees of resistance to Phytophthora are now being tested on MS medium (Murashige & Skoog 1962) with a range of toxin concentrations. The results are given in Table 3.

On MS medium with 2 mg l^{-1} 2,4-D, which results in rather loose callus growth, callus from the resistant source grew well in the presence of high toxin concentrations but turned black. In contrast, calli from the susceptible genotype grew poorly or not at all, depending on the toxin concentration. Tuber tissue from resistant clones also turned black in the presence of toxin (Foroughi-Wehr & Stolle 1985). From black and dying callus, pieces may survive which can be regenerated. Among the types that grew slowly, some mutated, grew normally, and developed into plants with resistance to the first phase of infection. Plants descended from surviving black callus should express a different type of resistance. It is probable that specific in vitro screening methods will be needed for each phase of the infection process.

Table 3. Influence of Phytophthora toxins on the increase of fresh weight (FW) and dry weight (DW) during three weeks of shoot tip culture on simple and double concentrated toxin (Foroughi-Wehr and Stolle 1985).

Clone	Toxin Concentration	FW (mg)	DW (mg)
15	Control	646 a	55 a
	1x	430 b	49 ab
	2x	338 b	43 b
17	Control	822 a	74 a
	1x	445 b	60 ab
	2x	359 b	52 b

Numbers followed by common letters are statistically identical at 5% level of significance

Screening for Fusarium resistance in protoplast populations

Since it is possible to regenerate plantlets from dihaploid potato protoplasts routinely, experiments were performed to screen for Fusarium resistance in potato. Protoplast populations of four different dihaploid clones: H 78.551/1 (M 3), H 75.1207/7 (M 4), AH 82.4521/6 (A 1), and AH 84.4568/6 (A 2) were grown on media containing exotoxin(s) produced by Fusarium sulphureum or F. solani var. coeruleum (hereafter F. coeruleum) cultured abiotically on defined media. All clones were susceptible to Fusarium under field conditions. Toxin concentration was standardized by determining the respiration activity of incubated potato cell-suspension cultures using an oxygen electrode. The relative concentration of the toxins is directly correlated with the inhibition of the respiratory activity of potato suspension cultures (Schuchmann 1985a). For screening for Fusarium resistance such standardized toxin solutions were used at a concentration which allows 1-2% of the protoplasts to form walls, divide and continue growth. From each of the four clones, protoplasts were isolated and about 1500 protoclones resistant to the toxin at the cellular level were regenerated. Because of the rapid passage through the callus phase, all clones at the same ploidy level were phenotypically quite uniform, indicative of low somaclonal variation. Retesting of secondary callus and leaves revealed a significant increase in tolerance of toxin, measured as increase in fresh weight. The reaction of the toxin on leaves was determined and compared with the controls (Figure 2A) and the fresh weight growth-rate of calli in the presence of the toxin was retested (Figure 2B). In all experiments the selected protoclones were less sensitive to the toxin treatment than the controls. The resistance was always higher when the first selection and the retest were performed with toxins from the same Fusarium species. When, for example, toxin from F. sulphureum was used for initial selection and toxins from F. coeruleum were used in the second screen, a resistance reaction was in most cases not significant, although a trend was clearly visible. This probably means that the toxin has both a specific and an unspecific action (Schuchmann 1985b).

How well the resistance of calli and leaves of selected protoclones to the toxin is correlated with the resistance of the tuber to the actual pathogen remains to be determined in next year's field-grown tuber generation. In addition to the different starting material used in the Fusarium system and the Phytophthora system i.e. protoplasts versus calli, another important difference lies in the procedures used. The callus is transferred several times to toxin-containing media, which means that escapes should be rare; the protoplast, however, has just one contact phase (7 weeks) with the toxins and a high rate of escapes is probable. For ensuring a reliable identification of resistant clones, the Phytophthora procedure is preferable. It induces too many abnormalities however, whereas the Fusarium selection technique results in more useful material, although a second screening is necessary.

Making use of combined resistance for selecting somatic hybrids

For applied breeding it would be highly desirable to combine resistances by exclusion of meiotic processes, particularly for quantitatively inherited characters. Phytophthora resistance is probably such a quantitatively inherited trait (Behnke 1979). After fusion, using polyethylene glycol, and cell wall regeneration, the Fusarium toxins are added, and when small colonies are formed the Phytophthora toxin is also added to the medium. It is expected that this procedure increases the frequency of fusion hybrids. Hybrid vigour is used as a general selection system for somatic hybrids. Scheider & Vasil (1980) demonstrated that the calli of

Figure 2. Effect of preselection with Fusarium toxins on a) leaves and b) callus, relative to unselected controls. On leaves the size of lesions produced by the toxin was measured, while in callus the respiration rate was determined.

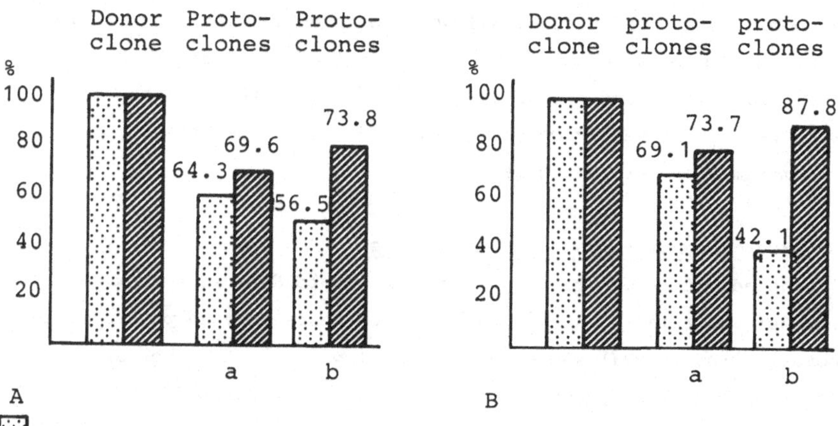

hybrid genotypes expressed heterotic growth in the petri dish. We make use of this fact as it has such a general application. The technique is not completely reliable as genotypes with unfavourable gene combinations will not be detected. However, genotypes with poor callus growth will probably not result in vigorous plants of economic value, even though they might be hybrids.

RAPID INCORPORATION OF VIRUS RESISTANCE THROUGH THE USE OF HAPLOIDS

Haploid induction

Potato microspores can be cultured either within (anther culture), or in isolation from, the anther. The method used for all anther culture experiments has been described by Wenzel & Foroughi-Wehr (1984). The most important features are summarized below. Flower buds 4-6 mm in length were harvested and precultured in dry test tubes for 2 days at 6°C. Then the anthers were dissected and plated on MS medium supplemented with 6% sucrose, 0.5% active charcoal and 1 mg l^{-1} 6-BAP. The macroscopic structures which developed within 3-4 weeks were transferred to MS medium supplemented with 0.5 mg l^{-1} zeatin and 3% sucrose. As soon as differentiation started, the tissue was transferred to MS medium supplemented with 0.5 mg l^{-1} 6-BAP and 3% sucrose. Uhrig (1985) demonstrated that the use of liquid medium for culturing potato anthers, instead of solid agar plates, improved the yield of embryoids. This method, together with two successive cycles of selection for good response to anther culture, allowed the induction and regeneration of plants from microspores on a large scale. When functional shoots were formed these were cut off and transferred as cuttings to the glasshouse. Ploidy levels were estimated by counting chloroplasts in the stomata, or by counting chromosomes of very young leaves.

Production of virus-resistant clones

The advantage of breeding at the monohaploid level is now well appreciated. Our experiments have concentrated on resistance to potato virus X (PVX), potato virus Y (PVY) and potato leafroll virus (PLRV). For all experiments with PVX and PVY, mechanical inoculation was carried out with mixtures of three different strains of the relevant virus. Retesting was performed on test plants and by ELISA. The presence of extreme resistance to PVX and PVY was tested by grafting scions on virus-infected tomato stocks. After 4 weeks the scions were retested on test plants and by ELISA.

Changes in resistance after reduction in the ploidy level have been found (Table 4). Susceptible donor clones gave rise to microspore-derived (androgenetic) clones with extreme resistance or field resistance. Also, as expected, numerous susceptible clones segregated.

Resistance to PVY, PVX and PLRV was also tested in the field. Tubers harvested in the glasshouse from androgenetic plants with low field resistance to PVY, PVX and/or PLRV were planted in the field in Cologne under high virus pressure. From 1957 clones tested in 1979, a sample of 230 was harvested and regrown in 1980. In the second year they were screened for field resistance to PVY and PLRV under natural conditions. This was repeated for 6 years. The results are given in Table 5.

Even after 7 years of field cultivation under natural virus infection, healthy clones were found. These clones are probably more resistant than the anther-donor material. This means that both monogenically and additively inherited resistance, and probably agronomic traits also, are retained during the haploidization process and that it is possible to select

Table 4. Inheritance of field resistance to PVX and PVY in androgenetic clones. The donor clones were 2x and the resistance was assessed after spontaneous doubling of the androgenetic clones.

	PVX resistance						PVY resistance					
	extreme			field			extreme			field		
number of clones	7			11			5			11		
	er	fr	s	er	fr	s	er	fr	s	er	fr	s
doubled androgenetic clones	1	1	5	1	1	9	2	1	2	1	2	8

er = extreme resistance, fr = field resistance, s = susceptible

Table 5. Freedom from virus infection (PVX, PVY, PLRV) in androgenetic clones derived from one field-resistant donor clone (H3-703), planted for 7 successive years.

Year	No of clones Planted	Healthy	% Healthy
1979	1957	230	12
1980	230	84	37
1981	84	54	64
1982	54	39	72
1983	39	26	66
1984	26	24	92
1985	24	23	96

haploids after meiotic segregation which express a higher level of field resistance than their donor clones. After genome doubling such resistances are present in the homozygous condition and consequently a quantitative trait such as field resistance can be transferred to the next generation like a qualitative character. This is probably the most important advantage in using doubled haploids.

CONCLUSIONS

The question to be answered is: Has disease resistance in potato been improved by *in vitro* systems? The answer is yes! Clones with reliable resistances have been produced via haploidization and it can clearly be demonstrated that *in vitro* systems can be used to screen for resistance to fungal diseases. We have followed two different lines; utilization of *in vitro*-induced variation and of natural variation following meiotic recombination. Both strategies resulted in new resistant clones. In general the natural variation expressed after meiosis following cross pollination is sufficient for selection purposes. Only if a desired character is not available, or if polyploid and vegetatively propagated crops have to be improved, does *in vitro* induced variation become a promising tool. In such situations it is, however, crucial to develop a selection system which allows screening of large populations of individuals in a small space, such as protoplasts in the petri dish. Then the natural mutation rate would be sufficient to allow the formation of a desired new genotype. Too drastic somaclonal variation is not beneficial, as too many characters are influenced, usually in a deleterious direction as shown by the potato/*Phytophthora* system. When, on the other hand, the vast majority of the clones regenerated from protoplasts are phenotypically normal, as in the potato/*Fusarium* system, chances of too many undesirable characters being created are greatly reduced.

The reduction of the ploidy level of potato in stages offers the advantage of a simple mode of inheritance and a better chance of combining quantitatively inherited characters in one clone. After reduction of the ploidy level different degrees of resistance to viruses could be detected. The data available demonstrate that field resistance to PLRV, PVX and PVY is maintained and expressed at all ploidy levels over a 7-year period, which means that both monogenically and polygenically inherited resistances are maintained during the successive haploidization processes. When a genome has passed through the monohaploid level, a homozygous additive character can be transferred efficiently to the next generation via hybridization in the same way that a monogenic character can be transferred. In the vegetatively propagated potato such a hybrid can be directly propagated as a new cultivar.

In our breeding scheme (Wenzel et al. 1979), when we found the female parent was a strong influence, we reduced the ploidy level to 2x by parthenogenesis. This was also true for the second step, where the parthenogenetic method works as well (van Breukelen et al. 1975). The second parthenogenetic step is, however, more laborious as no natural selection operates against 2x hybrids. As a consequence many more seeds and seedlings have to be screened for their ploidy level. Furthermore, there is no simple way of identifying amongst dihaploid S. tuberosum x S. phureja hybrids the spontaneously doubled monohaploid S. tuberosum clones, which are present in quite large quantities (Hermsen, personal communication), from the very few monohaploids (5 from 250 000 seeds, Wenzel 1979). Therefore, we concentrate on androgenetic monohaploid plant production. Experiments along these lines gave the following results: 1) The regeneration ability of microspores was dependent upon the genotype and could be transferred via sexual hybridization. 2) Homozygous resistant clones could be produced from hybrids carrying resistance in the heterozygous condition via anther culture, regardless of the mode of inheritance.

Differences between genotypes in regeneration ability are usually explained as being due to the effect of the medium, most probably to the phytohormone concentrations. This external influence is still not well understood, but both exogenous and endogenous phytohormone concentrations can influence the regeneration potential. Genotypes which are not too sensitive and which do not require highly balanced phytohormone levels are the ones which react more frequently. Others are irreversibly destroyed by the influence of the medium. Hence, we think that the chances of successful regeneration can be improved more by introducing tissue culture tolerance into genotypes, than by finding a specific medium for each genotype. This does not mean that genotypes exist which have no tissue culture ability at all; probably a specific medium can be successfully formulated for each genotype.

Finally it should be stressed that unconventional approaches to plant breeding can never stand on their own; they need to be integrated with classic breeding programmes.

REFERENCES

Behnke, M. (1979). Selection of potato callus for resistance to culture filtrates of Phytophthora infestans and regeneration of resistant plants. Theor. Appl. Genet., 55, 69-71.

Binding, H., Nehls, R., Schieder, O., Sopory, S.K. & Wenzel, G. (1978). Regeneration of mesophyll protoplasts isolated from dihaploid clones of Solanum tuberosum L. Physiol. Plant., 43, 52-4.

Breukelen, van E.W.M., Ramana, M.S. & Hermsen, J.G.Th. (1975). Monohaploids (n=x=12) from autotetraploid Solanum tuberosum (2n=4x=48) through two successive cycles of female parthenogenesis. Euphytica, 24, 567-74.

Foroughi-Wehr, B. & Stolle, K. (1985). Resistenzselektion in vitro am Beispiel des Systems Kartoffel/Phytophthora infestans (Mont.) de Bary. Nachrichtenbl. Deut. Pflanzenschutzd., 37, 170-3.

Murashige, T. & Skoog, F. (1962). A revised medium for rapid growth and bioassays with tobacco tissue culture. Physiol. Plant., 15, 473-97.

Schieder, O. & Vasil, I.K. (1980). Protoplast fusion and somatic hybridization. Rev. Cytol. Suppl., 11B, 21-46.

Schuchmann, R. (1985a). Methode zur relativen Konzentrationsbestimmung von Fusarium-Toxinen durch Messung der Respirationsrate. Nachrichtenbl. Deut. Pflanzenschutzd., 37, 81-4.

Schuchmann, R. (1985b). In vitro Selektion auf Fusarium-Resistenz bei der Kartoffel. Ph.D. Thesis, Technische Universität München.

Sree Ramulu, K., Dijkhuis, P. & Roest, S. (1983). Phenotypic variation and ploidy level of plants regenerated from protoplasts of tetraploid potato (Solanum tuberosum L. cv. Bintje). Theor. Appl. Genet., 65, 329-38.

Stolle, K. & Schöber, B. (1984). Wirkung eines Toxins von Phytophthora infestans (Mont.) de Bary auf Kartoffelgewebe. Potato Res., 27, 173-84.

Uhrig, H. (1985). Genetic selection and liquid medium conditions improve the yield of androgenetic plants from diploid potatoes. Theor. Appl. Genet., 71, 455-60.

Wenzel, G. (1979). Neue Wege in der Kartoffelzüchtung. Kartoffelbau, 30, 126-9.

Wenzel, G., Bapat, V.A. & Uhrig, H. (1982). New strategy to tackle breeding problems of potato. In Plant Cell Culture in Crop Improvement, eds. S.K. Sen & K.L. Giles, pp. 337-49. New York: Plenum Press.

Wenzel, G. & Foroughi-Wehr, B. (1984). Anther culture of Solanum tuberosum. In Cell Culture and Somatic Cell Genetics of Plants, ed. I.K. Vasil, pp. 293-301. New York: Academic Press.

Wenzel, G., Schieder, O., Przewozny, T., Sopory, S.K. & Melchers, G. (1979). Comparison of single cell culture derived Solanum tuberosum L. plants and a model for their application in breeding programs. Theor. Appl. Genet., 55, 49-55.

ANTHER CULTURE OF DIHAPLOID *SOLANUM TUBEROSUM* H_3703

N.P. Batty and J.M. Dunwell

INTRODUCTION

Temperature pretreatment of excised floral parts is a technique used to improve response to anther culture in many species. In some solanaceous species including *Solanum tuberosum* (Uhrig 1983) pretreatment has involved placing the excised buds at low temperatures, usually around 6°C, for 2-7 days prior to culture. However, more recent investigations on *Capsicum annuum* (Dumas de Vaulx *et al*. 1981) have shown that for this species, high temperature pretreatment during the first days of culture (i.e. preincubation) gives optimum results. The first report of preincubation at high temperatures (30°C, 2 days) for a tuberous *Solanum* species was by Cappodocia *et al*. (1984) on *S. chacoense* and various hybrids. However, no comparison was made between this treatment and the cold pretreatment described above. The present study involved the dihaploid *S. tuberosum* H_3703 and compared the effects of various pretreatments before and during culture and included the use of both high and low temperatures in two experiments.

MATERIALS AND METHODS

Scions of *S. tuberosum* cv H_3703 were grafted onto tomato rootstocks to promote and prolong flowering. For the first experiment plants were grown in growth cabinets at 20°C day, 16°C night, 18h day. Illumination was provided by fluorescent tubes and 25W tungsten filament lamps (5000-8000 lux). For the second experiment the same procedure was adopted except that plants were glasshouse grown. Buds were selected when 3.5-5.0mm in length. Cytological studies had shown that pollen from buds within this size range was predominantly at the late uninucleate stage.

In the first experiment random samples of buds were subjected to one of five treatments: pretreatment at 6°C; preincubation at 6°C; pretreatment at 30°C; preincubation at 30°C; cultured fresh, ie. without any pretreatment.

In the second experiment random samples of buds were subjected to preincubation at 6°C, 25°C or 30°C. Pretreated buds were placed in dry, sealed tubes and stored in darkness at 6°C or 30°C for 2 days. They were then surface sterilized and rinsed three times in sterile distilled water. The five anthers from each bud were then aseptically removed and plated on 10ml of culture medium (Wenzel <u>et al</u>. 1983) in a plastic petri dish (50 x 18mm). The plates were sealed and transferred to the incubation conditions described below. Preincubated buds were harvested, surface sterilized, rinsed, dissected and plated as described above.

Anthers were incubated at 25°C ± 1°C in the light (Grolux fluorescent tubes, intensity 30uEm^{-2} s^{-1}) with a 16h photoperiod. After 6 weeks incubation, anthers were examined and the percentage "responding" ie. producing macrostructures visible to the naked eye, was scored. The number of anthers producing plantlets and the number of plantlets produced by each anther were also recorded.

RESULTS

Table 1(a) shows that, as estimated by both the number of anthers responding and the number of plants produced per 100 anthers cultured,

Table 1. Effect of pretreatment and preincubation on anther response.

Treatment	No. anthers	% anthers responding	% anthers with pls*	% resp. anthers with pls*	Pls*/100 anthers cultured	Pls*/100 anthers responding
(a)						
Pretreat. 6°C	557	15.8	8.1	51.1	7.9	50.0
Preincub. 6°C	344	36.3	14.8	40.8	13.4	36.8
Pretreat. 30°C	390	3.3	1.3	38.5	1.8	53.9
Preincub. 30°C	470	56.0	36.4	65.0	49.2	87.8
Fresh	431	38.1	15.3	40.2	20.2	53.1
(b)						
Preincub. 6°C	210	18.6	9.3	51.3	4.3	23.1
Preincub. 25°C	240	33.8	8.8	11.3	11.3	33.3
Preincub. 30°C	260	30.0	20.8	69.2	18.1	60.3

*, pls = plantlets

preincubation was significantly better than pretreatment and that 30°C was significantly better than any other treatment. Comparison of the results with those for anthers cultured without pretreatment show that the use of either 6°C or 30°C pretreatment significantly reduced the percentage of anthers responding and thus the yield of plants per 100 anthers cultured. However, the proportion of the responding anthers which produced plants was not significantly reduced by pretreatment. Throughout, anthers preincubated at 6°C did not produce results that were significantly different from those obtained for fresh anthers. However, 30°C preincubation was shown to significantly increase both the number of anthers responding and also the quality of response as estimated by the proportion of responding anthers producing plants and the yield of plants per 100 anthers responding.

The results in Table 1(b) show that although overall yields were lower than in experiment one the relationship between the effects of the three preincubation temperatures remained constant. Compared with the two higher temperatures, 6°C preincubation resulted in reductions in both the number of anthers responding to culture and the proportion of the macro-structures produced which developed into plants. The combination of these two factors reduced the yield of plants per 100 anthers cultured for buds pre-incubated at 6°C to less than half that for buds preincubated at 25°C and to less than one quarter of the value obtained by preincubation at 30°C. Comparison of the results obtained from buds preincubated at 25°C and 30°C show that there were no significant differences between these treatments for the proportion of anthers cultured responding and the proportion of embryos produced becoming plants. The increase in yield of plants per 100 anthers cultured in the 30°C treatment was attributable to an increase in the number of embryos produced by each responding anther.

DISCUSSION

The results presented above show first that preincubation at high temperature (30°C) is significantly better than at low temperature (6°C) in terms of both the number of anthers responding to culture and, perhaps more importantly, the yield of plants per anther. Secondly, it is clear that far from improving yields, the use of low temperatures depressed yields when compared with buds cultured fresh. It would appear that in terms of pretreatment effects, the response of *S. tuberosum* is similar to its solanaceous relatives *S. melongena* (Dumas de Vaulx *et al*. 1981) and *C. annuum* (Dumas de Vaulx & Chambonnet 1982) which have been shown to respond well to high temperature preincubation.

REFERENCES

Cappadocia, M., Cheng, D.S.K. & Ludlum-Simonette, R. (1984). Plant regeneration from *in vitro* culture of anthers of *Solanum chacoense*. Bitt. and interspecific hybrids *S. tuberosum* x *S. chacoense* Bitt. Theor. Appl. Genet., 69, 139-43.

Dumas de Vaulx, R., Chambonnet, D. & Pochard, E. (1981). Culture *in vitro* d'antheres de piment (*Capsicum annuum* L.): amelioration des toux d'obtention de plantes chez differents genotypes par des traitements a + 35°C. Agronomie, 1, 859-64.

Dumas de Vaulx, R. & Chambonnet, D. (1982). Culture *in vitro* d'antheres d' aubergine (*Solanum melongena* L.): stimulation de la production de plantes au moyen de traitements a + 35°C associes a de faibles teneurs en substances de croissance. Agronomie, 2, 983-8.

Uhrig, H. (1983). Breeding for *Globodera pallida* resistance in potatoes. Z. Pflanzenzuchtg. 91, 211-8.

Wenzel, G., Bapat, V.A. & Uhrig, H. (1983). New strategy to tackle breeding problems of potato. *In* Plant Cell Culture in Crop Improvement, ed. S.K. Sen & K.L. Giles, pp. 337-49. New York: Plenum.

GENETIC MANIPULATION IN POTATO USING *AGROBACTERIUM*

G. Ooms

Improved potato varieties are obtained by bringing together desirable combinations of genes. Generally this is achieved by sexual hybridization followed by selection of plants with the desired characters. New techniques are now in existence that enable specific genes to be inserted into all the cells of a particular plant, leaving the residing somatic genome largely intact. In principle, such foreign genes can originate from any organism. They may be naturally existing genes, genetically engineered derivatives or may even consist of newly designed and chemically and/or biochemically synthesized DNA.

In this chapter three aspects of introducing specific genes into potato are discussed. Firstly the transformation of potato cells by direct infection of wounded shoot cultures of cultivars with *Agrobacterium*, the acquired growth properties of such transformed cells and the isolation of transformed plants. Secondly, the initial analysis of the expression of genes introduced into potato and finally implications for future genetic manipulations in potato, based on the known effects of genes already introduced.

Possible alternative ways of obtaining potato plants transformed with specific foreign DNA are omitted or only mentioned in passing. This is not to express a preference for direct *Agrobacterium* infections but simply because experimental progress in potato using other systems is less advanced. Other ways of transforming plant cells in tobacco using *Agrobacterium* include direct infections of wounded plant organs such as punched out leaf discs (Horsch *et al.* 1985), co-cultivation of *Agrobacterium* with cell-wall regenerating protoplasts (Marton *et al.* 1979), extended co-cultivation (transformation of small cell colonies with *Agrobacterium*; Muller *et al.* 1984 and Pollock *et al.* 1985) and fusion of plant protoplasts with *Agrobacterium* protoplasts, also called spheroplasts (Okada *et al.* 1985). Progress has also been made in transforming tobacco cells without using *Agrobacterium* by direct introduction of purified DNA into protoplasts (for review see Potrykus *et al.* 1985) and by fusion of protoplasts with DNA containing liposomes (Deshayes *et al.* 1985). The possibility of using viruses as vectors for the transfer of foreign genes into plants has been demonstrated by Brisson *et al.* (1984). Further prospects for altering the genetic information in potato by protoplast fusion are discussed by Jones (this volume).

NATURAL AGROBACTERIUM TRANSFORMATION

At the beginning of this century Smith and Townsend (1907) identified the bacterium *Agrobacterium tumefaciens* as the cause of plant tumour development (crown gall) on a wide range of dicotyledonous plants (Figure 1). The bacterium induces tumours only upon infection of plant wounds and early recognition processes are involved since effective attachment of the bacterium to the plant cell is required for tumour induction (Douglas *et al.* 1982). Nowadays much more is known about the molecular mechanism underlying crown gall formation. For comprehensive reviews and extensive lists of references the reader is referred to Nester *et al.* (1984), Hooykaas & Schilperoort (1983), Schell *et al.* (1984) and Binns (1984).

Virulent *Agrobacterium* contains, in addition to the chromosomal DNA, a large autonomously replicating DNA molecule called a plasmid. This so called tumour-inducing or Ti-plasmid is essential for virulence and is normally present in 1-2 copies per bacterium. It is generally 100-150 M

Figure 1. Tumours induced by infection with *Agrobacterium tumefaciens* strain T37. Wounded stems of *Solanum brevidens* (left), *S. tuberosum* cv. King Edward (middle), and monohaploid *S. tuberosum* line 7322 (right) were inoculated and grown *in vitro*. Approximately 6 weeks after infection, stem segments with tumours developing from the sites of infection were excised and photographed. Note the different responses of the three genotypes.

Dalton (150-225 kilo base pairs or kbp) in size, depending on the particular wild-type isolate. This equals approximately 3% of the total bacterial chromosome and corresponds to an estimated 150-200 bacterial genes of average size.

Recent evidence suggests that wounding of plant cells causes release of molecular factors of which one or several appear to induce expression of certain Ti-plasmid genes (Okker *et al*. 1985). These genes are probably located on a large section of the Ti-plasmid (encompassing at least six separate loci) which has been called the virulence or *Vir* region. *Vir* gene products play an important, albeit as yet little understood, role in the transfer of another distinct part of the Ti-plasmid into the plant cell soon after infection. This second segment of transferred DNA or T-DNA becomes covalently integrated into the nuclear genome of the plant. For the efficient transfer of T-DNA there is a further requirement for at least one of several short DNA fragments (Wang *et al*. 1984). These fragments are 25 bp long and they border both sides of the T-DNA region as unique direct imperfect repeats. It can be speculated that the repeats serve as recognition targets for *Vir* coded proteins involved in T-DNA transfer (Koukolikova-Nicola *et al*. 1985).

The genes located on the T-DNA are important in determining the characteristic properties of tumour cells. Natural *Agrobacterium* isolates may differ considerably in the number (from approximately five to 13) and functions of the genes that are located between the T-DNA borders. Furthermore, on a particular Ti-plasmid multiple T-DNA regions may exist but the T-DNA genes themselves have no effect on T-DNA transfer.

T-DNA in plant cells

The exact structure of T-DNA in the genome of a transformed plant cell may be diverse. It may be a simple copy of the T-DNA region(s) in the bacterium. Sometimes it is longer than the normal T-DNA and sometimes shorter. Multiple T-DNA copies may be present in tandem but not necessarily of the same length. So far no evidence exists for preferred sites or regions for T-DNA integration within the plant genome. Possibly natural T-DNA may even become integrated outside the nucleus, for example in chloroplast DNA (De Block *et al*. 1985).

T-DNA genes themselves are expressed in plant cells which means that they can be regarded as eukaryotic genes despite their immediate prokaryotic origin. The function of all T-DNA genes has not yet been elucidated but for some there are indications, and for others there is firm evidence, of what they do. Broadly speaking, four classes of T-DNA genes can currently be identified: (i) genes that cause biosynthesis of low molecular weight nitrogenous compounds, generally classified as "opines"; (ii) genes involved in active secretion of opines from transformed cells; (iii) T-DNA genes that cause changes in plant cell growth properties (plant oncogenes). These are best characterized by those that enhance cytokinin and auxin biosynthesis; (iv) unclassified genes that may belong to a previous category.

Opines

The finesse of transformation of plant cells by Agrobacterium is highlighted by the role played by opines. Opines are normally not found in plant tissues and they comprise a number of different compounds. The first opine identified was lysopine (N^2-D-1-carboxyethyl-L-lysine), a condensation product of the common α-keto acid pyruvate with the amino acid L-lysine (Biemann et al. 1960). Of a similar nature were the subsequently discovered opines, octopine (arginine plus pyruvate), octopinic acid (pyruvate plus ornithine) and histopine (histidine plus pyruvate). The biosynthesis in tumour cells of all these compounds, "the octopine family", is catalysed by a single enzyme, lysopine dehydrogenase or octopine synthase, which is a T-DNA coded monomeric enzyme of 358 amino acids and has a corresponding molecular weight of 38.733 D.

Members of the octopine family represent only a few of a range of opines coded for by different T-DNA genes located on plasmids from a variety of wild-type Agrobacterium isolates, but only the role of octopine will be discussed here.

After its biosynthesis in tumours, octopine is actively transported out of the transformed cells by another T-DNA coded protein, probably a permease (Messens et al. 1985). It is likely that the excreted octopine in natural tumours subsequently serves as a carbon, nitrogen and energy source specifically for tumour inducing agrobacteria. This is inferred from the presence of genes on the Ti-plasmid, located outside the T-DNA region that enable Agrobacterium to actively take up octopine from its surroundings and subsequently to degrade it back into arginine and pyruvate. Octopine also induces transmission of Ti-plasmids, by conjugation, from virulent Agrobacterium strains into avirulent, plasmid-free, Agrobacterium strains (Kerr et al. 1977). Thus, the plasmid can be regarded as a subcellular organism (Novick, 1980) that uses both plants and bacteria to create itself a niche for survival and multiplication.

The practical role for opines in current plant transformations is as an easily assayable marker for transformed cells. Furthermore, the flanking regions of opine synthase genes have been used in chimeric gene constructions to achieve expression in plants of other genes of bacterial origin, such as genes for hygromycin or amino glycoside-phosphotransferases, which give plant cells enhanced resistance to the antibiotics hygromycin B and kanamycin respectively (Waldron et al. 1985; Herera-Estrella et al. 1983).

T-DNA oncogenes

The particular function of the T-DNA cytokinin and auxin genes (a confusing variety of names exist for the particular loci) is important for our understanding of how crown galls develop. The T-DNA cytokinin gene codes for dimethylallylpyrophosphate, AMP transferase or cytokinin synthase (Akiyoshi et al. 1984) which enhances the first step in cytokinin biosynthesis. Two T-DNA genes code for enzymes of auxin biosynthesis. Almost certainly they convert tryptophan via indole-3-acetamide (probably by a mono oxygenase) to indole-3-acetic acid (by an amido hydrolase)(Schroder et al. 1984). In agreement with this, enhanced cytokinin and auxin levels have been detected in tobacco tumours (Akiyoshi et al. 1983) and these are thought to be the cause of cancerous growth. Because of the high endogenous hormone concentrations in tumour cells, cross feeding to neighbouring untransformed cells takes place. Tumours on a plant are, therefore, probably often mixtures of different populations, not only of transformed cells originating from independent transformations but also of untransformed cells (Sacristan & Melchers 1977; Ooms et al. 1982; Van Slogteren et al. 1983). It is unclear what, if anything, is important in controlling the ratio of the various kinds of cells in a tumour but it is very likely that considerable variation exists in the relative proportion of the untransformed cells to transformed cells in a crown gall.

The isolation of Agrobacterium mutants in which either a T-DNA auxin or cytokinin gene is inactivated, (by insertional inactivation through mutagenesis by a transposon or an IS element or by deletion through genetic engineering and microbial genetics) showed that such mutants caused tumours with different morphologies, particularly in tobacco (Ooms et al. 1981; Garfinkel et al. 1981). Inactivation of auxin genes gave tumours with shoots whilst inactivation of the cytokinin gene gave tumours with roots. These observations paralleled previous tissue culture experiments on exogenous hormone application with untransformed tobacco tissues. The addition of balanced amounts of auxins and cytokinins to the culture media caused undifferentiated (callus) growth, relatively high cytokinins in the medium caused shoot formation and relatively high auxin concentrations caused root formation (Skoog & Miller 1957). Therefore, instead of exogenous application of hormones, the T-DNA genes have brought about similar changes in differentiation of tissues by their endogenous application through genetic manipulation.

Tobacco transformed with Ti T-DNA

The shoots that originate from tobacco tumours are either transformed and contain Ti T-DNA (normally with the intact T-DNA cytokinin gene) or they are untransformed, in which case they are likely to have originated from the untransformed cells in the galls. Normally such transformed tobacco shoots have altered growth characteristics such as no root formation. Via grafting, however, they have been grown to maturity and have flowered and set seed from which plants with the same transformed characteristics have been grown (Wullems *et al.* 1981). These results showed that T-DNA genes and in particular the T-DNA cytokinin gene, caused specific changes in growth and development at the whole (transformed) plant level.

Potato transformed with Ti T-DNA

Because potato plants form tubers, it was of interest to study the changes the Ti T-DNA genes would bring about in the developmental process of tuber formation. In order to study this, similar Ti T-DNA transformed potato plants had to be isolated. By infecting *in vitro* grown potato plants with a shoot inducing *A. tumefaciens* strain and incubating for several months, it was possible to observe the development of tumours from which Ti T-DNA transformed shoots could be recovered (Ooms *et al.* 1983). Like the transformed tobacco plants described earlier the transformed shoots did not form roots but, via grafting, they developed into mature plants (Ooms & Lenton 1985). In the transformed potato plants the T-DNA cytokinin gene was expressed at low levels (on average only one or a few RNA molecules per cell (Burrell *et al.* 1985)). This caused approximately an eight-fold higher concentration of endogenous cytokinin (Ooms & Lenton 1985) which was considered directly responsible for changes in growth and development. Axillary buds developed into side-shoots and the transformed plants had a strong tendency to tuberize. The transformed tubers did not appear significantly different in morphology from normal tubers. After vernalization they sprouted and sufficient root growth occurred for the plant to grow to maturity and form second generation tubers (Ooms & Lenton 1985).

Vector development

The regeneration of potato plants transformed with Ti T-DNA is interesting in its own right. It has demonstrated the principle that transformed derivatives of potato cultivars may be reproducibly isolated and that specific changes in potato growth and development can be achieved. It is, however, important to isolate transformed plants without obvious phenotypic changes. Such plants would open the way to the introduction of other desirable genes without any unwanted "side effects".

Because *Agrobacterium* has an established record of efficiently and reproducibly delivering DNA into plant cells, several research groups set out to develop an *Agrobacterium* Ti-plasmid based vector system for plant transformation that would enable regeneration of phenotypically transformed normal plants. The general requirements were (i) maintenance of the natural DNA transfer efficiency (ii) removal of T-DNA onco genes and (iii) ease of manipulation for introduction of other genes into the plasmid vector. Furthermore, it would be preferable to have (iv) a selectable marker such as

Figure 2. Shoot cultures of potato line Mb1501B (left) and cv. Maris Bard (right). Line Mb1501B is a derivative of Maris Bard transformed with Ti T-DNA from the shoot-inducing *A. tumefaciens* strain LBA1501.

resistance to an antibiotic which would help selection for transformed cells and (v) a screenable marker, such as opine production, for early and easy confirmation that selected cells are indeed transformed. The aims of vector development have been achieved in various ways, as described below.

A brief summary of the current status of vector development inevitably does injustice to the subtle differences in the various constructions made, and to remaining questions on whether or not minor differences exist in their way of transforming plant cells. This is discussed in more detail elsewhere (Shaw 1984) and here only a few highlights are mentioned. Broadly speaking, two types of vectors have been constructed and both have been used successfully to obtain phenotypically normal plants transformed with other genes. This has mainly been carried out in *Nicotiana* and *Petunia* (De Block *et al*. 1984; Horsch *et al*. 1984). In one type of vector most of the DNA between the 25bp T-DNA borders on the Ti-plasmid was removed. The remaining T-DNA or newly inserted, unrelated, DNA functioned as an (artificial) target site for the introduction of other genes through homologous recombination (Horsch *et al*. 1984; De Blaere *et al*. 1985). The other genes were initially cloned in *E. coli* in a shuttle vector, a small plasmid also containing the target DNA. The shuttle plasmid plus its cloned gene was then transferred by conjugation into the *Agrobacterium* strain with the modified Ti-plasmid, where, by a homologous recombination event, the desired gene was incorporated into the modified T-DNA region. As a result, there was now a more or less straight forward replacement of new genes for the T-DNA oncogenes of the wild-type *Agrobacterium*. Using this technique, efficiency of plant cell transformation was apparently identical to that in wild-type *Agrobacterium*. In another type of vector the T-DNA region with its 25bp border sequences was cloned in a separate plasmid with a broad host range allowing the plasmid to replicate both in *E. coli* and in *Agrobacterium* (Hoekema *et al*. 1983; Bevan 1984). It was shown that an *Agrobacterium* strain with this broad host range plasmid plus a second modified Ti-plasmid, deleted for the entire T-DNA region (including the 25bp borders) but still containing the *Vir* region, was still able to transfer the T-DNA region from the first plasmid into plant cells. This approach enables cloning experiments on the broad host range plasmid (or so-called binary plasmid) *E. coli*, which is technically easier than in *Agrobacterium*. Once introduced into *Agrobacterium*, the cloned gene is immediately ready for plant transformation.

Potato plants transformed with nononcogenic T-DNA

As mentioned earlier both untransformed and transformed cells are present in natural crown galls. Infection of wounded potato plants *in vitro* with a mixed culture of a shoot-inducing *Agrobacterium* strain and a nononcogenic strain, manipulated for its T-DNA, should therefore give tumours with at least three kinds of cells: cells transformed with shoot inducing T-DNA, cells with T-DNA from the non-oncogenic *Agrobacterium* and untransformed cells. Furthermore, since both untransformed and high-cytokinin transformed plants have been recovered from potato tumours it was thought likely that if mixed infections were carried out, at least some of the cells in the developing tumour would spontaneously grow into plants transformed with the engineered (nononcogenic) T-DNA.

From such mixed infection experiments we have recovered a number of derivatives of potato cultivars Maris Bard and Desiree that produced opines and were transformed with nononcogenic T-DNA (unpublished). Further analysis showed that at least one transformed but morphologically and cytologically normal derivative of a potato cultivar has been isolated. The analysis showed however that most of the regenerated plants were morphologically different and aneuploid. There is therefore a further requirement to improve not only the efficiency of regenerating transformed plants but also to identify ways of minimizing the generation of additional changes in the existing genome. Such changes are known to be inherent in the process of plant regeneration from somatic cells of potato cultivars (Shepard *et al.* 1980; Shepard 1982; Karp *et al.* 1982; Sree Ramulu *et al.* 1983), but the extent to which this "somaclonal variation" (Larkin & Scowcroft 1981) occurs can be influenced by genotype and culture conditions (for review see Karp & Bright 1985).

EXPRESSION OF INTRODUCED GENES

From the perspective of potato breeding, the whole point of genetic manipulation is to introduce genes, preferably into cultivars or advanced breeding lines, that will be expressed and give desired changes in phenotypic characters. However, it is also easily recognized that taking a gene, any gene, and introducing it anywhere in the potato genome of a single cell and regenerating a transformed plant from this cell poses a range of questions as to how the introduced gene will ultimately be expressed in the different cells of the plant.

Whole organism transformations in animal cells with a variety of genes have shown that expression of many but not of all the introduced genes varied depending on the transformation event (see for mini reviews Palmiter & Brinster 1985; Maniatis 1985). On the other hand, from such analyses and also from similar experiments on gene expression with transformed tobacco plants, we know that tissue specificity and response to environmental stimuli such as light, are confined to the immediate flanking sequences, particularly those "upstream" of the structural coding region of the gene (Herrera-Estrella et al. 1984; Sengupta-Gopalan et al. 1985). By transferring sufficient of the flanking DNA together with the natural gene or the DNA of a marker gene into another plant, the introduced gene will be expressed in a more or less predictable manner.

In potato we have sought to extend our experience in whole plant transformation and to address further questions on gene expression by introducing the natural T-DNA from the bacterium Agrobacterium rhizogenes. This is a plant pathogen (Riker et al. 1930) related to A. tumefaciens and it also transforms plant cells with DNA from a large residing plasmid, an Ri or root-inducing plasmid. Ri T-DNA transformed cells grow as roots (hairy roots) and plants have been regenerated from cultures established from single roots (Tepfer 1984 and for potato; Ooms et al. 1985, 1986). As

Figure 3. Potato cv. Desiree (left) and Ri T-DNA transformed derivative D9X8a (right) in soil. The plantlets were grown from tissue culture shoots. Note that D9X8a established itself under these conditions faster than Desiree (from Ooms et al. 1985).

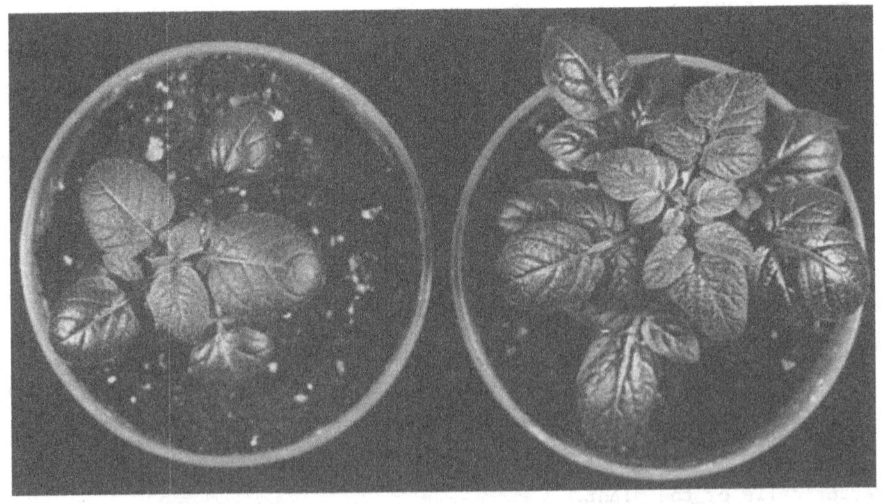

expected, the transformed potato plants have a number of changes in growth and development. Their leaves are crinkled, their roots have reduced geotropism and branch more frequently and their tubers are longer and have more pronounced eyes. These changes are T-DNA determined and not due to somaclonal variation. From the point of view of gene expression, the interest is that Ri T-DNA contains no less than nine closely linked genes with different, although largely unknown, molecular functions. Certain aspects of their expression can be analysed simultaneously and relatively easily. Recent experiments have shown that in different organs (leaf/stem, root, tuber) many of the genes directed synthesis of different concentrations of their corresponding RNAs (as judged by Northern blot analysis), suggesting some degree of organ-determined control of their expression (unpublished). Variation in expression in independently transformed plants remains to be investigated.

Ri T-DNA transformed potato plants have also contributed to questions on stability of expression during propagation of transformed

Figure 4. Field harvested tubers from Desiree (left) and its Ri T-DNA transformed derivative D9X8a (right). Note the characteristic and consistent change in tuber shape and the lower yield of D9X8a tubers (from Ooms *et al.* 1986).

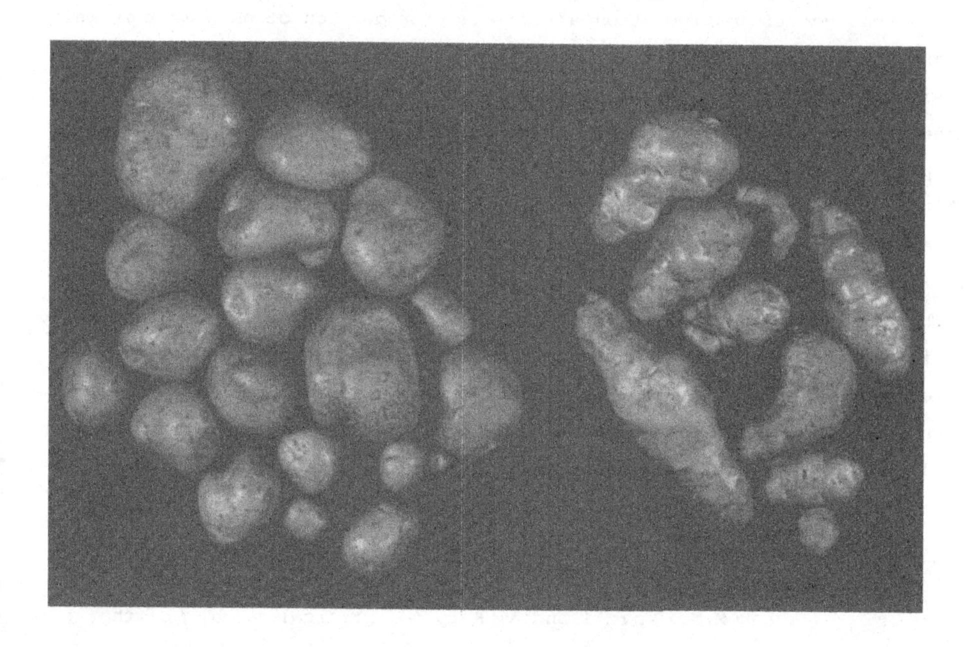

plants. The plants have been grown in a culture room and in the field over several tuber generations (Ooms et al. 1986). Invariably the plants showed their characteristic changes in habit, and even the shape of every single, newly developing tuber was characteristically different from that of untransformed tubers. This demonstrated a high degree of preservation of the changes induced by the transforming DNA in the potato plant and in tubers, at least in this particular case.

IMPLICATIONS OF T-DNA DETERMINED PHENOTYPIC CHANGES

The changes in potato development caused by Ti and Ri T-DNA described have implications extending beyond transformation and gene expression; they exemplify the research perspective for studying the molecular basis of specific changes in complex biological processes such as growth and development. Note that the observed effects are caused by specific major genes. As such, T-DNA genes themselves may well be among the genes that play a further role in this. However, many other naturally occurring genes are being, and will continue to be, sought after for such purposes. Examples include genes coding for various kinds of regulatory proteins, those that play a role in plant pathogen interactions or beneficial plant-microbe interactions, and genes involved in resistance to herbicides. The concept of genetic manipulation allows the introduction of an almost infinite number of molecular changes in the nucleotide sequence of such genes (for review, see Botstein & Shortle 1985). A corresponding range of modified genes can be introduced into plants and in doing so, in principle, a range of plants with slight differences in particular characteristics can be obtained. Of course, the questions on somaclonal variation, position effects of site(s) of integration and differences in structure and copy number of introduced DNA are important in this context. It is obvious that the current technology provides scope for more possibilities than actually can be handled. It is therefore inevitable that major challenges lie ahead in choosing those experiments that are the most informative, in continuing to search for techniques that enable early evaluation of results and in attempting the isolation of the most interesting genes for practical and scientific application of the genetic manipulation approach.

ACKNOWLEDGEMENTS

I thank M.M. Burrell and A. Karp for critical reading of the manuscript, members of the Photography Dept. at Rothamsted for preparation of the photographs and J.E. Hutchins for typing the manuscript.

REFERENCES

Akiyoshi, D.E., Morris, R.O., Hinz, R., Mischke, B.S., Kosuge, T., Garfinkel, D.J., Gordon, M.P. & Nester, E.W. (1983). Cytokinin/auxin balance in crown gall tumors is regulated by specific loci in the T-DNA. Proc. Natl. Acad. Sci. USA, 80, 407-11.

Akiyoshi, D.E., Klee, H., Amasino, R.M., Nester, E.W. & Gordon, M. . (1984). T-DNA of Agrobacterium tumefaciens encodes an enzyme of cytokinin biosynthesis. Proc. Natl. Acad. Sci. USA, 81, 5994-8.

Bevan, M. (1984). Binary Agrobacterium vectors for plant transformation. Nucl. Acid Res. 12, 8711-8721.

Biemann, K., Lioret, C., Asselimeau, K., Lederer, E. & Polonski, J. (1960). Sur la structure chimique de la lysopine, nouvel acide amine isole de tissu de crown-gall. Bull. Soc. Chim. Biol., 42, 979-91.

Binns, A.N. (1984). The biology and molecular biology of plant cells infected by Agrobacterium tumefaciens In Oxford Surveys of Plant Molecular and Cell Biology, 1, ed. B.J. Miflin pp. 133-61. Oxford: Oxford University Press.

Botstein, D. & Shortle, D. (1985). Strategies and applications of in vitro mutagenesis. Science, 229, 1193-201.

Brisson, N., Paszkowski, J., Penswick, J.R., Gronenborn, B., Potrykus, I. & Hohn, T. (1984). Expression of a bacterial gene in plants using a viral vector. Nature, 310, 511-14.

Burrell, M.M., Twell, D., Karp, A. & Ooms, G. (1985). Expression of Ti T-DNA in differentiated tissues of potato (Solanum tuberosum) cv. Maris Bard. Plant Mol. Biol., 5, 213-22.

De Blaere, R., Bytebier, B., de Greve, H., De Boeck, F., Schell, J., Van Montagu & Leemans, J. (1985). Efficient octopine Ti plasmid-derived vectors for Agrobacterium mediated gene transfer to plants. Nucl. Acid Res., 13, 4777-88.

De Block, M., Herrera-Estrella, L., Van Montagu, M., Schell, J. & Zambryski, P. (1984). Expression of foreign genes in regenerated plants and their progeny. EMBO J., 3, 1681-9.

De Block, M., Schell, J. & Van Montagu, M. (1985). Chloroplast transformation by Agrobacterium tumefaciens EMBO J., 4, 1367-72.

Deshayes, A., Herrera-Estrella & Caboche, M. (1985). Liposome-mediated transformation of tobacco mesophyll protoplasts by an Escherichia coli plasmid. EMBO J. 4, 2731-9.

Douglas, C.J., Halperin, W. & Nester, E.W. (1982). Agrobacterium tumefaciens mutants affected in attachment to plant cells. J. Bacteriol., 152, 1265-75.

Garfinkel, D.J., Simpson, R.B., Ream, L.W., White, F.F., Gordon, M.P. & Nester, E.W. (1981). Genetic analysis of crown gall: fine structure map of the T-DNA by site directed mutagenesis. Cell, 27, 143-54.

Herrera-Estrella, L., de Block, M., Messens, E., Hernalsteens, J.-P., Van Montagu, M. & Schell, J. (1983). Chimeric genes as dominant selectable markers in plant cells. EMBO J., 2, 987-95.

Herrera-Estrella, L., Van den Broeck, G., Maenhaut, R., Van Montagu, M., Schell, J., Timko, M. & Cashmore, A. (1984). Light-inducible and chloroplast-associated expression of a chimaeric gene introduced into Nicotiana tabacum using a Ti plasmid vector. Nature, 310, 115-20.

Hoekema, A., Van Haaren, M.J.J., Fellinger, A.J., Hooykaas, P.J.J. & Schilperoort, R.A. (1985). Non-oncogenic plant vectors for use in the Agrobacterium binary system. Plant Mol. Biol., 5, 86-9.

Hooykaas, P.J.J. & Schilperoort, R.A. (1983). The molecular genetics of crown gall tumorigenesis. Adv. Genet., 22, 210-83.

Horsch, R.B., Fraley, R.T., Rogers, S.G., Sanders, P.R., Lloyd, A. & Hoffman, N. (1984). Inheritance of functional foreign genes in plants. Science, 223, 496-8.

Horsch, R.B., Fry, S.E., Hoffman, N.L., Eichholtz, D., Rogers, S.G. & Fraley, R.T. (1985). A simple and general method for transferring genes into plants. Science, 277, 1229-31.

Karp, A., Nelson, R.S., Thomas, E. & Bright, S.W.J. (1982). Chromosome variation in protoplast-derived potato plants. Theor. Appl. Genet., 63, 265-72.

Karp, A. & Bright, S.W.J. (1985). On the causes and origins of somaclonal variation. In Oxford surveys of Plant Molecular and Cell Biology, 1, ed. B.J. Miflin, pp. 199-234. Oxford: Oxford University Press.

Kerr, A., Marigault, P. & Tempe, J. (1977). Transfer of virulence in vivo and in vitro in Agrobacterium. Nature, 265, 560-1.

Koukolikova-Nicola, Z., Shillito, R.D., Hohn, B., Wang, K., Van Montagu, M. & Zambryski, P. (1985). Involvement of circular intermediates in the transfer of T-DNA from Agrobacterium tumefaciens to plant cells. Nature, 313, 191-6.

Larkin, P.J. & Scowcroft, W.R. (1981). Somaclonal variation - a novel source of variability from cell cultures for plant improvement. Theor. Appl. Genet., 60, 197-214.

Maniatis, T. (1985). Targetting in mammalian cells. Nature, 317, 205-6.

Marton, L., Wullems, G.J., Molendijk, L. & Schilperoort, R.A. (1979). In vitro transformation of cultured cells from Nicotiana tabacum by Agrobacterium tumefaciens. Nature, 277, 129-31.

Messens, E., Lenaerts, A., Van Montagu, M. & Hedges, R.W. (1985). Genetic basis for opine secretion from crown gall tumor cells. Mol. Gen. Genet., 199, 344-8.

Muller, A., Manzora, T. & Lurquin, P.F. (1984). Crown gall transformation of tobacco callus by co cultivation with Agrobacterium tumefaciens. Biochim. Biophys. Res. Commun., 123, 458-62.

Nester, E.W., Gordon, M.P., Amasino, R.M. & Yanofski, M.F. (1984). Crown Gall - a molecular and physiological analysis. Ann. Rev. Plant Physiol., 35, 387-413.

Novick, R. (1980). Plasmids. Sci. Am., 243, 76-90.

Okada, K., Hasazawa, S., Syono, K. & Nagata, T. (1985). Further evidence for the transformation of Vinca rosea protoplasts by Agrobacterium tumefaciens spheroplasts. Plant Cell Rep., 4, 133-6.

Okker, R.J.H., Spaink, H., Hille, J., Van Brussel, T.A.N., Lugtenburg, B. & Schilperoort, R.A. (1984). Plant inducible virulence promotor of the Agrobacterium tumefaciens Ti-plasmid. Nature, 312, 564-6.

Ooms, G., Hooykaas, P.J.J., Moolenaar, G. & Schilperoort, R.A. (1981). Crown gall plant tumors of abnormal morphology, induced by Agrobacterium tumefaciens carrying mutated octopine Ti plasmids: analysis of T-DNA functions. Gene, 14, 33-50.

Ooms, G., Bakker, A., Molendijk, L., Wullems, G.J., Gordon, M.P., Nester, E.W. & Schilperoort, R.A. (1982). T-DNA organization in homogeneous and heterogeneous octopine type crown gall tissues of Nicotiana tabacum. Cell, 30, 589-97.

Ooms, G., Karp, A & Roberts, J. (1983). From tumour to tuber: tumour cell characteristics and chromosome numbers of crown gall-derived tetraploid potato plants (Solanum tuberosum cv. Maris Bard). Theor. Appl. Genet., 66, 169-72.

Ooms, G., Karp, A., Burrell, M.M., Twell, D. & Roberts, J. (1985). Genetic modification of potato development using Ri T-DNA. Theor. Appl. Genet., 70, 440-6.

Ooms, G. & Lenton, J.R. (1985). T-DNA genes to study plant development: precocious tuberisation and enhanced cytokinins in A. tumefaciens transformed potato. Plant Mol. Biol., 5, 205-12.

Ooms, G., Bossen, M.E., Burrell, M.M. & Karp, A. (1986). Genetic manipulation in potato using Agrobacterium rhizogenes. Potato Research. (In Press).

Palmiter, R.D. & Brinster, R.L. (1985). Transgenic mice. Cell, 41, 343-5.

Pollock, K., Barnfield, D.G., Robinson, S.J. & Shields, R. (1985). Transformation of protoplast-derived cell colonies and suspension cultures by Agrobacterium tumefaciens. Plant Cell Reports, 4, 202-5.

Potrykus, I., Shillito, R.D., Saul, M.W. & Paszkowski, J. (1985). Direct gene transfer: state of the art and future potential. Plant Mol. Reporter, 3, 117-28.

Riker, A.J., Barnfield, W.M., Wright, W.H., Keight, G.W. & Sagen, H.E. (1930). Studies on infectious hairy-root of nursery apple trees. J. Agric. Res., 41, 507-40.

Sacristan, M.D. & Melchers, G. (1977). Regeneration of plants from "habituated" and "Agrobacterium transformed" single cell clones of tobacco. Mol. Gen. Genet., 152, 111-7.

Schell, J., Herrera-Estrella, L., Zambryski, P., De Block, M., Joos, H., Willmitszer, L., Eckes, P., Rosalil, S. & Van Montagu, M. (1984). Genetic engineering of plants. In The Impact of Gene Transfer Techniques in Eukaryotic Cell Biology, ed. J.S. Schell & P. Starlinger, pp. 73-90. Berlin, Heidelberg, New York: Springer.

Schroder, G., Waffenschmidt, S., Weiler, E.W. & Schroder, J. (1984). The T-region of Ti-plasmids codes for an enzyme synthesizing indole-3-acetic acid. Eur. J. Biochem., 138, 3817.

Sengupta-Gopalan, C., Reichert, N.A., Barker, R.F., Hall, T.C. & Kemp, J.D. (1985). Developmentally regulated expression of the bean beta phaseolin gene in tobacco (Nicotiana tabacum cultivar Xanthi) seed. Proc. Natl. Acad. Sci. USA, 82, 3320-4.

Shaw, C.H. (1984). Ti-plasmid derived plant gene vectors. In Oxford Surveys of Plant Molecular and Cell Biology, 1, ed. B.J. Miflin, pp. 211-6. Oxford: Oxford University Press.

Shepard, J.F., Bidney, D. & Shahin, E. (1980). Potato protoplasts in crop improvement. Science, 208, 17-24.

Shepard, J.F. (1982). The regeneration of potato plants from leaf-cell protoplasts. Sci. Am., 246, 112-21.

Skoog, F. & Miller, C.O. (1957). Chemical regulation of growth and organ formation in plant tissues cultured in vitro. Symp. Soc. Exp. Biol., 11, 118-31.

Smith, E.F. & Townsend, C.O. (1907). A plant-tumor of bacterial origin. Science, 25, 671-3.

Van Slogteren, G.M.S., Hoge, J.H.C., Hooykaas, P.J.J. & Schilperoort, R.A. (1983). Clonal analysis of heterogeneous crown gall tumor tissues induced by wild-type and shooter mutant strains of Agrobacterium tumefaciens - expression of T-DNA genes. Plant Mol. Biol., 2, 321-33.

Sree Ramulu, K., Kijkhuis, P. & Roest, S. (1983). Phenotypic variation and ploidy level of plants regenerated from protoplasts of tetraploid potato, Solanum tuberosum L. cv. Bintje. Theor. Appl. Genet., 65, 329-38.

Tepfer, D. (1984). Transformation of several species of higher plants by Agrobacterium rhizogenes - sexual transmission of the transformed genotype and phenotype. Cell, 37, 959-67.

Waldron, C., Murphy, E.B., Roberts, J.L., Gustafson, G.D., Armour, S.L. & Malcolm, S.K. (1985). Resistance to hygromycin B: a new marker for plant transformation studies. Plant Mol. Biol., 5, 103-8.

Wang, K., Herrera-Estrella, L., Van Montagu, M. & Zambryski, P. (1984). Right 25 base pair terminus sequence of the nopaline T-DNA is essential for and determines direction of DNA transfer from Agrobacterium to the plant genome. Cell, 38, 455-62.

Wullems, G.J., Molendijk, L., Ooms, G. & Schilperoort, R.A. (1981). Retention of tumor markers in F1 progeny plants from in vitro induced octopine and nopaline tumor tissues. Cell, 24, 719-27.

PROSPECTS OF USING TUMOUR-INDUCING PLASMID-MEDIATED GENE TRANSFER FOR THE IMPROVEMENT OF POTATO VARIETIES

A. Blau, P. Eckes, J. Logemann, S. Rosahl, J. Sanchez-Serrano, R. Schmidt, J. Schell and L. Willmitzer

INTRODUCTION

Most dicotyledonous plants are susceptible to tumour formation as a result of infection of wounded sites by the Gram-negative bacterium *Agrobacterium tumefaciens* (for a general review see Caplan *et al*. 1983). The bacterium is necessary for tumour induction but not for tumour maintenance and growth. The ability of the bacteria to induce tumour formation is strictly limited to those Agrobacteria species harbouring a large extra-chromosomal DNA element, the so-called Ti-(tumour-inducing) plasmid, which has a size of 120-180 kb. The molecular basis underlying the neoplastic transformation is the transfer and stable integration of a well-defined part of the Ti-plasmid, the so-called T-DNA (T=Transfer or Tumour) into the plant nuclear DNA, leading to the formation of a gall. The Ti-plasmid therefore represents a natural gene vector to plant cells. As it is essential for the practical use of this system for gene transfer to obtain normal plants; mutants of the Ti-plasmid have been constructed with all the T-DNA genes leading to tumour formation deleted. Normal plants which are fertile and which sexually transmit the introduced genes as a single, dominant Mendelian locus (Zambryski *et al*. 1983) have been obtained.

The main limitation at the present time concerning the use of Ti-plasmid mediated gene transfer is the limited host range of *Agrobacterium* species. Thus whereas nearly all dicotyledonous plants are susceptible to *Agrobacterium* infection, monocotyledonous plants as a rule (despite certain exceptions reported recently, Hernalsteens *et al*. 1984; Hooykaas *et al*. 1984), and especially Gramineae, do not seem to be susceptible to *Agrobacterium* infection.

Of the ten most important crop plants, potato has a unique position in that it is on the one hand susceptible to *Agrobacterium tumefaciens* mediated gene transfer, and on the other hand its tissue culture system is well developed.

Here we describe experiments aimed at isolating genes from potato which are only expressed in a specific organ. The regulatory sequences of these genes should be of value in controlling the expression of foreign genes which might be beneficial to the potato plant.

EXPERIMENTAL

Isolation of organ-specific cDNA clones

Genes expressed only in leaves or tubers were isolated from potatoes. cDNA libraries were prepared by standard methods using poly-A+ RNA from leaves or tubers. About 5000 clones from each library were subsequently screened to identify organ-specific clones in a colony hybridization experiment using ^{32}P-labelled cDNA probes from various organs, i.e. roots, tubers, leaves and stems. Clones hybridizing only to the cDNAs from the homologous organ were further characterized.

Northern and dot-blot analysis demonstrates organ-specific expression of a number of cDNA clones

To verify the organ-specific expression of the cDNA clones obtained by the differential screening described above, the inserts were isolated from a number of leaf-specific cDNA clones (pcL600, pcL700 and pcL900) as well as from some tuber-specific clones (i.e. pcT700, pcT800 and pcT1500) and hybridized to RNAs of different organs isolated from field or glasshouse-grown potato plants. These experiments showed that clone pcL900 hybridizes strongly to a RNA of about 900 nucleotides which is predominantly present in leaves. A weak hybridization was also seen with stem A+ RNA whereas no hybridization was visible with RNA from either roots or tubers. A similar picture was seen for pcL600 and pcL700; a strong hybridization to RNA from leaves with about 600 and 700 nucleotides respectively, no hybridization to RNA from either root or tuber and a weaker though obvious hybridization with RNA from stem. With the different tuber cDNAs, in all cases hybridization was visible only with RNA from tubers, no signal was seen with RNA from leaves, roots or stems. Semi-quantitative experiments performed using dot- blot assays showed that the level of RNAs present in the homologous organs (i.e. leaf and tuber respectively) was at least several hundred-fold higher than the levels detected in heterologous organs.

Organ-specific expression is controlled at both the transcriptional and the post-transcriptional level. Organ-specific gene expression can result from transcriptional as well as post-transcriptional control. In order to distinguish between these possibilities nuclei were isolated from different organs of the potato plant under conditions retaining the transcriptional activity. After isolation and purification the nuclei were incubated to enable transcription to proceed in the presence of ^{32}P-labelled RNA precursors. After pulse treatment for 30 minutes, ^{32}P-labelled newly synthesized RNA was isolated and probed against single-stranded M13 clones of the different cDNAs immobilized on nitrocellulose filters. These experiments showed that in the case of leaf-specific cDNA clones, pcL600, pcL700 and pcL900, hybridization was seen with pulse-labelled RNA from leaf cell nuclei, whereas no signal was seen with pulse-labelled RNA from root or tuber cell nuclei. A reciprocal picture was obtained with the tuber-specific clones pcT700 and pcT1500 (i.e. transcriptional activity could be detected only in nuclei from tubers). These results indicate that the developmental (organ-) specific expression of these five clones is mainly (if not exclusively) controlled at the level of transcription.

A different picture emerges for the tuber-specific clone pcT800. Hybridization to this cDNA was seen not only with pulse-labelled RNA from tuber cell nuclei (the homologous organ) but also with pulse-labelled RNA from leaf cell nuclei. This, therefore, raises the possibility that the organ-specificity of the steady-state RNA homologous to pcT800 is due mainly to post-transcriptional processes involving either differential processing and/or stability. Run-off experiments performed in the presence of different concentrations of alpha-amanitin support the assumption that transcription of all six cDNA clones is performed by RNA polymerase II.

Proteins encoded by the different cDNA clones

Hybrid selection and *in vitro* translation experiments, as well as nucleotide sequence data, have identified the product of pcL900 as a small subunit of the ribulose-1,5-bisphosphate-carboxylase, that of pcT800 as the proteinase inhibitor II of potato tubers and that of pcT1500 as the major protein of potato tubers with a molecular weight of 40 K dalton, called "patatin". The products of pcL600, pcL700 and pcT700 have not yet been identified.

Analysis of genomic clones homologous to a leaf-specific (pcL700) or a tuber-specific clone (pcT1500).

A genomic library of a haploid potato line (HH 5793) was established using the lambda vector EMBL 4 and screened for genomic clones hybridizing to the different cDNA clones by plaque hybridization. Genomic clones corresponding to pcL700 and pcT1500 were used for further characterization.

Characterization of a genomic clone homologous to pcL700

The structure of the genomic clone pgL700 (Figure 1) was deduced from a comparison of the nucleotide sequences of cDNA and genomic DNA as well as by S1-type mapping experiments. pgL700 contains 4 introns with lengths varying from 96 to 838 nucleotides. The largest exons with 248 and 186 nucleotides are found at the 3' and 5'-termini respectively; the three internal exons are all rather small since they consist of less than 60 nucleotides each. About 20 nucleotides in front of the polyadenylation site the sequence GATAAA is found which probably functions as the polyadenylation signal. All intron-exon boundaries obey the GT-AG rule. pgL700 contains an open reading frame of 414 nucleotides bordered by a 32 bp long untranslated 5' leader sequence and a 167 bp long 3'-untranslated region. The amino acid sequence as deduced from nucleotide sequence data, indicates that the protein contains a fairly high amount of hydrophobic amino acids (70%).

cDNA cloning has indicated that the mRNA encoded by pgL700 is abundant (14 out of 5000 cDNA clones screened were homologous to pgL700). In order to see whether the cloned 4.5 kb genomic fragment contains all the necessary information for a high level expression, pgL700 was ear-marked by inserting a 470 bp DNA fragment derived from the coding region of the T-DNA gene 2 of the octopine-type plasmid pTiAch 5 (Gielen et al. 1984). This DNA fragment was inserted into the last exon as indicated in Figure 1 and transferred back to potato using Agrobacterium-derived vectors. If this gene is actively expressed in potato it should give rise to a chimeric RNA, 470

Figure 1. Structure of pgL700 (from *Solanum tuberosum*). Sizes and locations of the five exons (boxed) and the four intervening sequences (lined) are shown. To distinguish the activity of the endogenous pgL700 from the reintroduced pgL700, the last exon was modified by insertion of a 470 bp fragment of the T-DNA.

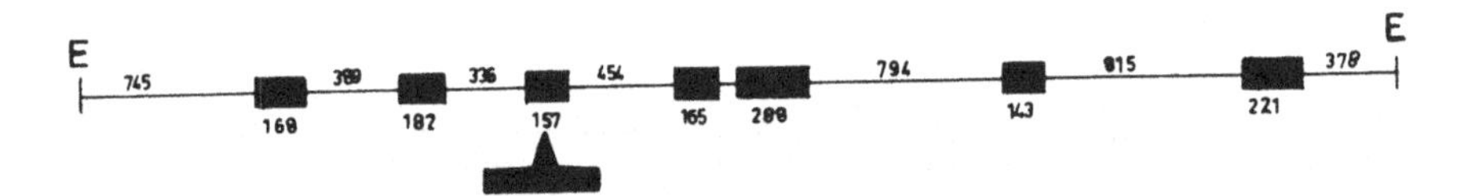

nucleotides longer than its normal counterpart. In addition, this RNA should specifically hybridize to the 470 bp fragment.

It was found that the reintroduced gene was transcribed giving rise to an RNA species of about 1200 nucleotides. Furthermore the amount of RNA made was approximately the same as the amount made from the endogenous gene. We consider this as evidence that the 4.5 kb genomic fragment pgL700 contains all the signals necessary for a high level of expression.

Characterization of a genomic clone (pgpat 1) homologous to pcT1500 (encoding patatin)

Figure 2 shows the structure of a genomic clone homologous to the patatin encoding cDNA clone pcT 1500. This structure was derived from a comparison of nucleotide sequences of different cDNA clones and pgpat 1. The gene is composed of six introns and seven exons of varying length. The gene contains an open reading frame coding for a protein of 386 amino acids. The exon-intron boundaries again obey the classical GT-AG rule. An AATAA sequence is located 20 nucleotides upstream of the polyadenylation site.

pgpat 1 was ear-marked by exon modification (cf. Figure 2) and was introduced into potato using Agrobacterium-derived vectors. The analysis of the activity of this modified gene however has to await the formation of tubers.

CONCLUSION

cDNA and genomic clones have been identified and characterized for two genes of potato which show an organ-specific expression. The further functional analysis of these genes by modification and Ti-plasmid mediated gene transfer into both potato and tobacco plants should lead to a deeper understanding of the mechanisms leading to developmentally controlled

Figure 2. Structure of pgT1500. The figure shows size and location of the seven exons (boxed) and six introns (line) of the patatin genomic clone pgT1500. The insertion of the 470 bp fragment into a HindIII site in the third exon is indicated.

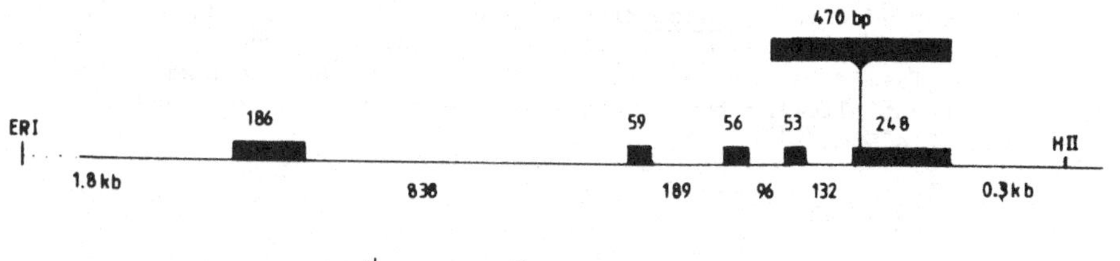

expression of plant genes. Our preliminary results indicate that 5' upstream sequences of these genes play an essential role in regulating the expression of these genes.

These genes and the corresponding regulatory regions might be of use to improve potato. Two examples of potential interest will be discussed in more detail.

Quality improvement

The protein content of potato tubers is about 2% of the total fresh weight. Forty per cent of the total protein is patatin. It is therefore conceivable to think of either increasing the protein content by increasing the gene dosage via the introduction of more genes using the *Agrobacterium tumefaciens* Ti-plasmid system and/or modifying the amino acid content by changing the coding region of the gene.

Unconventional resistance to pathogens

Genes, the products of which are toxic to plant pathogens, could be introduced into potato under the control of plant (potato) regulatory sequences. These "toxin" genes could be isolated from a wide variety of organisms, i.e. bacteria, plants, fungi, etc. Examples of potential candidates for such toxin genes are the endotoxin genes of different strains of *Bacillus thuringiensis*.

REFERENCES

Caplan, A., Herrera-Estrella, L., Inze, D., Van Haute, E., Van Montagu, M., Schell, J. & Zambryski, P. (1983). Introduction of genetic material into plant cells. Science, 222, 815-21.

Gielen, J., De Beuckeleer, M., Surinck, J., Deboeck, F., De Greve, H., Lemmers, M., Van Montagu, M. & Schell, J. (1984). The complete nucleotide sequence of the TL-DNA of the *Agrobacterium tumefaciens* plasmid pTiAch 5. EMBO J., 3, 835-46.

Hernalsteens, J.P., Thia-Toong, L., Schell, J. & Van Montagu, M. (1984). An Agrobacterium-transformed cell culture from the monocot Asparagus officinalis. EMBO J., 3, 3039-41.

Hooykaas-Van Slogteren, G.M.S., Hooykaas, P. & Schilperoort, R. (1984). Expression of Ti-plasmid genes in monocotyledonous plants infected with *Agrobacterium tumefaciens*. Nature, 311, 763-4.

Zambryski, P., Joos, H., Genetello, C., Van Montagu, M. & Schell, J. (1983). Ti-plasmid vector for the introduction of DNA into plant cells without altering their normal regeneration capacity. EMBO J., 2, 2143-50.

USE OF PROTOPLAST FUSION AND SOMACLONAL VARIATION IN POTATO BREEDING

M.G.K. Jones

INTRODUCTION

In common with applications of tissue culture to other crops, there is a fundamental divide between culture systems of potato in which the organization of meristems is maintained, and those where cultured cells pass through a disorganized callus phase. In tissue explants where meristem organization is maintained, and from which whole plants can be regenerated, such regenerated plants are normally uniform and true-to-type. Thus techniques such as micropropagation, virus elimination through meristem culture, embryo rescue, ploidy manipulation and germplasm storage, which employ this approach, are widely used by breeders as practical adjuncts to plant breeding programmes. However, when potato cells pass through a callus phase, the plants subsequently regenerated may well exhibit differences from the original genotype. Variation so produced has been termed somaclonal variation, and its nature needs to be outlined and taken account of in any manipulations that involve production of callus tissue.

In this article, manipulations of potato that involve callus culture are described, in particular techniques of protoplast fusion and somaclonal variation. The aim of work described here is to induce and exploit useful genetic changes in potato, and to evaluate the practical potential of these techniques in plant breeding.

PROTOPLAST CELL CULTURE AND CONSEQUENCES OF A CALLUS PHASE

Regeneration from explants

It is now well established that whole plants of potato can be regenerated from excised tissue segments such as leaf, stem or tuber. Although plants can be regenerated after culture on a single medium (Roest & Bokelmann 1976), our experience is that a simple two-stage procedure is applicable to a broad range of cultivars (Webb *et al.* 1983; Wheeler *et al.* 1985; Karp *et al.* 1985a). Surface-sterilized leaf discs are cultured on a medium containing auxin and cytokinin for about 2 weeks. During this time cell division is initiated at cut surfaces, particularly around leaf veins. The discs are then transferred to the second medium containing cytokinin and gibberellic acid. The discs develop further callus which becomes more nodular. Within the nodular regions meristems develop and shoots emerge by organogenesis after 3-4 weeks, their elongation being promoted by the gibberellic acid. Although there is some genotype dependence, such that some cultivars or lines may require modification to the standard hormonal regime, in most cases large numbers of shoots (>50) can be regenerated from a single disc. The shoots are then excised, rooted, transferred to soil in a propagator for about 1 week, and can be grown on to maturity.

Regeneration from protoplasts

The regeneration of plants from isolated wall-less cells, protoplasts, is considerably more involved than regeneration from explants. The genotype response is more marked, and more care is required both in growth of starting material so that the cells are in the correct physiological state for subsequent culture and division, and in specific media requirements for continued growth in culture and plant regeneration (Shepard & Totten 1977; Jones *et al.* 1983; Nelson *et al.* 1983; Haberlach *et al.* 1985). Protoplasts can be isolated from potato leaves of whole plants grown under carefully controlled conditions, or more conveniently from leaves or lines maintained as sterile shoot cultures. The steps involved - protoplast isolation, culture at an appropriate density, transfer to media that yield individual colonies, then induced differentiation and shoot emergence, have been outlined in a number of publications (Shepard & Totten 1977; Nelson *et al.* 1983; Sree Ramulu *et al.* 1983; Haberlach *et al.* 1985). A study of three important British cultivars (Maris Piper, Desiree and King Edward), illustrates some of the factors to be considered (Foulger & Jones 1986). Cv. Maris Piper can be regenerated relatively well by a range of methods (Gunn & Shepard 1981; Foulger & Jones 1986), but the same methods did not work for cvs. Desiree and King Edward. Improvements in protoplast yield, stability, division frequency, and subsequent culture and regeneration were obtained by altering the media used for shoot culture, increasing Ca^{2+} levels during protoplast isolation, using more complex additives (containing casein hydrolysate and coconut water), and modified shoot regeneration media. At present, it is possible to obtain colonies (c.1 month) from about 10% of isolated protoplasts, and 70-80% of these colonies will regenerate shoots. A point worth emphasising is the fact that the demonstration in principle that plants can be regenerated from a particular genotype cannot necessarily be equated with an ability to use that genotype in routine experiments, without further optimization of culture conditions.

The list of potato cultivars and wild species that can be regenerated from protoplasts is now extensive. It includes many European and American cultivars, advanced breeding lines, dihaploids derived from *Solanum tuberosum*, and wild species such as *S. brevidens*, *S. dulcamara*, *S. phureja*, *S. chacoense*, *S. nigrum*, *S. demissum*, *S. etuberosum* and *S. pennellii* (Jones 1985; Karp *et al.* 1985b; Haberlach *et al.* 1985, and references cited therein). For protoplast fusion, it is necessary to have reasonably efficient protoplast isolation and regeneration procedures for one, and preferably both, genotypes used.

Cytology of regenerated plants

In a series of studies at Rothamsted, cytological changes that occur in regenerated potato plants have been documented by Karp and coworkers (Creissen & Karp 1985; Karp *et al.* 1982, 1985a,b; Fish & Karp 1986). In summary, a significant proportion of the overall variation observed in regenerants can be ascribed to aneuploidy. (This aspect of variation is considered separately from somaclonal variation, which is discussed later). The extent of aneuploidy is greater in protoplast-derived plants than those derived from tissue explants, and aneuploidy is more pronounced in plants regenerated from the tetraploid level than plants regenerated from lower ploidy levels. In the latter instances there is a greater tendency to an increase in ploidy level. Thus plants regenerated from tetraploid explants are 90-95% euploid, with aneuploids having 48 ± 1-2 chromosomes. Of plants regenerated from monohaploid ($2n=x=12$) explants, the majority double in chromosome number with isolated regenerants remaining monohaploid or doubling again to tetraploid. From dihaploid explants about 40% of regenerants remain at the original ploidy level, 60% double to the tetraploid level, and few are are aneuploid (Karp *et al.* 1985a). The situation regarding protoplast-derived plants obtained from tetraploid leaves is more complicated, and there is evidence that culture conditions can affect the degree of aneuploidy in regenerants. For example, early work with cv. Maris Bard indicated only 5% euploid regenerants, with the majority of plants apparently having doubled to the octoploid level, then lost some chromosomes (Karp *et al.* 1982). However, recent work with Maris Bard, using four different early culture conditions (different hormone, salt and sugar components), has yielded plants that are 60-70% euploids, about 8% aneuploids around the tetraploid level, and about 27% aneuploids at the octoploid level (Fish & Karp 1986). Protoplast-derived regenerants from other tetraploid cultivars have yielded a similar pattern of euploidy and aneuploidy (Creissen & Karp 1985). From their general appearance in shoot culture, and tendency to delayed rooting of shoots, the majority of aneuploids can in fact be recognized and discarded for practical applications.

In addition to aneuploidy, a number of other structural changes to chromosomes have been observed, involving translocations, deletions and amplifications (Creissen & Karp 1985; Fish & Karp 1986).

PROTOPLAST FUSION

The isolation of protoplasts and their ability to regenerate whole plants renders them amenable to fusion, and this provides an alternative method for production of hybrid plants. Possible applications include the opportunity to form hybrids that cannot be produced sexually, the asexual combination of complete or partial genomes and the possibility of transferring cytoplasmically encoded traits. At present, molecular events or genes underlying useful characters, such as disease resistance, are completely uncharacterized. Protoplast fusion offers a means of transferring such traits without a requirement for any detailed molecular knowledge, as long as the characters can be identified in some way.

Fusion technique

Both chemical and electrical techniques are now available to fuse protoplasts together. Chemical procedures, using polyethylene glycol (PEG) to agglutinate protoplasts followed by treatment with high pH/Ca^{2+} ions, have been used successfully for some years with various modifications (e.g. Evans 1983). More recently, a physical technique using electrical pulses has become available (Zimmermann 1982). This involves alignment of protoplasts in a solution of low conductivity (e.g. 9% mannitol) between electrodes by applying a high frequency alternating field (1 MHz). Protoplasts act like dipoles, and move by 'dielectrophoresis' to form chains. Once in contact, fusion is induced by applying a brief (50 μsec) direct current pulse of sufficient amplitude (1-3 KV cm^{-1}) to cause transient, reversible, membrane breakdown and pore formation. Neighbouring protoplast plasma membranes begin to fuse, and heterokaryons are then obtained. The procedure can be controlled to direct fusion between different protoplast populations, and can be carried out on a preparative scale (Tempelaar & Jones 1985a,b).

Heterokaryon selection

Probably the most difficult aspect of fusion is selection of fused heterokaryons from nonfused or self-fused parental protoplasts. Much effort has gone into selection early after fusion but, in fact, in most successful experiments resulting in hybrid plants, identification of hybrid material has been made at a later stage, such as on regenerated shoots or whole plants (based on phenotype or isozymes). Early selection methods include mechanical isolation of heterokaryons 1-3 days after fusion, complementation selection of different mutant lines or different culture lines are available but the possibility now exists of adding selectable traits by transformation (Ooms, this volume), that can be used for selection of hybrid material after fusion. Identification of hybrid material can now also employ molecular techniques, such as species or cultivar-specific DNA probes, that will allow rapid screening of large numbers of regenerated colonies to pick out those with DNA from both partners (Saul & Potrykus 1984).

Whatever method is employed to identify hybrids, the aim is to regenerate whole plants, as from single protoplasts. However, after fusion, the frequency of heterokaryons that divide and can subsequently regenerate plants is reduced.

PRACTICAL APPLICATIONS AND RESULTS

For potato, the potential of protoplast fusion is in (1) combining selected dihaploids, or dihaploids and diploid species, to produce tetraploid hybrids, (2) the transfer of useful traits from a donor to correct identified defects in a tetraploid receptor cultivar and (3) transfer of cytoplasmically encoded traits, such as herbicide resistance, by movement of chloroplasts or mitochondria into selected lines. It is worth emphasizing that, even if a sexual cross can be achieved between two species, either by using a bridging species or by other breeding techniques, hybridization by fusion can yield different combinations of genetic material, and is therefore still worth attempting for such cases. The range of genotypes that can be combined may also be extended by protoplast fusion.

Fusion at the dihaploid level

The potential application of dihaploid potato lines in breeding is well established (Hermesen & Ramanna 1981; Hermundstand & Peloquin, this volume) because of simplified genetics and improved crossability with diploid wild species. The combination of selected dihaploid potato lines exhibiting, for example, different disease resistance traits can be achieved by fusion to yield tetraploids. However, it is necessary to establish that tetraploid regenerants produced in this way are hybrids, and not derived from spontaneous chromosome doubling during culture. This has been achieved using Aec (aminoethyl cysteine) and 5MT (5-methyl tryptophan) resistant dihaploid lines following electrofusion of protoplasts, with the demonstration of hybrid callus (but without plant regeneration)(de Vries *et al*. 1985). Dihaploid lines to be fused may also be characterized by isozyme or DNA restriction fragment length polymorphism analyses, or by any suitable phenotypic markers. The report by Austin *et al*. (1985b) of the successful synthesis of tetraploid plants from fusion of dihaploid potato protoplasts, shows that this approach is practical. The tetraploid fusion progeny they obtained showed increased vigour, and in general other characters were intermediate, although pigmentation was enhanced in tubers and anthers. There are a number of groups working in this area and many more successful reports can be expected.

Fusion with other species

Hybrid plants produced by fusion of potato protoplasts with six different Solanaceous species have been reported. Melchers *et al*. (1978) fused protoplasts of a yellow-green mutant of tomato ($2n=2x=24$) with protoplasts from suspension cultures of dihaploid potato (nominally $2n=2x=24$). Hybrid shoots were selected using culture differences and morphology. Analysis of hybrids indicated the nuclear components from both parents were present, but that chloroplasts originated from only one parent (Melchers *et al*. 1978; Schiller *et al*. 1982). Shepard *et al*. (1983) also obtained hybrids after fusing tetraploid potato somaclone protoplasts with those from two tomato cultivars. Butenko & Kuchko (1980) recovered a hybrid plant from fusion of tetraploid potato and *S. chacoense* protoplasts.

One combination that has received particular attention is between *S. tuberosum* and *S. brevidens*. The latter species is reported to exhibit some cold tolerance and resistance to potato leafroll virus (PLRV), but along with other members of the *Etuberosa*, which are often sexually incompatible, can only be crossed with difficulty to *S. tuberosum* (Johnston & Hanneman 1982). Barsby *et al.* (1984) fused protoplasts of *S. brevidens* and an albino clone of cv. Russet Burbank and isolated 32 hybrid plants that displayed morphological features intermediate between the two parents. Of six plants analysed in detail, two had the additive number of 72 chromosomes. The hybrids contained the nuclear encoded RuBPCase small subunit polypeptides from both parents, but chloroplasts from *S. brevidens*. Austin *et al.* (1985a) produced tetraploid somatic hybrids following fusion of protoplasts of *S. brevidens* with a diploid tuber-bearing line from *S. phureja*. Hybrids were similarly intermediate in morphology between the parents. They demonstrated transfer of resistance to PLRV, and the hybrids produced fertile pollen and eggs. Field experiments with hybrid plants have been carried out. At Rothamsted, hybrid plants have also been obtained by fusing protoplasts of tetraploid potato with suspension-culture protoplasts of *S. brevidens* (D. Foulger, N. Fish, unpublished). As with the work of Austin *et al.* (1985a), hybrid plants differed from each other, and some showed vigorous growth. In this case the hybrid plants were probably aneuploid, since the chromosome complement of the *S. brevidens* suspension-culture protoplasts was probably not 24 at the time of fusion.

Cytoplasmic traits

Transfer of chloroplasts and mitochondria has already been successfully achieved in rape (*Brassica napus*) by protoplast fusion (Pelletier *et al.* 1983). Similar work on potato has been limited to an attempt to transfer atrazine (herbicide) resistant chloroplasts from *S. nigrum* into potato. Somewhat perversely, the result was the transfer of atrazine-sensitive chloroplasts from potato to *S. nigrum* (Binding *et al.* 1982).

Partial genome transfer

Work aimed at transfer of a limited number of characters from irradiated donor protoplasts to tetraploid potato receptor protoplasts has also been limited. Attempts to transfer a well-defined trait - resistance to the amino-acid analogue aminoethyl cysteine (Aec^R) - from *Nicotiana sylvestris* into potato have been made (Foulger *et al.* 1985). In this case *N. sylvestris* was very vigorous in culture, and even after irradiation, heterokaryons formed by fusion yielded essentially *N. sylvestris* shoots, apparently due to repair mechanisms operating in the heterokaryon. Similarly, from heterokaryons isolated without prior irradiation, the N. sylvestris eliminated the potato chromosomes. In this case, the choice of N. sylvestris as a fusion partner was unlucky, and the same problem is not expected in current experiments with different S. tuberosum cultivars as donors and acceptors.

General comments on fusion

The transfer of nuclear-encoded genetic information by protoplast fusion cannot provide the same degree of precision as may be obtained in the future by genetic engineering techniques. However, it has the potential for total combination or partial transfer of characters with unknown or complex genetic bases. Asexual transfer of complex traits such as disease resistance is, at least in the short term, more likely to be achieved using protoplast fusion than by genetic engineering using recombinant DNA technology, as has been demonstrated by Austin et al. (1985a). Fertility of hybrids is important to allow introgression of useful introduced characters. The synthesis of tetraploids from selected dihaploids by fusion also promises to be of immediate practical benefit. Because in general non-selective methods of gene transfer are used, and cytological changes can occur as a result of culture manipulations, for applied results it will be necessary to produce many hybrid plants and search through them for the desired combination of traits, screening out aneuploid plants at an early stage. With improved understanding of the basis of culture-induced variation, it should be possible to reduce the number of aberrant types and to make sensible choices of parental lines to fuse.

SOMACLONAL VARIATION

Somaclonal variation in potato has been described at length in a number of publications (e.g. Shepard et al. 1980; Karp et al. 1985b; Jones 1985; Karp & Bright 1985) and space here limits a detailed discussion. Comments will be confined to our experiences at Rothamsted (D. Foulger, S.W.J. Bright) and Dept. of Agriculture of Northern Ireland (N. Evans), and discussion of whether culture-induced variability is of use to plant breeders.

That plants regenerated from tissues following a callus phase can differ from parental tissues is clearly established. Some phenotypic variation, usually the most pronounced, is a result of aneuploidy and this has already been discussed. Nevertheless, variant plants can still be obtained which have apparently normal cytology (Wheeler et al. 1985). These are of more interest, and are referred to here as somaclonal variants. Although there has been no direct comparison of variants produced from protoplasts and those produced from explants, it appears that the same range of variants can be generated. Therefore, because fewer abnormal plants are produced from explants, it is recommended that this technically simpler technique is adopted for those interested in practical assessment of somaclonal variation.

Assessment of variation

Clones of cv. Desiree, regenerated from explants, have been analysed in field experiments over three tuber generations for tuber characteristics (Bright *et al.* 1985; Evans *et al.* 1986). Tubers were assessed for general impression, shape, eye depth, skin colour, common scab (*Streptomyces* spp.) and flesh colour (on respective scales of 1-5) and tuber number and weight. Specific gravity, dry matter and resistance to *Globodera pallida* were also assessed. In all cases the range of observed values for the regenerant population was much wider than that of the control groups, and statistically significant (1% confidence level) differences were evident between control and regenerant populations. Clones with significantly higher or lower scores were observed for most of the characters.

There was a general decrease in yield, with no clones significantly greater than the control. However, Evans *et al.* (1986) found an interesting change in size distribution resulting in an increase in ware yield in 45 clones (out of 78 clones examined in detail), as a result of a decrease in oversize tubers (>80 mm diameter) and an increase in the 40-60 mm fraction.

The incidence of common scab was also reduced in specific regenerants compared with controls in each of the 3 test years. However, no clones resistant to *G. pallida* were found in regenerants.

As an overall measure of variation of more than 30 characters scored in field experiments, principal component analysis indicated that at least half of the regenerants differed from the parental genotype in some way (Bright, pers. comm.; Jones *et al.* 1984).

General comments on somaclonal variation

It is thus possible to obtain stable changes in useful agronomic characters of potato by somaclonal variation. There are, as yet, a number of unanswered questions concerning its basis, stability of particular characters, and the range of alterations that can be obtained. "Hot spots" of change (e.g. red to white pigmentation) can be observed, and also traits where no change has been found. These may reflect the underlying genetics of the traits. Somaclonal variation has provoked some emotion, but a sensible approach is to keep an open mind and to use the technique if it is appropriate. A conventional breeding programme takes 10-12 years to produce a variety. It is still early days for newer techniques of genetic manipulation. It would appear that somaclonal variation does offer a method of up-grading established cultivars, and although it is too early to make a firm pronouncement, the signs are that new, useful varieties will emerge from somaclonal programmes.

CONCLUSION

Close contact needs to be maintained between laboratory-based scientists and breeders, to develop and evaluate techniques such as protoplast fusion, somaclonal variation and transformation. Breeders can then decide how much effort to put into such techniques relative to conventional approaches. There will of course be no substitute for field testing new lines, however they may be generated.

REFERENCES

Austin, S., Baer, A. & Helgeson, J.P. (1985a). Transfer of resistance to potato leaf roll virus from *Solanum brevidens* into *Solanum tuberosum* by somatic fusion. Plant Science, 39, 75-82.

Austin, S., Baer, A., Ehlenfeldt, M., Kazmierczak, P.J. & Helgeson, J.P. (1985b). Intraspecific fusions of *Solanum tuberosum*. Theor. Appl. Biol., in press.

Barsby, T.L., Shepard, J.F., Kemble, R.J. & Wong, R. (1984). Somatic hybridization in the genus *Solanum*: *S. tuberosum* and *S. brevidens*. Plant Cell Reports, 3, 165-7.

Binding, H., Jain, S.M., Finger, J., Mordhurst, G., Nehls, R. & Gressel, J. (1982). Somatic hybridization of atrazine resistant biotype of *Solanum nigrum* with *Solanum tuberosum*. Part 1: Clonal variation in morphology and in atrazine sensitivity. Theor. Appl. Genet., 63, 273-7.

Bright, S.W.J., Ooms, G., Foulger, D. & Karp, A. (1985). Mutation and tissue culture. *In* Plant Tissue Culture and Its Agricultural Applications, ed. L.A. Withers & P.G. Alderson. London: Butterworths, in press.

Butenko, R.G. & Kuchko, A.A. (1980). Somatic hybridization of *Solanum tuberosum* L. and *Solanum chacoense* Bitt. by protoplast fusion. Soviet Pl. Physiol. 24, 541-6.

Creissen, G.P. & Karp, A. (1985). Karyotypic changes in potato plants regenerated from protoplasts. Plant Cell Tissue and Organ Culture, 4, 171-82.

Evans, D.A. (1983). Protoplast fusion. *In* Handbook of Plant Cell Culture Vol. 1: Techniques for Propagation and Breeding, eds. D.A. Evans *et al.*, pp. 291-321. New York: Macmillan.

Evans, N.E., Foulger, D., Farrer, L. & Bright, S.W.J. (1986). Somaclonal variation in explant-derived potato clones over three tuber generations. In press.

Fish, N. & Karp, A. (1986). Improvements in regeneration from protoplasts of potato cv. Maris Bard and studies on chromosome stability. 1. The effect of initial culture media. Theor. & Appl. Genet., in press.

Foulger, D. & Jones, M.G.K. (1986). Improved efficiency of regeneration from protoplasts of important potato cultivars. Plant Cell Reports (in press).

Foulger, D., Fish, N., Bright, S.W.J. & Jones, M.G.K. (1985). Studies on the transfer of genetic information between tobacco and potato by protoplast fusion. In Genetic Engineering of Plants and Microorganisms Important for Agriculture, eds. E. Magnien & D. de Nettancourt, pp. 154-5. Dordrecht: Nijhoff/Junk.

Gunn, R.E. & Shepard, J.F. (1981). Regeneration of plants from mesophyll-derived protoplasts of British potato (Solanum tuberosum L.) cultivars. Plant Science Letters, 22, 97-101.

Haberlach, G.T., Cohen, B.A., Reichert, N.A., Baer, M.A., Towill, L.E. & Helgeson, J.P. (1985). Isolation, culture and regeneration of protoplasts from potato and several related Solanum species. Plant Science, 39, 67-74.

Hermesen, J.G.Th. & Ramanna, M.S. (1981). Haploidy and plant breeding. Phil. Trans. Roy. Soc. London B, 292, 499-507.

Johnston, S.A. & Hanneman, R.E. (1982). Manipulations of endosperm balance number overcome crossing barriers between diploid Solanum species. Science, 217, 446-8.

Jones, M.G.K. (1985). Protoplast and somaclonal research on potato. In Innovative Methods for Propagating Potatoes (Report of XXVIII Planning Conference), ed. O.T. Page, pp. 305-18. Lima: International Potato Center.

Jones, M.G.K., Bright, S.W.J., Nelson, R.S., Foulger, D., Creissen, G.P., Karp, A. & Ooms, G. (1983). Variation in plants regenerated from protoplasts and complex explants of potato. In Proceedings of 6th International Protoplast Symposium, Experientia, 45 (Suppl), 150-1.

Jones, M.G.K., Maddock, S.E., Karp, A., Nelson, R.S., Creisson, G.P., Foulger, D. & Bright, S.W.J. (1984). Tissue and protoplast culture - a novel way to new crop varieties. In The World Biotech Report 1984, Volume 1: Europe, pp. 443-54. Pinner: Online.

Karp, A. & Bright, S.W.J. (1985). On the causes and origins of somaclonal variation. In Oxford Surveys of Plant Molecular & Cell Biology, 2, ed. B.J. Miflin, pp. 199-234. Oxford: Oxford University Press.

Karp, A., Nelson, R.S., Thomas, E. & Bright, S.W.J. (1982). Chromosome variation in protoplast-derived potato plants. Theor. & Appl. Genet., 63, 265-72.

Karp, A., Risiott, R., Jones, M.G.K. & Bright, S.W.J. (1985a). Chromosome doubling in monohaploid and dihaploid potatoes by regeneration from cultured leaf explants. Plant Cell Tissue & Organ Culture, 3, 363-73.

Karp, A., Jones, M.G.K., Ooms, G. & Bright, S.W.J. (1985b). Potato protoplasts and tissue culture. In Plant Breeding Progress Reviews, ed. G.E. Russell, in press.

Melchers, G., Sacristan, M.D. & Holder, A.A. (1978). Somatic hybrid plants of potato and tomato regenerated from fused protoplasts. Carlsberg Res. Commun., 43, 203-18.

Nelson, R.S., Creissen, G.P. & Bright, S.W.J. (1983). Plant regeneration from protoplasts of Solanum brevidens. Plant Sci. Letts., 30, 355-67.

Pelletier, G., Primard, C. & Vedel, F. (1983). Intergeneric cytoplasm hybridization in Cruciferae by protoplast fusion. Proceedings of the 6th International Protoplast Symposium, Basel, Experientia (Suppl), 45, 286-7.

Roest, S. & Bokelmann, G.S. (1976). Vegetative propagation of Solanum tuberosum L. in vitro. Potato Res. 19, 173-8.

Saul, M.W. & Potrykus, I. (1984). Species-specific repetitive DNA used to identify interspecific somatic hybrids. Plant Cell Reports, 3, 65-7.

Schiller, B., Herrmann, R.G. & Melchers, G. (1982). Restriction endonuclease analysis of plastid DNA from tomato, potato and some of their somatic hybrids. Mol. & Gen. Genet., 186, 453-9.

Shepard, J.F. & Totten, R.E. (1977). Mesophyll cell protoplasts of potato: isolation, proliferation and plant regeneration. Plant Physiol., 60, 313-6.

Shepard, J.F., Bidney, D. & Shahin, E. (1980). Potato protoplasts in crop improvement. Science, 208, 17-24.

Shepard, J.F., Bidney, D., Barsby, T. & Kemble, R. (1983). Genetic transfer in plants through interspecific protoplast fusion. Science, 219, 683-8.

Sree Ramulu, K., Dijkhuis, P. & Roest, S. (1983). Phenotypic variation and ploidy level of plants regenerated from protoplasts of tetraploid potato (Solanum tuberosum L. cv. 'Bintje'). Theor. & Appl. Genet., 65, 329-38.

Tempelaar, M.J. & Jones, M.G.K. (1985a). Fusion characteristics of plant protoplasts in electric fields. Planta, 165, 205-16.

Tempelaar, M.J. & Jones, M.G.K. (1985b). Directed electrofusion between protoplasts with different responses in a mass fusion system. Plant Cell Reports, 4, 92-5.

de Vries, S.E., Jacobsen, E., Jones, M.G.K. & Tempelaar, M.J. (1985). Somatic cell genetics of potato III - electrofusion of two amino-acid analogue resistance cell-lines. Proceedings Eucarpia Symposium, Berlin, p.90.

Webb, J.K., Osifo, E.O. & Henshaw, G.G. (1983). Shoot regeneration from leaflet discs of six cultivars of potato (Solanum tuberosum subsp. tuberosum). Plant Sci. Letts., 30, 1-8.

Wheeler, V.A., Evans, N.E., Foulger, D., Webb, K.J., Karp, A., Franklin, J. & Bright, S.W.J. (1985). Shoot formation from explant cultures of fourteen potato cultivars and studies of the cytology and morphology of the regenerated plants. Ann. Bot., in press.

Zimmermann, U. (1982). Electric field-mediated fusion and related electrical phenomena. Biochim. Biophys. Acta, 694, 227-77.

THE POTENTIAL VALUE OF SOMACLONAL VARIANTS IN POTATO IMPROVEMENT

A.J. Thomson

INTRODUCTION

The regeneration of potato plants from somatic tissues such as leaf mesophyll protoplasts is now well established (Gunn 1982). Examination of the somaclones raised from these tissues has revealed variation for a range of characters, some such as resistance to common scab (*Streptomyces* spp.) being of economic importance in potato breeding (Gunn *et al.* 1985). However it has been demonstrated that the variation occurs only for a limited number of characters and in many respects the somaclones are identical to the parental variety from which they were derived (Secor & Shepard 1981). This suggests that it may be possible to upgrade existing varieties by selecting amongst somaclones for improvements in those characters in which the parental variety performs poorly while maintaining the same level of character expression for those characters in which the parental variety performs well (Thomson *et al.* 1986). Some recent data obtained to explore this possibility are reported here.

MATERIALS AND METHODS

The data were recorded from field trials of somaclones derived from either Feltwell or Maris Piper which are varieties bred at the Plant Breeding Institute, Cambridge. The trials comprised either two or three replicates of six-tuber plots grown in 1984 and 1985. Total harvested yield and general tuber appearance were recorded. In 1985 the produce from those somaclones selected from the 1984 trials was riddled to obtain the different size grades of the tubers. All the somaclones trialled so far have been derived from protoplasts but over 400 somaclones derived from leaf discs of Maris Piper will enter trials for the first time in 1986.

RESULTS

There were large differences between somaclones in total harvested yield. In 1984, six out of 197 Feltwell somaclones and two out of 229 Maris Piper somaclones outyield their respective parental variety although the difference was significant (p <0.05) for only one Maris Piper somaclone (Table 1). Fifty two Feltwell and 49 Maris Piper somaclones were selected from the 1984 trials, on the basis of yield and general tuber appearance, for retrial in 1985. Nine of the Maris Piper somaclones had a higher yield than Maris Piper itself but this was only significant (p <0.05) for two of them. Although these nine somaclones included the two which out-yielded Maris Piper in 1984, there was a significant years x somaclones interaction (p <0.001). No Feltwell somaclones outyielded Feltwell itself.

The general appearance of the somaclones was agronomically acceptable and most had as good as or better tuber shape and uniformity than the parental variety although the tuber size was smaller. This was confirmed when the produce from the 1985 trials was graded into a small ware (40-60mm) and a large ware (60-80mm) fraction. Whereas Maris Piper had 57% of its yield in the small ware fraction the corresponding figure for its somaclones was 84%. In Feltwell the difference between the grades was less, the somaclones having 63% and Feltwell having 67% in the small ware fraction. Nevertheless, 13 Maris Piper somaclones and no Feltwell somaclones have been selected for further trial in 1986.

A new group of 184 Maris Piper somaclones was grown in field trials for the first time in 1985, after stringent rejection of somaclones showing poor vigour during the culturing stages. Ten of these somaclones significantly outyielded Maris Piper (p <0.05) and had good overall tuber characteristics. Sixty were selected for retrialling in 1986.

Table 1. Number of somaclones in different yield classes relative to the parental variety.

	Feltwell somaclones		Maris Piper somaclones	
Yield class	1984	1985	1984	1985
>110%	1	0	1*	2*+1
100-110%	5	0	1	6
90-99%	38	4	7	5
<90%	153	48	220	35
Total	197	52	229	49

* Significantly higher yielding than Maris Piper (p <0.05).

DISCUSSION

A yield at least as good as that of existing varieties is required in a new variety for it to become successful. Earlier trial results had indicated that relatively large proportions of high-yielding somaclones could be obtained (Thomson et al. 1986). In the present trials 13 somaclones, or 3.1% of the population first trialled in 1984, have been selected after two years' trialling. This selection rate compares reasonably favourably with selection rates in conventional potato breeding programmes (e.g. Howard et al. 1978), especially if one considers that the somaclones were derived from high-yielding, well-adapted varieties. However little is yet known about the performance of these somaclones for all the other factors, including most disease resistances and quality characteristics, which contribute to the success of a potato variety.

It was disappointing that such a big proportion of the yield of the somaclones in 1985 was in the small ware fraction. It is possible that the absence of large tubers is an indication of late maturity which could lend support to the contention of Sanford et al. (1984) that somaclones are merely bolters. Whatever the reason for a high proportion of small tubers, the size distribution is crucial to the success of varieties and the chances of selecting a commercially acceptable somaclone from a variety such as Maris Piper will be severely limited if larger tubers can not be produced.

Useful variation amongst somaclones has been found for some disease resistances (e.g. Gunn et al. 1985). Improved selection rates and a better size distribution of tubers may result from the use of more complex explants such as leaf discs which result in fewer obviously aberrant individuals than protoplasts. Such somaclones will be in trial for the first time in 1986. However until a somaclone with good agronomic characteristics and stable performance has been selected, doubts remain about the value of the technique in applied potato breeding.

REFERENCES

Gunn, R.E. (1982). Breeding new potato varieties from protoplasts. 8th Long Ashton Symposium, 12-15 Sept., 1982.

Gunn, R.E., Jellis, G.J. & Starling, N.C. (1985). Improved resistance to common scab (Streptomyces scabies) in protoplast-derived potato somaclones previously selected for high yield. Tests of Agrochem. and Cult., No. 6, (Ann. Appl. Biol., 106, Suppl.), pp. 162-3.

Howard, H.W., Cole, C.S., Fuller, J.M., Jellis, G.J. & Thomson, A.J. (1978). Potato breeding problems with special reference to selecting progeny of the cross of Pentland Crown x Maris Piper. Plant Breeding Institute, Cambridge, Ann. Rep. for 1977, pp. 22-50.

Sanford, J.C., Weeden, N.F. & Chyi, Y.S. (1984). Regarding the novelty and breeding value of protoplast-derived variants of Russet Burbank (Solanum tuberosum L.). Euphytica, 33, 709-15.

Secor, G.A. & Shepard, J.F. (1981). Variability of protoplast-derived potato clones. Crop Sci., 21, 102-5.

Thomson, A.J., Gunn, R.E., Jellis, G.J., Boulton, R.E. & Lacey, C.N.D. (1986). The evaluation of potato somaclones. In Somaclonal Variations and Crop Improvement, ed. J. Semal, pp. 236-43. Dordrecht: Martinus Nijhoff.

USE OF _IN VITRO_ CULTURE OF _SOLANUM TUBEROSUM_ IN POTATO BREEDING

B. Colin, F. Lammin and Y. Dattee

The potato is an autotetraploid plant, multiplied vegetatively in most instances, with many clones being male sterile, or even completely sterile. The potato was brought into France by only two introductions of a limited number of clones, and was subsequently severely selected to change photoperiod sensitivity. Genetic variability is thus restricted, and this is serious since there is evidence of a correlation between heterozygosity and vigour.

CURRENT SELECTION PRACTICES

Potato breeding practice is in part dictated by the peculiarities of the species. Selection is applied to the products of crosses between plants, and while there are data on the progeny, they are of limited value given the incomplete genetic information for the species. Selection is thus largely ineffective. The sheer mass of vegetative material imposes drastic selection in the early stages, with few representatives per clone. The risk of discarding worthwhile material is high. Hougas and Peloquin (1958) proposed using diploids (dihaploids) as a means of overcoming these difficulties.

USE OF DIPLOIDS IN BREEDING PROGRAMMES

Production of diploids

The first diploids were obtained by Hougas and Peloquin (1957) using _Solanum phureja_ to provoke parthenogenesis. Dunwell and Sunderland (1973) obtained diploids by anther culture and Irikura (1972) cultured anthers of diploids to produce haploids. It is worth investigating several methods of haploid production because, as San (1977) has shown for barley, plants produced by androgenesis, gynogenesis and interspecific crosses can be different from one another.

Possibilities offered by diploids

Diploids provide access to the genetic resources of the wild diploid species known to be resistant to a number of diseases. They facilitate a rapid appreciation of the variation available in a tetraploid parent, and simplify its genetic analysis. The genotype is more readily evaluated and the accumulation of favourable genes is faster.

Problems posed by the use of diploids

Dihaploids of *S. tuberosum* are almost all sterile. The wild diploid species are agronomically unacceptable (very long stolons, small tubers, photoperiod sensitivity, etc.). However, haploid crosses between *S. tuberosum* and wild species give progeny which are vigorous and fertile.

Recovery of tetraploidy

This step is essential since tetraploids are more vigorous and productive than diploids. Three techniques are available.

1. Doubling with colchicine, which gives duplex (inbred) plants lacking vigour.
2. Unreduced gametes: their production is under genetic control and integrating the appropriate genes into selected populations is not easy. Furthermore, the maintenance of diploid populations carrying such genes would pose problems.
3. Protoplast fusion: this technique is very promising.

Use of protoplast fusion in potato breeding

Wenzel (1979) proposed a scheme combining haploidization and protoplast fusion. However his interest was restricted to monofactorial characters with genes acting cumulatively. His model is of theoretical interest but is not adapted to the constraints of potato breeding. Most characters of interest such as yield and disease resistance are controlled by several genes and inter-locus interactions are important.

Proposed scheme

We thus propose a selection scheme which involves the use of diploid species and *in vitro* culture techniques, while seeking to meet the particular requirements of potato breeding (see Figure 1). At present, the application of the whole scheme is not possible. The genetics of the potato are not well known, and it will be important to study the correlations between diploids and tetraploids carrying the same genes so that diploid populations can be selected for their use in fusion.

Protoplast fusion

To date we know of no successful protoplast fusion involving only S. tuberosum. Furthermore, fusion is not necessarily total and there

Figure 1. Use of diploids and protoplast fusion in potato breeding.

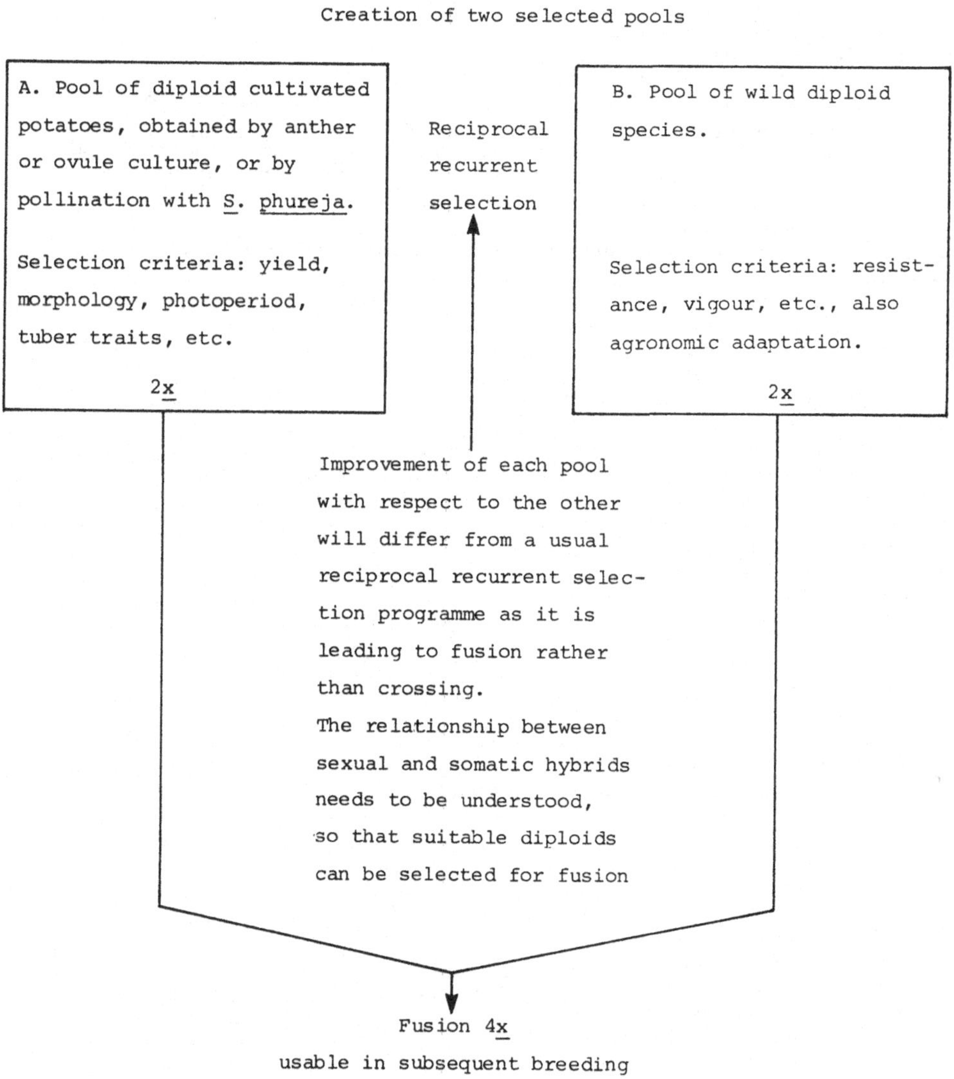

Creation of two selected pools

A. Pool of diploid cultivated potatoes, obtained by anther or ovule culture, or by pollination with S. phureja.

Selection criteria: yield, morphology, photoperiod, tuber traits, etc.

2x

Reciprocal recurrent selection

B. Pool of wild diploid species.

Selection criteria: resistance, vigour, etc., also agronomic adaptation.

2x

Improvement of each pool with respect to the other will differ from a usual reciprocal recurrent selection programme as it is leading to fusion rather than crossing.
The relationship between sexual and somatic hybrids needs to be understood, so that suitable diploids can be selected for fusion

Fusion 4x
usable in subsequent breeding

will remain the problem of screening fusion products, even if partial fusion products constitute very valuable breeding material. Nevertheless, this scheme does have the advantage of being applicable in stages, each stage improving breeding efficiency. It is only the lack of development of a fusion technique which restricts progress. We are currently working on this problem.

REFERENCES

Dunwell, J.M. & Sunderland, N. (1973). Anther culture of *Solanum tuberosum* L. Euphytica, *22*, 317-23.

Irikura, Y. & Sakagucki, S. (1972). Induction of 12 chromosomes plants from anther culture in a tuberous solanum. Potato Res., *15*, 170-3.

Hougas, R.W. & Peloquin, S.J. (1957). A haploid plant of potato variety Katahdin. Nature, *180*, 1209.

Hougas, R.W. & Peloquin, S.J. (1958). The potential of potato haploid in breeding and genetic research. Amer. Potato. J., *35*, 701-7.

San, L.J. & Demarly, Y. (1983). Gynogenesis *in vitro* and biometric studies of doubled haploids obtained by three techniques in *Hordeum vulgare* L. Proc. 10th Cong. EUCARPIA, Wageningen, The Netherlands, p. 347.

Wenzel, G. (1979). New directions in potato breeding. Kartoffelbau, *30*, 160-2.

COMMENTARY

A PRACTICAL BREEDER'S VIEW OF THE CURRENT STATE OF POTATO BREEDING AND EVALUATION

A.J. Thomson

INTRODUCTION

Potatoes differ from other arable crops grown in temperate countries in a number of ways which influence the conduct and logistics of breeding programmes and compound the difficulties of variety assessment. These differences can be summarized as follows.

1. The potato crop is propagated vegetatively from tubers and varieties exist as clones with the exception of crops grown from true potato seed (TPS), mainly in the tropics and subtropics. In temperate regions, however, the consequences of clonal methods include the establishment of a diversity of schemes for seed tuber multiplication, maintenance and certification of healthy stocks and of the importance given to breeding for resistance, especially to virus diseases.

2. There are more characters of economic importance in potatoes than in any other temperate arable crop. This means that the chances of selecting a variety with good performance for all the characters of importance are very small. All varieties are, therefore, compromises and perform well or less well for the range of important characters. This has contributed to the relatively slow progress made in potato breeding and partly explains why the replacement rate of old varieties by new ones is less rapid in potatoes than in other arable crops.

3. Because of the autotetraploid nature of potatoes, the F_1 population raised from intercrossing two tetraploid parents is highly heterozygous. Progeny testing as a means of evaluating parental performance is laborious and has seldom been applied routinely to breeding programmes. Breeding at the diploid level has the advantages of greater diversity in choice of parents and a broader genetic base allied with a better understanding of the inheritance of specific characters. However it is generally accepted that the improved performance of tetraploids over diploids is a consequence of multiallelism so that a return to the tetraploid state is

required for the production of finished varieties.

4. The uses to which potatoes can be put are very diverse, ranging from domestic and table use to the production of French fries, crisps (chips), dehydrated products, industrial starch and alcohol. It is not surprising, therefore, that different sectors of the potato industry attach different priorities to the qualities required in potatoes. For example, the grower for the domestic market is concerned with yield and tuber appearance to a far greater extent than the potato processor who requires high dry matter and low reducing sugar content as top priorities. This divergence of requirements compounds the problems of the variety assessor and potato breeder who have to rank their objectives according to the market being considered.

BREEDING OBJECTIVES

Breeding objectives can include any of the large range of characters which contribute to the success of a new variety. However the inclusion of a multiplicity of breeding objectives in a single programme creates obvious problems in producing the segregating population containing the desired individuals. If such a population were produced the even greater problem exists of identifying the elite individuals. Breeders therefore usually adopt a more realistic approach by selecting positively for only a few characters of prime importance and rely largely on chance that amongst the selected sub-group there are clones which also perform well enough for all the other secondary characters which are required for a variety to be successful. In practice the knowledge and skill of the breeder in choosing the right parents to achieve these ends is very important.

The main selection criteria in any breeding programme are marketable yield and quality. These characters must be at acceptable levels, first of all to the grower of both seed and ware tubers and secondly to the end users. Marketable yield is relatively easily defined as the yield of sound potatoes without defects. In certain countries, tubers must also attain a specified size. Quality is a much more difficult character to define as it depends not only on the end use but also on a large element of subjectivity, particularly when flavour and taste are considered. However varieties which are clearly inferior and do not match the normal quality standards for the particular end use will not be widely grown.

Although yield and quality are essential attributes of a new variety, disease resistance is much more contentious because of the number and type of diseases as well as the different priorities that can be attached to them. These priorities usually reflect national concerns although there are certain criteria in common between countries. Different countries vary in their attitude to breeding objectives, ranging from the pragmatism of the Netherlands (van Loon) where the objective was stated as "would I as a farmer like to grow the clone on a large scale" to the much more formal listing of objectives in Poland (Swiezynski), FRG (Munzert), GDR (Scholtz) and the USA (Martin).

It is therefore impossible to define a uniform set of breeding objectives because of these national differences and priorities and also because of the differences in the organisation of the breeding work. Some countries, however, are very definite in their requirements. For example, in the FRG resistance to the *Globodera rostochiensis* species of potato cyst nematode is absolutely essential and clones lacking this resistance will not be grown (Munzert). Virus resistance is also universally sought along with resistance to late blight (*Phytophthora infestans*), *Erwinia* and storage diseases. Resistance to common scab (*Streptomyces* spp.) is also given widespread consideration.

Although the emphases given to the various objectives depends upon national conditions it is generally true that unless environmental considerations override simple economics, high levels of disease resistance will seldom if ever compensate for inferior yield and quality. Only when disease control by chemical means becomes unacceptable because of environmental risks or cost will genetic disease control through resistant varieties begin to compensate for low yield.

VARIETY ASSESSMENT

The question has been raised as to whether or not variety assessment is at all feasible or meaningful using relatively small plots and the limited trials carried out by breeders and agronomists. In attempting to answer this question, one has to distinguish between the aims of the breeder and those of the agronomist charged with conducting official trials for certification or National List purposes.

Selection procedures in breeding programmes have been largely governed by the complexity and polygenic nature of the inheritance of most characters. Consequently in the early stages of a breeding programme when tubers are few and the number of clones present is large, assessment has been carried out on characters which are relatively easy to measure e.g.

stolon length, eye depth, tuber shape, etc. Selection has, in effect, been negative in the sense that the worst clones have been rejected rather than positive in which the best clones are retained. This emphasis on characters which are usually of secondary importance only will inevitably result in the rejection of valuable clones before they have been screened for those characters of primary importance.

It is therefore very important that early generation testing methods be devised which reflect the real priorities of the breeding programme. These methods will be easy to develop for major gene characters such as resistance to *G. rostochiensis* and potato virus Y as comparatively few tubers are required for such tests. The introduction of multiplex parents into breeding programmes for these characters has obvious benefits (MacKay) but this is not possible for all characters. However even for polygenically inherited characters progress in increasing selection efficiency is being made by improving early generation testing methods (Brown; Lacey *et al.*; Wenzel *et al.*). Nevertheless the accurate assessment of characters such as yield requires many tubers and here the problems of the breeder merge with those of the agronomist responsible for carrying out official trials.

The difficulties of official variety assessment described by Talbot can be seen by the lack of standardization and the various schemes adopted in the different European countries (Table 1). The variation between the countries in what is considered appropriate for number of years, sites, replicates and plants per plot is marked. This is not surprising, however,

Table 1. Official trial schemes in European countries.

	France	Netherlands	Poland	FRG	UK
Years	2	OP 4 T 3	3	3	NL 2 RL 3+2
Sites	15-20	OP 25-40 T 2-7	28-38	5-12	NL 3 RL 6-20
Replicates	5	OP 1 T 3	4	3	NL 2 RL 3
Plants/plot	60-100	OP 16-36 T 50-70	60	60	NL 100 RL 40-100
Seed origin	Yr1-breeder Yr2-common	OP breeder T common	common	breeder	NL breeder RL common
Varieties	10-20 10-20	OP 200-20 T 10-20	10-20	40-50	NL 15-20 RL 6-12

OP = Observation plot; T = Replicated trial;
NL = National List; RL = Recommended List

and merely reinforces the argument that potato trialling is more complex than that of other arable crops e.g. the potato crop has a bigger percentage of its total variation associated with environmental interaction than other arable crops (Talbot).

Despite the difficulties of standardization of potato trials, it is very important that official testing is seen as a means of augmenting the efforts of breeders and not, as is more commonly the case, frustrating them. There is still a great tendency for official testers to assume absolute authority and not to collaborate and cooperate with breeders as much as they might. One can speculate that many highly successful varieties which were bred a number of years ago but which are still in demand today would probably not have passed the standards of several of the present-day evaluation and selection schemes. A degree of flexibility is therefore required in official testing and in this regard it is interesting to compare the rigidity of the French scoring system for new varieties (Perennec) with the pragmatism and concern for profitability of the Dutch system (van der Woude). Potatoes are much more central to the agricultural economy of the Netherlands than they are to that of France and this may be reflected in the difference between the evaluation schemes of the two countries.

BREEDING METHODS

There is great diversity in the potato breeding methods being proposed and, in certain cases, adopted in different countries. The traditional method involves clonal selection at the tetraploid level where progress continues to be made. Almost all the widely-grown varieties in present day agriculture in temperate countries have been produced in this way. Yet there is widespread concern, at least among potato geneticists and researchers, about the slow rate of improvement in new varieties and the absence of any significant increase in performance. The evidence most commonly cited in support of this is the continued popularity of some very old varieties. Although there is acknowledgement that this is partly due to some of the pecularities of the potato crop already discussed, there is a firmly held view that variation within the Solanum tuberosum ssp. tuberosum (Tuberosum) gene pool is limited and inhibiting breeding progress. However practical breeders involved in selection have introduced wild species into their programmes over the decades and have been able to demonstrate that variation exists for most of the characters of importance. There are, of course, exceptions such as the lack of good sources of resistance to leaf roll virus and possibly also to the Erwinia complex of pathogens. Nevertheless, most practical breeders contend that it is not the lack of variation which is the main reason for the slow genetic advance but rather the complexities of the number of characters to be considered and of combining good performance for a large number of characters into a single clone.

The problems of the narrowness of the Tuberosum gene pool can not, however, be ignored and it is claimed that genetic advance could be quicker by adopting new techniques and breeding methods which are better able to exploit diverse germplasm. Breeding at the diploid level has a number of advantages over tetraploids, including crossability with wild diploid species, better understanding of inheritance and ease of elimination of undesirable genes (Colin et al.). Great emphasis has been put on the use of diploids as the source of genetic variation by many authors, exemplified by Hermsen who advocated that evaluation at the diploid level is required for the characters of most interest. This requires a sufficient number of accessions and a big enough population size to assess the extent of the variation present in any species, followed by "prebreeding" to upgrade the level of polygenic traits and select for combinations of genes. Once this has been done it is then necessary to convert the diploids to tetraploids and it is generally accepted that the best method is by 2n gametes.

The mode of 2n pollen formation is important because if the pollen is produced by first division restitution (FDR) with crossing over, about 80 per cent of the heterozygosity of the 2x parent is transmitted whereas without crossing over this rises to 100 per cent (Hermundstad & Peloquin). Thus gametes obtained by FDR without crossing over have identical genotypes and in 4x-2x crosses all the variation in the 4x progeny should arise from the 4x parent only. Even 2x clones which produce gametes by FDR with crossing over have an improved transmission of genes to the progeny compared with clones which do not have the FDR characteristic. However 4x progeny testing is still necessary to determine the parental value for yield of 2x clones as there is no significant correlation between yield of the 2x parent and the 4x progenies derived from 4x-2x crosses (Masson & Peloquin).

Nevertheless by breeding at the 2x level and exploiting these methods of gamete formation to return to the tetraploid level, the breeder now has powerful tools at his disposal for accessing the enormous variation present in diploid species and for transferring this variation almost intact to tetraploid genotypes. It is therefore not surprising that the use of unreduced gametes, which has been pioneered by Peloquin's group in Wisconsin, has attracted widespread interest. However unreduced gametes have yet to make a major impact on the provision of new improved varieties in temperate regions, perhaps because the "prebreeding" stage at the diploid level has not been extensive enough and too few characters have been considered, although significant progress in the provision of new germplasm for the tropics and subtropics has been reported.

The problem of widening the range of variation available to breeders has been tackled in a different way by the use of Neotuberosum germplasm produced by selecting amongst a set of S. tuberosum ssp. andigena (Andigena) gentoypes over a number of generations and eventually intercrossing with Tuberosum germplasm (Plaisted; MacKay). The programme at Cornell University has reached the stage where rapid improvement is predicted in adapted temperate varieties. Already one variety has been released and heterosis in crosses with Tuberosum clones has been obtained. Combining ability is good with big increases in the yields of hybrids between the Andigena material and Tuberosum varieties although this is due to a greater number of tubers rather than to an increase in tuber size. The Neotuberosum material also contains valuable sources of resistance to major diseases such as late blight and potato viruses Y and X. It is expected that the proportion of Tuberosum x Neotuberosum progenies that are good enough to be selected will gradually increase as the Neotuberosum material is further improved through selection (Plaisted).

The breeding method used at the International Potato Center in Peru (CIP) has been conditioned by its philosophy not to release varieties but to produce improved germplasm for distribution to developing countries (Mendoza). This approach allows greater freedom in incorporating variation from diverse, wild sources because the final step in the breeding programme, which is concerned with producing improved varieties adapted to local conditions, is not of prime consideration. CIP decided, therefore, to exploit additive genetic variation by conducting a population breeding programme using phenotypic recurrent selection followed by progeny testing in the later generations. The selection of elite parents is given high priority because not only has it advantages in the provision of good progenies for developing countries, it is also of major importance in the TPS breeding programme in which parents with high general combining ability for yield and tuber uniformity are essential. The population breeding programme has now reached the stage where good germplasm for the hot, humid tropics is available. It is doubtful whether this would have been considered possible a few years ago or whether it could ever have been achieved using the conventional clonal selection methods used by breeders in northern, temperate countries.

At CIP much effort has also been put into the use of TPS for potato production. It is now generally assumed that hybrid TPS populations will be superior to open pollinated progenies because of their greater heterozygosity and it is advocated that the parents of the hybrids should have good combining ability for yield and be complementary for important disease and agronomic characters. However because a certain degree of uniformity of performance for the array of important traits in the progeny is necessary, an inbreeding strategy has also been proposed. This strategy runs counter to the accepted beliefs that phenotypic advantage in outbreeders depends on heterozygosity. The debate has been reviewed by Jackson and is, as yet, unresolved. Although TPS is already being used successfully in tropical and subtropical countries, its potential in northern, temperate regions could be greatly enhanced if the inbreeding approach were found to be feasible.

Returning to the original theme, the problem of variation and the number of characters to be considered, a potential solution is offered by the exploitation of somaclonal variation. It is well known that potato somaclones raised from protoplasts or more complex explants such as leaf discs vary for only some of the important characters. This differs from the progeny from a sexual cross in which the resulting population segregates for virtually all characters. Somaclonal variation could therefore be a means of improving the performance of existing varieties (Jones; Thomson). Evidence to date on the evaluation of potato somaclones shows that restricted variation for a number of economically important traits has been found but that there are some unresolved problems, such as the propensity to produce small tubers, which create doubts about the feasibility of the technique for producing new varieties (Thomson). However the potential rewards which the technique could bring by upgrading the performance of varieties which are already popular and widely grown justify continued efforts.

There are other means by which existing cultivars can be upgraded. The rapid progress that has been made in biotechnology and genetic engineering in recent years has provided the capability of introducing specific genes into all the cells of a potato cultivar while leaving the somatic genome intact (Flavell; Ooms). This research is still in its infancy and has been the subject of much speculation and promotion. Many breeders are of the opinion that the benefits which could accrue from such work have been grossly exaggerated. Nevertheless the insertion of new genes into potatoes will become routine in the near future once the genes which control economically important traits have been characterized. The absence of easily identified genes at the molecular level is probably the major impediment to progress in genetic engineering but advances in this area are now being made and cDNA clones have been identified and characterized for potato genes showing organ-specific expression (Blau *et al.*). Up to now progress in genetic engineering has been made in the laboratory only and it is becoming increasingly apparent that the consequences of DNA transformation in terms of correlated responses for other characters or deleterious side effects have to be evaluated under practical field conditions. This requires the active cooperation of breeders and a commitment to new technology which has not always been in evidence.

The precise modification of plants by DNA transformation is not the only novel technique which is beginning to make an impact on potato breeding. Protoplast fusion also has exciting possibilities, especially as it enables characters to be combined without the detailed knowledge of molecular biology required in DNA transformation (Jones). However one of the main drawbacks of protoplast fusion is the difficulty of identifying the fused heterokaryons, but here molecular biology in the form of DNA probes can assist. The main advantages of protoplast fusion for potato breeders are its ability to overcome crossing barriers between different species and genotypes and to transfer cytoplasmically inherited characters.

These new techniques of DNA transformation and protoplast fusion are not yet at the stage of general applicability although the use of cDNA probes for the detection of specific viruses is gradually replacing serological methods in breeding programmes (Boulton *et al.*). The advances made in tissue culture and micropropagation, however, are readily applicable and have enabled radical approaches to breeding to be introduced. The progressive reduction in the pioidy level has provided an elegant solution to the problem of combining quantitatively inherited characters in a single clone (Wenzel *et al.*). The results obtained for field resistance to PVY, PVX and potato leafroll virus after 7 years of field trials are encouraging, especially as these quantitatively inherited traits, which are in the homozygous state after chromosome doubling, can be transferred to the next generation as if they were qualitative characters. The work on *in vitro* screening for resistance to *Phytophthora* and *Fusarium* is also of great interest to breeders. *In vitro* screening permits the use of much larger populations than could be handled conventionally but requires that regeneration to whole plants does not introduce too much variation to nullify this advantage.

CONCLUSIONS

The discussion on breeding objectives is mainly based on the evidence from state-funded organisations in Europe. It is interesting to compare these objectives, which include a large element of what is perceived to be in the particular national interest, with the much more commercially oriented objectives usually set by privately funded breeders e.g. in the UK, Dunnett described his principal aim as the improvement of the cultivar Desiree and he was motivated by the need to sell seed tubers worldwide and create a thriving export business. This is broadly similar to the aims of much of the highly successful Dutch potato industry. It is therefore worth remembering that although state funding of breeding programmes can allow strategic decisions to be taken on breeding objectives, it may also lead to a loss of commercialism. Clearly a balance has to be achieved between these differing viewpoints so that objectives can be set realistically and in tune with growers' and consumers' needs.

This is also true of potato breeding methods which are undergoing a fundamental reappraisal. Many new ideas and techniques associated with molecular biology and tissue culture are being developed. There has usually been only minimal contact between traditional breeders involved in clonal selection and the new generation of genetic engineers and tissue culturists. The contrasts in experience and outlook between those working on whole plants in the field and those engaged in laboratory experiments have led to a degree of mutual suspicion. In particular, the practical breeders have not been impressed by some of the wild claims made by molecular biologists. It now seems that these attitudes are moderating, however, and conventional breeding is being increasingly considered as a necessary adjunct for the translation of the new science into agricultural practice. It has been repeatedly emphasised that scientists working on unconventional techniques must be closely associated with agronomists and breeders involved in field work and selection programmes (Flavell; Jones; Wenzel et al.). It is to the benefit of all who work on potatoes that mutual understanding of problems and objectives is fostered and the conference on which this book is based has provided an excellent forum from which to develop these relationships.

REFERENCES

(All references are to chapters in this volume)

Blau, A., Eckes, P., Logemann, J., Rosahl, S., Sanchez-Serrano, J., Schell, Schmidt, R., Schell, J. & Willmitzer, L. Prospects of using Ti-plasmid mediated gene transfer for the improvement of potato varieties.

Boulton, R.E., Jellis, G.J. & Squire, A.M. Breeding for resistance to potato viruses with special reference to cDNA probes.

Brown, J. The efficiency of early generation selection.

Colin, B., Lammin, F. & Dattee, Y. Use of '*in vitro*' culture of *Solanum tuberosum* in potato breeding.

Dunnett, J.M. Private potato breeding in the UK.

Flavell, R.B. Recent progress in molecular biology and its possible impact on potato breeding. An overview.

Hermsen, J.G.Th. Efficient utilization of wild and primitive species in potato breeding.

Hermundstad, S.A. & Peloquin, S.J. Breeding at the 2*x* level and sexual polyploidization.

Jackson, M.T. Breeding strategies for true potato seed.

Jones, M.G.K. Use of protoplast fusion and somaclonal variation in potato breeding.

Lacey, C.N.D., Jellis, G.J., Starling, N.C. & Currell, S.B. A joint cyst nematode/late blight test for early generation screening of potato clones.

Mackay, G.R. Potato breeding in the United Kingdom.

Martin, M.W. Breeding multi-resistant potato germplasm.

Masson, M.F. & Peloquin, S.J. Heterosis for tuber yields and total solid content with 4*x*-2*x* FDR-CO crosses in potato.

Mendoza, H.A. Advances in population breeding and its potential impact on the efficiency of breeding potatoes for developing countries.

Munzert, M. Potato breeding strategy in the German Federal Republic.

Ooms, G. Genetic manipulation in potato using *Agrobacterium.*

Perennec, P. Variety assessment in France.

Plaisted, R.L. Advances and limitations in the uitlization of Neotuberosum in potato breeding.

Scholtz, M. Potato breeding strategy in the German Democratic Republic.

Swiezynski, K.M. Potato breeding strategy in Poland.

Talbot, M. Establishing standards in variety assessment.

Thomson, A.J. The potential value of somaclonal variants in potato improvement.

van der Woude, K. Variety assessment in the Netherlands.

van Loon, J.P. Potato breeding strategy in the Netherlands.

Wenzel, G., Debnath, S.C., Schuchmann, R. & Foroughi-Wehr, B. Combined application of classical and unconventional techniques in breeding for disease resistant potatoes.

INDEX